Upscaling of Single- and Two-Phase Flow in Reservoir Engineering

Upscaling of Single- and Two-Phase Flow in Reservoir Engineering

Hans Bruining

Civil Engineering and Geosciences, Geoscience & Engineering, Section Reservoir Engineering, Delft University of Technology (TU Delft), Delft, The Netherlands

CRC Press
Taylor & Francis Group
Boca Raton London New York

CRC Press is an imprint of the
Taylor & Francis Group, an **informa** business

CRC Press/Balkema is an imprint of the Taylor & Francis Group, an informa business

Library of Congress Cataloging-in-Publication Data
A catalog record has been requested for this book
Published by: CRC Press/Balkema
        Schipholweg 107C, 2316 XC Leiden, The Netherlands
        e-mail: enquiries@taylorandfrancis.com
        www.routledge.com – www.taylorandfrancis.com

ISBN: 978-0-367-76743-3 (hbk)
ISBN: 978-0-367-76744-0 (pbk)
ISBN: 978-1-003-16838-6 (ebk)

DOI: 10.1201/9781003168386

Typeset in Times New Roman
by codeMantra

Contents

Preface

This book,[1] dedicated to Prof. D.N. Dietz, benefited from numerous suggestions of Ph-D students[2] and discussions with staff[3] and specialists.[4] The emphasis in the book is on physical modeling and upscaling related to one-phase and two-phase flow in porous media models that are currently used in reservoir engineering and reservoir simulation and shows where these models use approximations. We only need to be aware of the approximations in modeling to understand its shortcomings, not to avoid them. All models use approximations, which is succinctly summarized in a statement by Box [37], "all models are false, some models are useful", a statement frequently quoted by Prof. W.R. Rossen. The consequences of the limitations due to approximations are largely left to engineering judgment. The book compiles many of the remarks of leading specialists in the fifties and sixties, who were still there when I was as an assistant professor teaching at the TU-Delft (1979-present). I would like to mention Professor De Josselin de Jong, who argued that generally in Darcy's law, the Darcy velocity (specific discharge) could not be represented as being proportional to the gradient of a single scalar. This is so because, in general, the density depends on the space coordinates. Writing the Darcy velocity as being proportional to the gradient of potential leads to an erroneous formulation in the case of natural convection flows or saltwater intrusion when extracting drinking water from the dunes. I would like to mention Prof Dietz who taught me the importance of simple modeling of processes that today carry his name, viz., the Dietz tongue, the Dietz shape factors and the semi-steady-state simplification of transient flow near wells. Indeed, water displacement in a layer can be represented between two extremes: the completely dispersed phase Buckley–Leverett model and the completely segregated interface model, where an interface separates an upper region where only the oleic phase flows from a lower region where only the aqueous phase flows. Indeed, the interface model is more realistic than the Buckley–Leverett model, which assumes complete dispersion of the phases (no interface). Together with Ruud Schotting and Hans van Duijn, we used an interface model to describe bottom water coning in an oil reservoir [47], which showed the strength of interface modeling. By disregarding viscous forces in the X-dip direction, any model between a completely dispersed model and an interface model can be obtained. However, implicit in the so-called Vertical (X-dip) equilibrium model is that equilibrium between capillary and gravity forces in the X-dip direction is much faster than saturation development in the dip direction; this ignores the occasionally long times required to reach X-dip equilibrium. Disregarding capillary forces ignores the presence of a transition zone between an aqueous phase-dominated region and an oleic phase-dominated region, thus leading to unrealistic results. At a Mobility ratio > Gravity number +1 (M > G+1), a shock solution is obtained, which can be represented by a stationary

---

[1]author: Hans Bruining, Delft University of Technology, Department of Geotechnology, Stevinweg 1, 2628 CN, Delft, The Netherlands.

[2]Diederik van Batenburg, Raymond Godderij, Ewout Biezen, Claes Palmgren, Alies Bartelds, Buulong Nguyen, Nikolai Siemons, Willem Jan Plug, Saikat Mazumder, Patrick van Hemert, Hamidreza Salimi, Rouhi Farajzadeh, Ehsan Eftekhari, Geert van der Kraan, Elisa Battistutta, Rahul Thorat, Mohammed Chahardowli, Roozbeh Khosrokhavar, Amin Ameri Ghasrodashti, Negar Khoshnevis Gargar, Anas Hassan.

[3]Pacelli Zitha, Harry Ronde.

[4]Willem Heijnen, Adri Maljaars. Many intricate Latex syntax problems have been solved as part of my collaboration with Bernard Meulenbroek.

interface, which has a constant inclination with respect to the layer direction. At $M < G+1$, a rarefaction (x/t) solution (of the Buckley–Leverett type) is obtained.

The mathematical background required to understand the Buckley–Leverett theory is nontrivial. In the engineering literature [63], the Welge tangent [247] construction was explained as a consequence of a mass balance consideration as opposed to as a consequence of a positive capillary diffusion function, a corollary of an entropy condition [30]. Prof. Dr. C.J. Van Duijn, with whom I intensively collaborated, made me acquainted with the mathematical theory of conservation laws, which was developed by Oleinik (1962) and also Krushkov (1980), who showed that the Welge tangent (1953) construction followed naturally by representing the shock condition as part of a traveling wave solution. The traveling wave formulation resolved an issue of closure for the application of the extended Buckley Leverett theory to a system of conservation laws that described the displacement of oil by steam [46]. The relevant knowledge on hyperbolic conservation laws was further put on a solid foundation during my numerous visits to Prof. Dr. Dan Marchesin at the Instituto Nacional de Matemática Pura e Aplicada in Rio de Janeiro where we extended the application domain to oil combustion and geochemistry. Alexei Mailybaev showed for combustion that solutions could be obtained under conditions for which I never considered it feasible to obtain an analytical solution. Pavel Bedrikovetskii introduced me to percolation theory [20,204].

Theo Olsthoorn [171] presented in the 1980s the possibility of using EXCEL[5] to illustrate almost any relevant numerical method. Based on this, it is possible to show that EXCEL can be used straightforwardly to obtain the numerical solution of a Buckley–Leverett model, considering the simultaneous flow of oil/gas/water. This shows again that EXCEL is a powerful educational tool. Teaching as a guest professor at the University of Leuven allowed to perfect the EXCEL method for Buckley-Leverett type of problems by a suggestion of one of the students for proper initialization. As a consequence, illustration with EXCEL is pervasive in this book.

As to geostatistical aspects, EXCEL can also be used to generate lognormally distributed random fields, with given variance and average, which allows to illustrate numerous methods that calculate upscaled permeabilities. In this, I greatly benefited from a sabbatical year that I spent, at the advice of John Waggoner, with Prof. Lake at the University of Texas at Austin. This also formed the basis of Chapter 5 dealing with mixing (dispersion) in porous media. The word dispersion is confusing for specialists dealing with wave phenomena, where it means the frequency wave vector relation. In the groundwater and petroleum engineering literature, it refers to a diffusional mechanism for mixing and the dispersion coefficient has units $[m^2/s]$. After a suggestion of prof. Van Duijn, it was possible to show quantitatively that the Peclet number dependence of longitudinal dispersion can be reasonably accurately obtained with homogenization [42].

Chapter 1 shows why recovery of hydrocarbons is still relevant, and how in doing so we can, accepting that we cannot abolish fossil fuels overnight, optimally reduce the carbon footprint. It also provides some basic examples to refute over-simplistic descriptions corroborating or denying greenhouse gas problems. It also shows how to carry out an Exergy Return on Exergy Investment (ERoEI) analysis [110], which may help to assess the importance of measures that alleviate global warming problems. Chapter 2 treats incompressible one-phase flow. It is shown that the relation $-\mathbf{u} = \mathbf{grad}(p + \rho gz)$ can only be used under ideal circumstances and which formulation to use when circumstances are no longer ideal. Subsequently, it gives an overview of upscaling techniques for one-phase flow and briefly explains homogenization. Chapter 3 describes the compressible flow in porous media. It uses Laplace transformation to obtain equations for pressure

---

[5]Excel definition: a software program created by Microsoft that uses spreadsheets to organize numbers and data with formulas and functions. Excel analysis is ubiquitous around the world and used by businesses of all sizes to perform financial analysis.

decline and pressure build-up. Chapter 4 treats two-phase flow. The first part consists of defining constitutive relations, such as capillary pressure and relative permeabilities. For relative permeabilities, it emphasizes wetting effects and shows that low endpoint permeabilities are expected for water-wet media at the expense of high residual oil saturation. In treating Buckley Leverett flow, we show why the Welge tangent condition can be used to find the correct shock condition. The interface modeling emphasizes using the Dietz-Dupuit approximation. Chapter 5 discusses dispersion and shows the difference between molecular diffusion and macroscopic (core level) hydrodynamic longitudinal and transverse hydrodynamic dispersion. If the scale extends to the reservoir scale, instead of macroscopic dispersion sometimes the word gigascopic dispersion is used. Chapter 5 also shows that the dispersion coefficient can be related to the integral of the velocity autocorrelation function. This can be used to relate the dispersion coefficient to the product of a correlation length and the variance of the logarithm of the permeability, which can be verified with numerical simulation.

OBJECTIVE OF BOOK

To enable the reader, based on the treatment of simplified models of both stationary and transient flow in porous media

- To set up simple models of one-phase and two-phase flow and find the solution of the ensuing model equations,
- to apply modifications of Darcy's law, e.g., for high flow rates, anisotropic permeabilities, etc.,
- to calculate average permeabilities,
- to describe both capillary driven phase mixing and diffusion-driven component mixing,
- to use the models for partial validation of reservoir simulation results,
- recognition of salient features in reservoir simulation results.

We claim that understanding of two-phase flow in porous media can only be acquired by modeling. In other words, we try to make sense of flow in porous media by modeling, i.e., simplification of the real world. A quote I once noted down about modeling is "Our minds are not going to be kept empty of them and those who have not decent ideas about modelling in their heads will have bad". I cannot retrieve its origin.

About the author

Hans Bruining is professor emeritus in geoenvironmental engineering of the Technical University of Delft, which is ranked 10 as one of the top technical institutes worldwide in Engineering. He holds a PhD degree from the University of Amsterdam. He is the founder of the Dietz-De Josselin de Jong laboratory. His special interests are the environmental aspects of fossil fuel recovery, enhanced oil recovery and theory and experiments of complex flow processes in porous media. He is the review chairman of the *Society of Petroleum Engineering Journal* (SPEJ). He is the recipient of the SPE Distinguished Achievement Award for Petroleum Engineering Faculty (2012). The international award recognizes superiority in classroom teaching, excellence in research, significant contributions to the petroleum engineering profession and/or special effectiveness in advising and guiding students.

1 Dutch and Worldwide Energy Recovery; Exergy Return on Exergy Invested

OBJECTIVE OF THIS CHAPTER

To give the course participant, based on generally available data, a background in the problems associated with replacing fossil fuels[1] by something else [159].

- It is useful to construct a figure like Figure 1.3 with your preferred energy strategy,
- make a division of the national and worldwide energy supply in terms of the present requirement of gas, oil, coal, renewable and nuclear,
- make also a division of the energy consumption of various sectors, such as industry, agricultural services, traffic, household, electricity /heat, refineries and others,
- make a division [159] of your liking for distributing between "renewable" sources [159] such as wind (2.0 MW /km^2 (on shore) - 3.0 MW /km^2 (off shore), solar PV panels (5 MW/km^2), concentrating solar power (15 MW/km^2, tidal pools (3 MW/km^2, tidal streams (6 MW/km^2, biofuel (0.5 MW/km^2), energy saving, clean zero carbon footprint fossil fuel using CO_2 storage, etc., on the one hand and fossil fuel (gas oil/coal) and nuclear on the other hand,
- to show that renewables suffer from an extremely low energy density. For wind, it is typically the power of a bicycle lamp per square meter.

We adopt the approach of MacKay [159] in suggesting that a proposed alternative strategy must be based on arithmetics. Here we keep in mind that for the Dutch situation, we use 110 GW and worldwide we use 15000 GW-18000 GW. Possibly, such amounts are not necessary or can be partly replaced by renewables, but the ideas to decrease them must "add up" and be backed up by arithmetics.

INTRODUCTION

The introduction of an energy recovery process poses four questions. (a) What is the impact on the total energy budget? (b) Is there a net gain, i.e., is the energy required to recover the energy less than the energy produced? (c) What is the net carbon footprint? (d) Is there a gap between the optimum recovery and the current state of the art?

Conversion to renewable energy is a challenge. Moreover, renewable energy sources are sparse [242]; for instance, on shore wind energy delivers 2.0 MW/km^2 [159], meaning that the entire surface area 40,000 km^2 of the Netherlands is required to supply its total energy by wind energy. Moreover, renewable energy sources are intermittent due to differences between day and night, seasonal fluctuations and weather effects, which implies that efficient storage methods have to be implemented [74]. It is convenient to express the energy in terms of exergy, i.e., the

---

[1]to reduce carbon dioxide emission.

DOI: 10.1201/9781003168386-1

energy that can be converted into work as otherwise energies of different quality are put on a single heap. A crude Exergy Return on Exergy Invested (ERoEI) analysis is not difficult and is of great help to show whether the use of the selected exergy source is a viable option. The ERoEI analysis [87, 90, 114] calculates the recovered exergy and compares it to the exergy costs that need to be sacrificed (drilling costs, fluid circulation costs, etc.) to make the recovery process possible. Such an analysis will also provide an estimate of the carbon footprint. Another advantage of this analysis is that it shows where the improvement of the recovery process is useful. We have also added some facts, which the reader may find useful, to put the exergy recovery process in a more general perspective. For fossil fuel recovery, it is important to know how much the worldwide temperature increases with increasing carbon dioxide in the atmosphere. Information regarding drilling costs is provided such that the reader can easily carry out his own approximate ERoEI analysis and possibly find inspiration to introduce renewable energy sources much faster than the current trends suggest.

1.1 FRACTION FOSSIL IN CURRENT ENERGY MIX

Figure 1.1 shows the fraction of fossil fuel in the Dutch and worldwide energy mix. Indeed, according to [100], in 2015, the Netherlands (world) uses 38.7 (4331.3) Mtoe[2] oil, 28.6 (3135.2) Mtoe natural gas, 10.6 (3839.9) Mtoe coal, 0.9 (583.1) nuclear energy, 0.0 (892.9) Mtoe hydroelectricity, and 2.7 (364.9) Mtoe renewable energy. The primary energy consumption in the Netherlands in 2015 was 81.6 (13147.3) Mtoe. This means that in 2015, 95.5% of the primary energy consumption in the Netherlands is fossil fuel-based, compared to 86.0% worldwide [100]. These unfavorable numbers in the Netherlands reflect the availability of methane in a huge gas field. The World Bank quotes 91.36% as opposed to 95.5% for the Netherlands,[3] showing typical discrepancies of global energy-related data in the literature. If one studies this subject more closely, one finds that the large uncertainties stem from all kinds of reasons. It is outside the scope of the introduction into this matter to state anything else than that the fossil fuel fraction in the Dutch situation is around 90%. The reason for the smaller relative contribution of

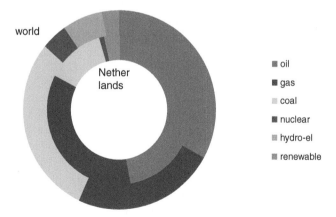

Figure 1.1 Fraction of fossil energy with respect to hydroelectric, nuclear and renewable energy. The inner disk represents the situation in the Netherlands and the outside disk represents the situation worldwide.

---

[2]one ton oil equivalent $\approx$ 42 GJ/ton.

[3]https://data.worldbank.org/indicator/EG.USE.COMM.FO.ZS.

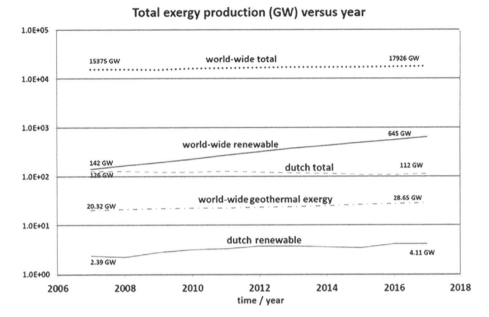

Figure 1.2 Worldwide and Dutch primary and renewable exergy recovery developments [100]. The total worldwide (Dutch) exergy consumption/production increased between 2007 and 2017 from 15375 GW (126 GW) to 17926 GW (112 GW). The renewable part increased from 142 GW (2.39 GW) to 645 GW (4.11 GW). The part of geothermal exergy increased from 20.32 GW to 28.65 GW.

fossil fuel worldwide than in the Netherlands is that worldwide nuclear energy and hydroelectricity play a larger role (Figure 1.1). The same is true about the worldwide figures, stating that the fossil fuel fraction is about 80%–90%. The fraction of the fossil fuel fraction in the electricity production in the EU depends largely on the acceptance of nuclear energy production (see "https://www.reddit.com/r/europe/comments/btxb63/power generation by source in eu countries"). This shows that one possible route [32] to reduce fossil fuel consumption is by massively introducing nuclear energy [132, 166], accepting a 60–80 month construction time. Data regarding energy consumption can be found in BP statistical yearly review of world energy (we used the June 2017 edition) [100]. Figure 1.2 shows that worldwide energy consumption increased from 15,400 to 17,900 GW between 2007 and 2017. The renewable energy generation climbed in 10 years from 142 GW in 2007 to 646 GW, almost a fivefold increase. The Dutch energy supply/consumption decreased from 126.4 to 112.1 GW in the decade from 2007 to 2017.

The introduction of renewables is slow [83, 200]. Across all 13 available scenarios, net zero GHG emissions are reached around 2055–2075 (rounded to the nearest 5 years). Net zero CO_2 emissions are reached earlier. Figure 1.2 shows the trends in total worldwide energy consumption, worldwide renewable, the Dutch energy consumption, worldwide geothermal energy supply, and the Dutch renewable energy supply. It is unavoidable that the Oil and Gas industry will play a major role in the transition toward the sustainable society [223]. Figure 1.3 shows how in the Netherlands (drawn curve) and worldwide (dotted curve) the share of renewables is steadily increasing. The extrapolation shows that around 2028[4] we may expect that around 10% of our energy supply will consist of renewables, which roughly corresponds to predictions of many

---

[4]it is possible that new technology allows that this occurs much sooner.

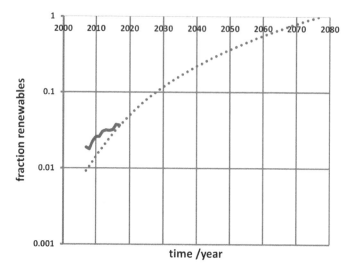

Figure 1.3 Fraction renewable [100] implemented worldwide. Extrapolation gives the order of magnitude in most predictions [100]; if the present trend persists, it can be expected that by 2038, 20% of the energy supply will be by renewables. The drawn curve shows the Dutch situation.

scenario analyses quoted in the literature. It is therefore also important to assess the role of a fossil fuel industry [223] toward a sustainable society.

1.2 POSSIBLE NEW DEVELOPMENTS

Figure 1.4 shows the current (2008) partition (fossil and renewable/nuclear) of the Dutch energy supply and Dutch energy consumption in various activities. Civilian energy consumption is by electricity heat (9.7 GW), household (14.1 GW), and traffic (15.5 GW), around 40% of the energy consumption. Figure 1.4 also shows a possible division into energy sources that have a low carbon footprint. The implementation of renewable sources is hampered by the intermittent supply of solar and wind energy. The conversion of fossil fuel to pumping power [87] goes at an efficiency of

$$\eta_{fuel \to pump} = \eta_{elec} \times \eta_{driver} \times \eta_{pump} = 0.45 \times 0.9 \times 0.8 \approx 0.325 \qquad (1.1)$$

Figure 1.4 shows a possible partition of the Dutch energy consumption choosing fuels with a low carbon footprint. Such an alternative would largely rely on solar and wind energy. For other renewables, MacKay [159] suggests to use tidal energy using the North seas as a reservoir. None of this has been implemented yet on a large scale in the Netherlands, of which the energy supply uses 3.3% for renewables.

MacKay [159] has a number of suggestions to personally save energy by

- using a wooly jumper so that one can put the thermostat at 15°C,
- avoiding single as opposed to shared usage of a car and when possible walk, bike or use public transport,
- reading your meters and take action,
- eating vegetarian during 6 days per week,
- not buying clutter,
- using double-glazed windows.

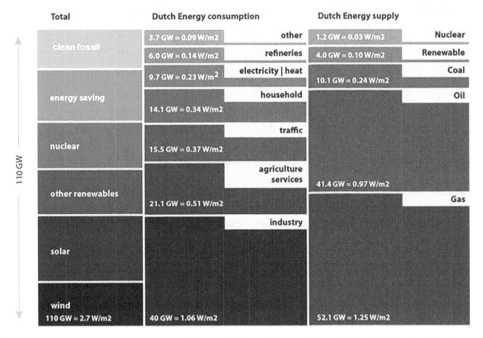

Figure 1.4 Division of energy use related to activities and a possible partition (2008) when Dutch energy need is to be distributed according to low carbon footprint emitting energy sources.

These saving options according to [159] show what an average citizen may contribute to reduce the energy requirement. It may lead to a reduction of 10%–30%, if these ideas were generally adopted. Numerous scenarios have been developed to predict the fossil fuel contribution in the future.[5]

More efficient [57, 90, 106] fossil fuel production may also help reduce carbon emissions. As about 17% of the energy [182] from fossil fuels is used for their production, it is to be expected that efforts to reduce production costs will on the short term be effective in reducing greenhouse gas production. Unfortunately, it can be expected that developing countries will increase their fossil fuel use annihilating this positive effect.

Figure 1.5 plots the radiative forcing ($\ln(CO_2/CO_{2,initial})$) versus the measured temperature increase. It shows a steep increase until 1940, after which the temperature remains more or less constant until 1980 in spite of increasing radiative forcing. After 1980, the temperature increases, an increase that persists until our time. The constant temperature behavior until 1980 may be attributed to aerosols [244].

However, there can be no doubt [189] (pp 218 ff) that carbon dioxide plays an important role in global warming. One of the main issues to be aware of, which pleads against the idea that CO_2 is a minor player with respect to H_2O, is that Lambert-Beer does not describe the behavior of the outgoing long wave radiation (OLR) as not only adsorption but also emission of radiation must be taken into account as is done in the Schwarzschild equation [189], an integral equation.

A summary of the mechanism shows that in spite of the presence of water, a much stronger greenhouse gas than carbon dioxide, carbon dioxide has still an important influence on the

---

[5]Prediction is difficult especially the future, The Yale Book of Quotations by Fred R. Shapiro, Section Niels Bohr, Quote Page 92, Yale University Press, New Haven.

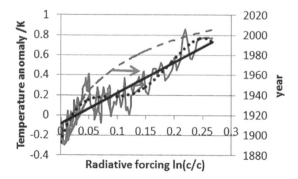

Figure 1.5 Temperature versus radiative forcing over the last 120 years. The radiative forcing is defined by the natural logarithm of the carbon dioxide concentration/concentration at a reference point in this case in the year 1901.5. The dashed curve shows the radiative forcing versus year. The drawn curve shows the average radiative forcing versus temperature, whereas the dotted curve shows radiative forcing versus temperature. Between 1940 and 1975, in spite of the increase in radiative forcing, no significant temperature increase was observed. This is attributed to postwar sulfuric aerosol emission by the industry and natural volcanic eruptions.

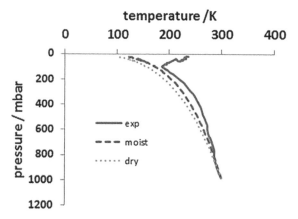

Figure 1.6 Pressure versus temperature calculated from the adiabatic assumption, i.e., a parcel of humid air expands adiabatically, and the air is everywhere saturated with water vapor. At the upper boundary of the troposphere (Atmospheric layer adjacent to the Earth surface) the temperature starts again to increase with height.

outgoing long wave radiation. First one needs to realize that the temperature from the Earth surface decreases due to adiabatic expansion (Figure 1.6). For dry air, it can be shown that the temperature decreases with height according to $g/C_p \approx 9.81[\mathrm{m/s^2}]/1000[\mathrm{J/kg}]$ approximately 1 deg/100m. Given its simplicity, the formula is remarkably accurate with respect to the measured temperature decline (it can be improved by taking moist air into account). The Schwarzschild equation takes both adsorption and emission into account. Lambert-Beer is only valid when the radiation is far away from Planckian thermal equilibrium and emission can be disregarded. If the temperature in the troposphere (the atmospheric layer closest to the Earth surface) was constant, the net adsorption + emission would add up to zero. Therefore, the radiation in the troposphere

passes through a number of radiation equilibria [189] at decreasing temperatures. The troposphere extends from 8 km at the poles to 16 km at the equator. Beyond the troposphere, we have
the stratosphere where the atmospheric density becomes so low that radiation processes rather
than adiabatic cooling determine the temperature, and the temperature starts to increase (Figure
1.6) in the stratosphere. In the stratosphere, other greenhouse gases like ozon leads to an increase
in the temperature with height. Water as a greenhouse gas only plays a role until its vapor pressure
becomes so low so that it no longer contributes to the absorption of radiation. The spectrum of
the outgoing radiation is characterized by its Planckian equilibrium. Above a certain height only
at the characteristic frequencies for carbon dioxide, radiation is adsorbed. In the upper part of the
troposphere, the outgoing radiation is determined by the carbon dioxide concentration ([189]).
Only considering the adsorption of radiation as used in the Lambert-Beer law gives us a lower
limit of the OLR. The effect of increasing amounts of carbon dioxide can be only understood
by considering the Schwarzschild equation, which includes emission, in an adiabatically cooled
atmosphere.

1.3 EXERGY

We will introduce the concept of exergy to enable the reader to make a crude estimate of the
viability of a recovery process of interest. Exergy is the part of the energy that can be converted
to work. Nicholas Sadi Carnot (1796–1832) [53] analyzed steam engines and found that a body
at elevated absolute temperature T in an environment of a temperature T_0 can deliver

$$Ex = C_w \Delta T \left(1 - \frac{T_0}{T} \right) \tag{1.2}$$

of work. This amount of work is also equal to the exergy. For instance, one cubic meter of water
($C_w \approx 4.184E6$ [J/m^3/K]) at $T_{res} = 353$ K in an environment of $T_{ref} = 313$ K has an exergy
of $C_w(T_{res} - T_{ref}) \times (1 - T_{ref}/T_{res}) = 167.4 \times 10^6 \times 0.113 \approx 19 \times 10^6$ J, where $(1 - T_0/T) =
0.113)$ is the Carnot factor.

1.4 EXERGY RETURN ON EXERGY INVESTED (ERoEI) ANALYSIS

We need to spend energy in order to recover energy [111]. In many processes, the amount of
energy for recovery exceeds the recovered energy. Therefore, it is important to make quick estimates of the recovered energy and the energy required to recover the energy. The ERoEI gives
an efficient procedure to implement such an analysis.

   The main components of an approximate ERoEI analysis are

Exergy recovery   •   for hydrocarbons, this is approximately (a) its combustion energy or
        (b) 10.7 kWh/liter liquid hydrocarbon,[6] or (c) 50 MJ/kg-CH$_4$ (methane), whatever is
        more convenient.
   •   for geothermal exergy, this is the Carnot factor $((1 - T_o/T)$ see also Eq. (1.2)) ×
        thermal energy production rate $\approx \chi[m^3/s]C_w(T - T_{ref})[J/m^3]$ with respect to a con-
        veniently chosen reference temperature, e.g., $T_{ref} \approx 40^oC$ or for supercritical condi-
        tions the enthalpy production $\chi(H - H_0)$ with respect to the enthalpy H_0 at a reference
        state.
   •   for wind energy, it is 2 W/m^2, and for solar energy, it is 10–20 W/m^2, which can also
        be used to power oil platforms [109].

---

[6]Mackay [159] uses the estimate of 10.7 kWh /liter, which differs from 41.87 GJ/ton of oil.

Daily handling and operation costs about 17% ([182] p. 49, [90]) of the produced fossil fuel energy is used for recovery, transportation, and processing. we assume also 17% of the fuel energy costs for handling and operation of other energy sources. This means that a higher efficiency in field operations is able to reduce the greenhouse gas emission [85].

Circulation costs the fluid circulation costs are $Q\Delta\Phi$, where Q is the flow rate and Φ is the mechanical energy difference between the injection and production well. As the kinetic energy is usually negligible, the mechanical energy difference $\Delta\Phi \approx \Delta P$, i.e., approximately equals the pressure difference for horizontal flow. Following Muskat [169], pp 177, we can estimate that the pumping power P required to circulate the oil at a rate $\chi\,[\text{m}^3/\text{s}]$ from injection to production well is given by

$$P = 3\chi \left(P_{wi} - P_{wp}\right) = \frac{3\chi^2 \left(\mu_{inj} + \mu_{prod}\right)/2}{2\pi k \zeta} \left(\ln \frac{D_{ip}}{r_w} + s_i + s_p\right), \tag{1.3}$$

where s_i, s_p are the skin factors for the injection and production well, respectively. Note that the power is quadratically dependent on the flow rate χ. The pressures at the injection well and production well with radius $r_w = 0.1$m are denoted by P_{wi}, P_{wp}, respectively. Here it is assumed that the two wells are located in an infinite layer of constant thickness ζ at a distance D_{ip} from each other. For the viscosity, we take its average between the produced oil and injected water viscosity, $\mu_{av} = \left(\mu_w + \mu_{oil}\right)/2$ in [Pa.s]. The skin factors s_i, s_p (see [64], Dake 1978)) describe the additional resistance near the wells. The power needed for pumping is $P = 3\,\chi\Delta P$. The factor of 3 arises to account for the conversion of fossil fuel energy to electricity and the efficiency of the pump [87] (see also Eq. (1.1))

$$\eta_{mp} = \text{Efficiency (methane } to \text{ electricity } to \text{ pump)} \approx 1/3. \tag{1.4}$$

Consider one m^3 of water at 80°C to release its heat toward 40°C. Its energy content amounts to $40[K] \times 4.184 \times 10^6 \text{J/m}^3/\text{K} = 167.3\,\text{MJ}$, while its exergy content amounts to $40[\text{K}] \times 4.184 \times 10^6 \text{J/m}^3/\text{K} \times CF(0.11) \approx 18.4\,\text{MJ}$. Here CF = 0.11 denotes the Carnot factor. The circulation energy to produce the one m^3 against 100 bar is $10^7/0.3 \approx 33.3\,\text{MJ}$, where the factor 0.3 (see Eq. (1.4)) represents the efficiency for conversion of fossil fuel to electricity and then to pumping energy. In this case, geothermal energy leads to a net loss of energy unless one argues that the CF should be ignored as geothermal energy is only used for heating purposes. However, it is possible to use a heat pump with a coefficient of performance well above one [105]. Indeed, for a heat pump, a COP value of 4 means that the addition of 1 kW of electric energy is needed to have a release of 4 kW of heat at the condenser.

Energy for tubing/casing We assume steel casing and tubing pipes and a depth of 2.333 km. The weight of the casing [14], is typically 30 lbs/ft = 44.7 kg/m = 44.7 ton/km. For the tubing, it would be 6.5 lbs/ft = 9.7 kg/m = 9.7 ton/km.

The density of concrete is typically 2,400 kg per cubic meter, while the density of steel is 8,050 kg per cubic meter. Cement costs are typically 1.5 GJ/ton [74] as opposed to 35.8 GJ/ton for steel. This amount must be divided by the lifetime of say a geothermal project (30 years) = 9.47e8 s. Exergy of casing and tubing = (44.7+9.7) [ton/km] ×depth[2.333 km] × 2[wells] × 35.8[GJ/tonsteel] lifetime = 9087.12/lifetime = 9087.12 GJ/9.47 e8 s) = 9598.5 W. Consequently, the costs of concrete and steel can be disregarded with respect to the energy produced in a geothermal doublet.

A geothermal reservoir produces say 100 m^3 hot water /hr of 80°C, which can be cooled down to 40°C. This corresponds to $100[\text{m}^3/\text{h}] \times 4.184e6[\text{J/K/m}^3] \times 40[\text{K}]/3600[\text{s/h}] = 4.65E6[\text{W}]$

Drilling costs it can be expected that the energy of drilling, in order of magnitude, corresponds to the crushing energy (third theory Bond equation), [33, 34, 167] of the formation,

$$W = W_i \left(\frac{10}{\sqrt{P(mm)}} - \frac{10}{\sqrt{F(mm)}} \right), \tag{1.5}$$

where W_i is of the order of 50 MJ/ton and where P is the 80% passing size for the product and F is the 80% passing size for the feed. The size of drill cuttings is of the order of $P = 100\,\mu m - 10\,mm$, which would typically make the crushing energy

$$W_V = W\rho_s \approx 1581[\text{MJ/ton}] \times 2.35[\text{ton/m}^3] \approx 3,700\text{MJ/m}^3 \tag{1.6}$$

for chips of 0.1 mm, using that $\sqrt{F} >> \sqrt{P}$ and assuming a rock density of $\rho_s = 2350$ kg/m$^3 = 2.35$ ton/m^3. We note that $50[\text{MJ/m}] \times 10/\sqrt{(0.1)} = 1,581$ MJ/ton.

The crushing energy Ξ can be used to estimate the drilling energy [11, 14, 150, 158, 173, 196, 226]. For a drilled hole with a radius of $r_w = 0.1$ m, the drilling energy would be

$$\Xi = W_V \Delta\psi = 116\text{MJ/m} \quad (cf. \quad 100\text{MJ/m}]), \tag{1.7}$$

where $\Delta\psi = \pi r_w^2 \, [\text{m}^3/\text{m}]$.

We assume the same efficiency reduction factors as we used for the circulation costs, i.e., conversion of fossil fuel to pumping power goes at an efficiency of $\eta_{elec} \times \eta_{driver} \times \eta_{pump} = 0.45 \times 0.9 \times 0.8 \approx 0.325$ (see Eq. (1.4)). All the same, drilling costs are negligible with respect to the produced oil energy or produced geothermal energy. Indeed, drilling costs are of the order of 100 MJ/m, (estimated from applied torque). Lifetime cumulative oil productions of wells are of the order of 50,000–100,000 bbl/well [39]. About 17% ([182], pp 49) of the produced fossil fuel energy is a ballpark percentage used for recovery, transportation, and processing, showing that recovery costs are not negligible. Assuming a combustion energy of 10.7 kWh /L [159], hundred thousand barrel corresponds to $0.17[-] \times 10.7[\text{kWh/L}] \times 10^5[bbl] \times [159l/bbl] \times 3.6 \times 10^6[\text{J/kWh}] = 10^5$ GJ. Drilling one hole to a depth of 2,333 m costs 2333 m $\times$ 100 MJ/m depth $\approx$ 233 GJ, being much smaller than energy required for recovery, transportation, and processing. Indeed, the drilling costs are approximately 233/1e5 $\approx$ 0.2% of the recovered hydrocarbon energy.

Let us assume that the lifetime cumulative water production of low enthalpy geothermal energy for a reservoir of choice is 26 million m^3. The produced energy of 26×10^6 m^3 of water, i.e., over the lifetime of the well would be 26×10^6 m$^3 \times 40[\text{K}] \times 4.183 \times 10^6[\text{J/m}^3/\text{K}] \times CF(0.11) = 4.78 \times 10^5$GJ, where CF is the Carnot factor ($= (80-40)/353 = 0.11$). The drilling costs for drilling two holes to a depth of 2333 [m], i.e., 2×2333 m $\times$ 100 MJ/m amounts to 466 GJ/0.3 = 1550 GJ, where 0.3 approximates the efficiency factor. Indeed, the drilling costs must be divided by 0.3 to account for the efficiency (See Eq. (1.4)). Therefore, the drilling costs are less than 0.3% of the recovered geothermal energy (Table 1.1).

In practice, drilling costs [104] in dollar or euro's are larger because ecological [220] aspects and specialist knowledge of drilling and petroleum engineers are required. The arguments of Granowski [104] of the much larger exergetic costs have been nowhere adopted by us.

Sequestration costs In addition, we need to add the sequestration costs of carbon dioxide and the separation costs of carbon dioxide from nitrogen. Capture costs of CO_2 from the combustion mixture with mono-ethanol-amine and sequestration costs (Eftekhari et al. (2012) and Oyenekan and Gary (2006)) are between 4.3 and 8.3 MJ/kg-CO_2 = 11.83–22.83 MJ/ kg-CH_4, subtracting 11.83/50–22.83/50 = 24%–46% from the LHV (50 MJ/kg) of methane. For

Table 1.1

Input Data Geothermal Reservoirs

Quantity	Value	Symbol	Unit
Well distance	1000	A	m
Layer thickness	25	ζ	m
Production rate	100	Q	m^3/h
Depth	2333	H	m
Permeability	0.987×10^{-13}	k	m^2
Geothermal gradient	31.3	Tz	°C/km
Viscosity injection water	1.14×10^{-3}	$\mu_{inj}(T=15°C)$	Pa.s
Viscosity production water	0.35×10^{-3}	$\mu_{prod}(T=80°C)$	Pa.s
Heat capacity of water	4.184×10^6	C_w	$J/m^3/K$
Heat capacity layer and bounding rocks	2.3×10^6	C_{res}	$J/m^3/K$
Thermal conductivity	2.0	κ	W/m/K
Reservoir temperature	353	T_{res}	K
Injection temperature	288	T_{inj}	K
Well radius	0.1	r_w	m

the average capture + storage costs, we will assume that it is 5 MJ/kg-CO_2 [87]. Indeed 16 g of methane will produce about 44 g of CO_2. We will disregard the 17% costs, mentioned above, for recovery, transportation and processing of the methane combined with separation and sequestration of CO_2. Following the example with methane with a lower heating value (LHV) of (50.0 MJ/kg) producing 44/16 = 2.75 kg CO_2, captured and stored at 5 MJ/kg), tells us that 27.5% of the methane energy needs to be used for a single storage step. We have to store also the carbon dioxide resulting from compensating for the 27.5% exergy for capture and storage. This means in the cumulative sense that we have to multiply by $\Sigma(1+0.275+0.0756+\cdots) = 1/(1-0.275) = 1.38$ to account for the capture and storage of CO_2 when methane is used. This means that 38% of the produced fossil energy will be used for storage in the case of methane combustion. For oil, i.e., we use the homologous series of alkanes as model, where the additional (CH_2) has a LHV of 43.5 (MJ/kg) producing 1 kg $\times M_{CO_2}/M_{CH_2}$ 44/14 = 3.14 kg CO_2 to be stored at 5 MJ/kg; therefore, $3.14 \times 5[MJ/kg]/43.5[MJ/kg] = 36.092\%$ of the oil energy accounts for a single storage step. Again, we have to compensate the energy loss for capture and storage, and we have to multiply by $\Sigma(1+0.36092 + 0.130263 + \ldots) = 1/(1-0.36092) = 1.5647$ to account for CO_2 storage when oil is used, i.e., 56.47% of the produced fossil energy will be used for storage. For benzene $(CH)_n$ with n = 6 as model for coal with a lower heating value of 40.17 MJ/kg, we need producing (molar weight of CO_2/ molar weight of CH) = 44/13 = 3.38 kg CO_2/ kg-coal to be captured and stored at 5 MJ/kg-CO_2 [87]; now $3.38[-] \times 5[MJ/kg]/40.17[MJ/kg] = 42.13\%$ of the coal energy accounts for a single storage step. Again, we have to compensate the energy loss for storage and we have to multiply by $\Sigma(1+0.4213+0.1775+\ldots) = 1/(1-0.4213) = 1.728$ to account for CO_2 storage when coal is used, i.e., 72.8% of the produced fossil energy will be used for capture and storage (Table 1.2).

Energy for cutting removal We illustrate this aspect by an example. A hole with a diameter of twice the well radius $d = 2r_w$ of 0.2 m has a volume of $V_{spec} = \pi d^2/4 = 0.031$ m^3/m. The drilling energy per unit length (see also Eq. (1.6)), would be $E_{dr} = W_V V_{spec} = 3,700 MJ/m^3 \times 0.031 \approx 114.7 MJ/m$. The drilling hole has a depth of $H = 2,333$ m. The rock density ρ_s is 2,350 [kg/m^3]. The mass of cuttings would correspond to $m = \pi(d^2/4)H[m^3]\rho_s[kg/m^3] \approx 1.722e5$ kg or a volume of $\xi = \pi d^2 H/4 \approx 73.3 m^3$. This corresponds to a potential energy

Table 1.2

Footprint of Carbon Dioxide MJ/kg-fuel

Energy per mass of fuel E_k	f_p = kg carbon dioxide per MJ exergy (energy) [kg/MJ]
50 [MJ/kg-CH4]	2.75 kg CO_2/ 50 MJ methane = 55.0 g CO_2/ MJ methane
43.5 [MJ/kg-CH2]	3.14 kg CO_2/ 43.5 MJ oil = 72.1. g CO_2/ MJ oil
40.17 [MJ/kg-CH]	3.38 kg CO_2/ 40.17 MJ coal = 84.1 g CO_2/ MJ coal
167.3 [MJ/m^3 (ton)-water]	5.5 kg CO_2/167.3 MJ water energy = 3.29 g CO_2/ MJ water
18.4 [MJ/m^3 (ton)-water]	5.5 kg CO_2/18.4 [MJ/m^3-water] = 29.9 g CO_2/ MJ water(exergy)

(assuming that the average position of a particle is in the middle of the height of a well) of $\rho_s \xi g H /2 \approx 1.971$ GJ or about 1.971GJ/2,333 = 0.844 MJ/m. A typical load of cuttings in the drilling mud is $\eta = 4.5$ v/v% (3-6 v/v%) [Wikipedia]. This means that we need to circulate of the order of $\xi/\eta = 1,628$ m^3 drilling fluid. Assuming a viscous pressure drop of 100 bar, this would correspond to $(1628 \text{m}^3 \times 100[bar] \times 10^5 [\text{Pa/bar}]/2,333[\text{m}] = 16 \text{GJ}/2,333[\text{m}] \approx 6.98$ MJ/m. This means that the hole cleaning would account for 6.08% of the drilling energy, i.e., 6.98 MJ/m/114.7 MJ/m.

Conversion from financial costs to exergetic costs It is advisable to calculate exergetic costs directly from exergy balances. This, however, is not always possible due to lack of data [74]. For the European Union, inclusive electricity prices for household consumers - bi-annual data (from 2007 onward) ec.europa.eu/eurostat/statistics-explained/index.php/Electricity-price-statistics, 0.1267 €/kWh + 0.0883 €/kWh (estimated from a bar graph, TAX + VAT) = 0.215 €/kWh = 59.7 €/GJ. Therefore the conversion factor from GJ/€ is

$$0.215 €/\text{kWh} = 59.7 €/\text{GJ} \tag{1.8}$$

By way of example, we can show that this number (59.7 €/ GJ) in Eq. (1.8) is very useful for order of magnitude estimates of exergy requirements by estimating the cost of a car. Embodied energy of a car [159] is 76000 kWh = $76,000 \times 3.6e6 = 273.6$GJ $\approx 16,334$ €. This is indeed the order of magnitude of the price of a car. Lower numbers are quoted, but they usually assume that the car consists of recycled material. Use that one liter of hydrocarbons corresponds approximately to 10.7 kWh. Fuel costs are typically 0.8 kWh/km (13.4 km/L) [159]. A car driving a total of 200,000 km, spends for fuel 160,000 kWh. Therefore, roughly one-third of the exergy costs of a car is due to manufacturing and two-thirds due to fuel costs. This example emphasizes the importance of accounting for materials energy.

As already indicated above, there is a large margin of uncertainty in the amount of exergy that corresponds to one €, i.e., 1/59.7 GJ/€, but it can be used to get an order of magnitude indication of the exergetic costs. If they are much smaller than other costs, it is fair to disregard their contribution in exergetic calculations. If they are far larger, then it is useful to find the reason for the discrepancy, e.g., expensive specialist knowledge is required as in drilling.

Labor and investment costs Ferroni and Hopkirk [92] mention the exergy costs of labor and investment costs. However, in industrial processes, "physical work" is, per se, numerically negligible with respect to the much higher power of the machines and processes that this work commands [208]).

Other costs other costs in terms of Joules are usually relatively small. More accurate calculations are often a waste of effort.

1.4.1 EXERCISE ERoEI

About an Exergy Return on Exergy Invested (ERoEI) analysis.

A field segment is producing oil between an injection well and a producing well according to

$$\chi = \chi_0 \exp(-t/\ ), \tag{1.9}$$

where $\chi_0 = 1000$ bbl / day and is two years. The pressure drop stays constant at 100 bar, and the total fluid production (water + oil) also stays constant at 1000 bbl/day. Tubing and casing have each a weight of 100 kg/m, with a steel energy cost of 32 GJ/ton. The reservoir is at a depth of 3000m. Energy costs of concrete can be neglected. Drilling costs are 100 MJ/m for each well. Efficiency from fuel costs to electricity is one-third (See Eq. (1.4)). The produced oil has a lower heating value of 10.7 kWh/liter [159].

1. Calculate the so-called zero time t_{zero}, i.e., the time when the fossil energy produced equals the effective circulation energy $3\chi_0\Delta P$ (see Eq. (1.4)).
2. what is the initial energy production?
3. what are the drilling costs [14, 158, 196] divided by t_{zero}?
4. disregarding the costs for concrete what are the well costs divided by t_{zero}?
5. what is the energy for the cutting removal divided by $t_{zero} \neq 0$?
6. what is the percentage of produced fossil fuel energy spent on wells, drilling, circulation and cutting removal?

1.4.2 ANTHROPOGENIC EMISSIONS VERSUS NATURAL SEQUESTRATION

Worldwide Emissions of Carbon Dioxide

Worldwide emissions of carbon dioxide are around 39.18 Gton/year in 2014 and possibly 51 Gton in 2020 [95]. The total mass of gases in the atmosphere is 5.1e6 Gton ≈ 1.013 [kg/cm^2] × Earth surface [5.1 e18 cm^2 = 5.1 e8 km^2] . The CO_2 mass increase is (Figure 1.7) 2.56×10^{-6} [molwt$_{CO_2}$/molwt$_{air}$(44/28.96] × 5.10e6 Gton = 13.16 Gton, meaning that $39.18 - 13.16 = 26.02$ Gton/year is sequestered naturally. Of this amount [162] 7.3 Gton/year [246] is stored in the ocean. Therefore, the remaining 18.7 Gton/year is stored on land. Per unit surface area storage on land is more effective than storage in oceans.

This indicates that enhancing natural sequestration or removal from the atmosphere may help significantly (Workman et al. (2011)) reduce the carbon dioxide concentration in the atmosphere.

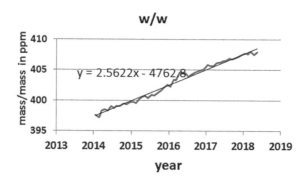

Figure 1.7 Weight fraction concentration of carbon dioxide in atmosphere.

However, removing from the atmosphere will be costly due to the upgrading of low concentrations. Transfer rates $\phi"_A$ are determined by the square root of the diffusion coefficient D times the first-order reaction rate k of carbon dioxide with the dissolved minerals [22] pp. 273, Eq. IV-57 × the surface concentration, [179].

$$\phi"_A = \sqrt{kD}c_A \qquad (1.10)$$

The surface concentration c_A becomes less [9, 10] with increasing ionic strength in the aqueous phase due to the dissolved minerals. Indeed the solubility of carbon dioxide becomes less as salt dissolves in the water. In addition, the dissolved minerals increase the viscosity and thus decrease the diffusion coefficient. The reaction rate k is with aluminosilicate transported by rivers.

Energy Use in an Air Plane

An air plane uses 53 ± 16 kWh $\approx 191 \pm 47$ MJ per 100 passenger kilometer [159]. As explained above, 43.5 MJ of gasoline produces 3.14 kg of carbon dioxide or 191 MJ produces 3.14 [kg] × 191 / 43.5 = 13.8 kg CO_2/ 100 passenger km. A round trip to Brazil (20,000 km) produces 13.8 × 200 = 2.76 ton carbon dioxide per passenger. This corresponds to 2.76 [ton] / 6.6 [ton/ha] = 0.42 ha of yearly crop growth. As indicated below crops can store 660 ± 440 ton -CO_2 /km^2/year = 6.6 ± 4.4 ton -CO_2 /ha / year. Multiplied by the earth land surface ($0.3 \times 5.1e8$km^2 × 550 = 84.15 Gton/year, meaning that part of the land area does not naturally sequester as much CO_2 as the average crop.

Use of Bioenergy to Relieve the Carbon Footprint

The following is only added to get an impression of order of magnitude estimates of bio-carbon storage. Bioenergy (grass) crops have the potential to store 1631 Tg-C/year = 1.631 Gton-C/year ≈ 6.0 Gton CO_2 /year worldwide. Storage capacities range [130] from 4.5 ton C/km^2/year to 40 ton C/km^2/year in temperate grassland soils. In the Netherlands (40,000 km^2), this would correspond to 0.18–1.6 Mton/year. Typical sequestration rates, in tonnes of carbon per hectare per year, are: 0.8–2.4 tonnes in boreal forests, 0.7–7.5 tonnes in temperate regions and 3.2–10 tonnes in the tropics [41, 42].

This is about half of the amount of naturally sequestered carbon dioxide, i.e., 13.15 Gton CO_2 /year.

Atmospheric storage [123] is outside the scope of this monograph. Crops can store at a rate of 0.6–3.0 ton-C/ha/yr or as quoted by Yang [254], i.e., 0.6 ton-C/ha/year. This is equivalent to 0.6 ton ×100 × 44/12 = 220 ton-CO_2/km^2/year. For the entire area in the Netherlands 40,000 km^2, this would correspond to 8.8 Mton CO_2/year. This would be only a fraction of Dutch greenhouse gas emission, i.e., The Netherlands emits 165.317 Mton CO_2 / year ($\approx 0.17Gt$/ year). The sequestration in "city forests" is of the order of 277 ton-C/km^2/year, [161]. If the Netherlands (40,000 km^2) were completely covered with forests, it could sequester about 10 Mton-C/year ≈ 36.7 Mton—CO_2 per year. The transfer capacity is 390 ton C km^{-2} year^{-1} on high activity mineral soils, but only 190 ton C km^{-2} year^{-1} on sandy soils ([217]).

1.4.3   EXERCISE: TREES TO COMPENSATE FOR INTERCONTINENTAL FLIGHTS

About the possibility of planting trees to compensate for the emission of carbon dioxide during intercontinentights

Given that an airplane uses 53 ± 16 kWh $\approx 191 \pm 47$ MJ per 100 passenger kilometer [159] and that the distance from Amsterdam to Rio de Janeiro and back is about $2 \times 10,000 = 20,000$

km, discuss the possibility of tree plant in Rio de Janeiro or in Amsterdam (0.8–2.4 tonnes in boreal forests, 0.7–7.5 tonnes in temperate regions and 3.2–10 tonnes in the tropics [41]) to compensate for the CO_2 emission from the intercontinental flights.

Mineral Storage

Mineral storage capacity in wollastonite ($CaSiO_3$) is 0.17 kg-CO_2/ (kg wollastonite) [186]. World reserves of wollastonite (see Wikipedia) are estimated 100 million ton and is thus unable to sequester significant amounts of carbon dioxide with respect to its yearly increase ($\sim$13.2 Gton/year CO_2).

There is sufficient olivine to store the yearly CO_2 mass increase of 13.2 Gton/year.

2 One-Phase Flow

OBJECTIVE OF THIS CHAPTER

To enable the course participant, to understand Darcy's law in porous media inclusive its limitations:

- To distinguish between the volumetric flux (u) [m/s], the interstitial velocity (v) [m/s] and the true velocity $v \times$ tortuosity as shown in Figure 2.1.
- understand that the Darcy velocity, in general, cannot be described as a gradient of a potential but $\mathbf{u} = -k(x,y,z)/\mu(x,y,z)(\mathbf{grad}\,p + \rho(x,y,z)g\mathbf{e_z})$,
- to understand when Darcy's law is not valid and how the amended Darcy's law can be written in case of inertia, slip and anisotropy,
- to enable understanding the Carman-Kozeny equation, which indicates that the permeability is quadratically dependent on grain size and strongly dependent on the porosity $\varphi^3/(1-\varphi)^2$,
- to enable understanding upscaling methods for permeability,
- to generate uncorrelated stochastic random fields with EXCEL,
- to enable understanding the notions R(epresentative) E(lementary) V(olume), and P(eriodic) U(nit) C(ell),
- enable understanding of arithmetic, harmonic, geometric averaging,
- understanding the limitations of the vertically (z)-averaged problem in two space dimensions,
- understanding the effective medium approximation,
- to derive the result of homogenization as averaged permeability of a subdomain using periodic boundary conditions,
- to numerically calculate flow velocities in a heterogeneous permeability field, for instance with EXCEL,
- to appreciate that a correctly averaged permeability can lead to erroneous forecasts of production of miscible and immiscible displacement descriptions.

INTRODUCTION

Single-phase flow in porous media sets the fundamental basis to understand a subset of aspects of porous media flow that plays an important part in all its applications. We confine our interest to incompressible flow. The most important building blocks are the mass conservation (Section 2.1) and Darcy's law (Section 2.2), which relates the flow of fluids to the applied pressure gradient and gravity forces. We will combine with the mass balance equation to give some examples that allow an analytical solution (see Section 2.3). We will also discuss the Carman-Kozeny relation as it is one of the most powerful estimators of the permeability of unconsolidated sand packs. We will address two aspects, viz., (a) the conditions of validity of Darcy's law and the modified versions (Section 2.4) that can be used in case that Darcy's law is not valid and (b) methods of upscaling of single-phase permeability from the core scale to reservoir scale. In Section 2.5, we will discuss some methods to generate very simple heterogeneity fields.

Ad (a): The flow velocity can be expressed in terms of the gradient of the potential $(p + \rho g z)$ that incorporates both gravity forces and the pressure, only when the density is not position-dependent. It is, therefore, necessary to reformulate the law for position-dependent densities.

Moreover, it is necessary to adapt the law for high flow rates or gas flow in tight media. In layered media, the law must be extended to incorporate a permeability tensor, which allows flow to occur in an oblique direction with respect to the applied potential gradient, i.e., if the layering assumes an arbitrary angle with respect to a Cartesian axis system of reference. Both mass conservation and Darcy's law assume from the outset the existence of a Representative Elementary Volume (REV), which allows the meaningful definition of averaged properties in a volume that is small with respect to the system length. The concept of an REV is clear for a quantity like porosity, but for permeability, unlike porosity, not only the permeability values but also the distribution of the permeability heterogeneity determines the average value.

Ad (b): In the early twentieth century, upscaling (Section 2.6) was mainly performed by using various averages, e.g., the arithmetic average, the geometric average and the harmonic average. The average permeability value for flow perpendicular to the layer direction requires a harmonic average, which is lower than the average for flow parallel to the layer direction, which requires an arithmetic average. These averages set the so-called Wiener bounds [250], meaning that no average can be lower than the harmonic average or higher than the arithmetic average. In statistically homogeneous media, these averages are even confined by the more restrictive Hashin-Shtrikman bounds [113]. In almost all cases, it is necessary to upscale the permeabilities, meaning that a meaningful average must be determined from a distribution of permeabilities (see Section 2.7). For educational purposes, we will use EXCEL (Section 2.A) to solve the numerical equations and to obtain average permeabilities from an arbitrary arrangement of permeabilities. The harmonic, arithmetic and geometric average was popular upscaling methods in the pre-computer era because they require simple calculations that can be performed on a hand calculator. It is tempting to use arithmetic averaging perpendicular to the layer direction to convert a 3-D flow problem to a 2-D flow problem (see Subsection 2.6.2). However, this simplification procedure is only valid under some conditions. In the nineties, more sophisticated upscaling procedures started to enjoy widespread use. There is the effective medium approximation (see Appendix 2.C), which can be handled analytically. The effective medium approximation, in spite of its simplicity, shows the fundamental concepts of heterogeneous flow. Other sophisticated upscaling methods, such as volume averaging or homogenization (see Appendix 2.D), lead to numerical procedures that allow to determine the upscaled averages. For a statistically homogeneous heterogeneity structure, it is possible to show that the averages do not depend on the boundary conditions, i.e., Dirichlet, Neumann or periodic, provided that separation of scales is possible [35].

2.1   MASS CONSERVATION

All isothermal flow problems can be described using the mass conservation equation and the momentum balance equation. The momentum balance equation in porous media is Darcy's law or some modification of it and will be described below. The mass conservation equation states that the accumulation of mass $\frac{dm}{dt}$ equals the net inflow of mass across the boundaries

$$\frac{dm}{dt} = - \oint_S \rho \mathbf{u}.\mathbf{n}dS. \tag{2.1}$$

In Eq. (2.1), $\mathbf{u}$ denotes the specific discharge, also called Darcy's velocity, which denotes the volumetric flux [m^3/ m^2/s]. For stationary situations, which we mainly address in this chapter, we take $dm/dt = 0$. The application of the divergence theorem [93], Chapter 2, (Integral theorem of Gauss) leads to

$$\oint_S \rho \mathbf{u}.\mathbf{n}dS = \int_V \mathbf{div}\,(\rho\,\mathbf{u})\,dV = 0. \tag{2.2}$$

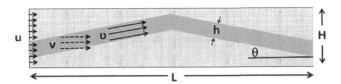

Figure 2.1 Volumetric flux (Darcy velocity), interstitial velocity and true velocity for creeping Stokes flow.

Eq. (2.4) holds for a representative elementary volume V. If $\mathbf{u}$ is defined as average over this representative elementary volume we may write

$$\mathbf{div}\,(\rho\,\mathbf{u}) = 0. \tag{2.3}$$

The mass balance equation is also derived by averaging over a volume that is much larger than the volume shown in Figure 2.1, i.e.,

$$\oint_S \rho\mathbf{u}.\mathbf{n}dS = \int_V \mathbf{div}\,(\rho\,\mathbf{u})\,dV = 0. \tag{2.4}$$

Eq. (2.4) holds for a representative elementary volume V. If $\mathbf{u}$ is defined as average over this representative elementary volume, we may write

$$\mathbf{div}\,(\rho\,\mathbf{u}) = 0. \tag{2.5}$$

2.2 DARCY'S LAW OF FLOW IN POROUS MEDIA

Darcy conducted experiments in a setup similar to the one shown in Figure 2.2. The drawing is based on a similar drawing by [18]. A tube filled with sand of length L and with cross-section A is connected to two vessels. We take an arbitrary datum level, which is drawn here below the cell. The water level in measurement tubes with respect to the datum level is called the total head. As the pressure in the measurement tube is hydrostatic, we find that at the upstream boundary of the sand pack and at elevation z_1, the pressure p_1 divided by the product of the density ρ and the acceleration due to gravity g, i.e., $p_1/\rho g$ is equal to the height of the water level in the upstream measurement tube with respect to z_1. In the same way, we observe that $p_2/\rho g$ is equal to the height of the water level in the upstream measurement tube with respect to z_2. Therefore, we can define the total head upstream as $h_1 = p_1/\rho g + z_1$ and the total head downstream as $h_2 = p_2/\rho g + z_2$.

   Darcy found that the total flow rate Q [m³/s] is proportional to difference between the total heads $h_2 - h_1$ divided by the length L [m] of the sand pack and proportional to the cross-section A [m²]. The proportionality constant K is called the hydraulic conductivity and has units [m/s]. We obtain

$$Q = KA\frac{h_1 - h_2}{L} = KA\frac{(p_1/(\rho g) + z_1) - (p_2/(\rho g) + z_2)}{L}, \tag{2.6}$$

where p is the pressure, ρ the density of the fluid and g the acceleration due to gravity.

2.2.1 DEFINITIONS USED IN HYDROLOGY AND PETROLEUM ENGINEERING

It has always been useful for a petroleum engineer to understand the hydrology literature. In the same way, it is useful for the hydrologist to understand the petroleum engineering literature. Agricultural engineers use again a different notation, where the flow potential [163] is defined

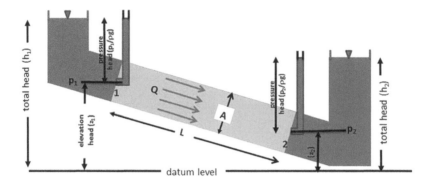

Figure 2.2   Definition of total head, elevation head, pressure head.

Table 2.1
Petroleum (right) and Hydrology (left) Nomenclature and Symbols

h	piezometric head (total head)	[m]	$\Leftrightarrow$	Φ	potential	[Pa]
$\frac{h_2-h_1}{L}$	hydraulic gradient	[−]	$\Leftrightarrow$	$\Phi_2 - \Phi_1$	potential gradient	[Pa/m]
$p/(\rho g)$	pressure head	[m]	$\Leftrightarrow$	p	pressure	[Pa]
$v^2/(2g)$	velocity head	[m]	$\Leftrightarrow$	$\frac{1}{2}\rho v^2$	kinetic energy	$\left[\mathrm{J}/\mathrm{m}^3\right]$
z	elevation head	[m]	$\Leftrightarrow$	$\rho g z$	potential energy	$\left[\mathrm{J}/\mathrm{m}^3\right]$
Q	total discharge	$\left[\mathrm{m}^3/s\right]$	$\Leftrightarrow$	Q	flow rate	$\left[\mathrm{m}^3/s\right]$
$q = Q/A$	specific discharge	$\left[\mathrm{m}^3/\mathrm{m}^2/s\right]$	$\Leftrightarrow$	$u = Q/A$	Darcy velocity	$\left[\mathrm{m}^3/\mathrm{m}^2/s\right]$
$v = q/n$	pore velocity	[m/s]	$\Leftrightarrow$	$v = u/\varphi$	pore velocity	[m/s]
n	porosity	[−]	$\Leftrightarrow$	φ	porosity	[−]

by $\psi = hg$ and the conductivity $\mathcal{K}$ in units of s (seconds) as $Q[\mathrm{m}^3/\mathrm{s}] = -\mathcal{K}[\mathrm{s}]A[\mathrm{m}^2]\,(\partial_l \psi)\,[\mathrm{m}/\mathrm{s}^2]$ but we leave this as being outside the interest of this book. Indeed, the three disciplines use completely different symbols, forms of equations and nomenclature, but agricultural engineers frequently also adopt the hydrology nomenclature. In many cases, Petroleum Engineers and Geo-hydrologists try to solve the same problems, which are hidden behind different nomenclatures. Table 2.1 summarizes some nomenclatures used in hydrology (left) and the equivalent nomenclature used in petroleum engineering (right).

For constant density, the formulations are completely equivalent. The formulation used by Petroleum Engineers can be straightforwardly extended if the density is position-dependent. The formulation used by Hydrologists, in terms of total heads, pressure heads and elevation heads cannot be easily extended to a form where the density is position-dependent. Indeed, if we start with Eq. (2.6) and use that ρg = constant we find

$$q = \frac{-K}{\rho g}\frac{\mathrm{d}\,(p+\rho g z)}{\mathrm{d}l} . \tag{2.7}$$

We choose the z-coordinate as pointing upward, i.e., in the opposite direction as the acceleration due to gravity. Petroleum engineers need to deal with fluids of different viscosities (oil, gas and water), and thus the total discharge Q is inversely proportional to the viscosity μ [Pa s].

Table 2.2
Typical Values of Hydraulic Conductivity or Permeabilities

Type	Hydraulic conductivity [m/s]	permeability [*Darcy*]
Gravel	$3 \times 10^{-4} - 3 \times 10^{-2}$	$30 - 3000$ *Darcy*
Coarse sand	$9 \times 10^{-4} - 6 \times 10^{-3}$	$90 - 600$ *Darcy*
Medium sand	$9 \times 10^{-7} - 5 \times 10^{-4}$	90 mD $- 50$ *Darcy*
Fine sand	$2 \times 10^{-7} - 3 \times 10^{-4}$	20 mD $- 30$ *Darcy*
Clay	$1 \times 10^{-11} - 4.7 \times 10^{-9}$	1μ Darcy $- 0.47$ mD
Sandstone	$3 \times 10^{-10} - 6 \times 10^{-6}$	30μ Darcy $- 600$ mD
shale	$1 \times 10^{-13} - 2 \times 10^{-9}$	10 nD - 0.2 mD

It turns out that the resistance force in the direction of flow described by the elongational viscosity is also proportional to the shear viscosity μ [93]. Therefore, they replace the hydraulic conductivity with its components, i.e., $K = k\rho g/\mu$. Here the permeability k is expressed in [m^2], where we can use that one Darcy is [$0.987 \ 10^{-12}$ m^2]. The viscosity μ of water is 10^{-3} [Pa s] = 1 cP, roughly the value at 20°C, whereas ρ is the density, which for water is approximately 1000 [kg/m^3]. Finally, the acceleration due to gravity g is 9.81 [m/s^2]. Therefore, Darcy's law for petroleum Engineers at constant density reads

$$u = \frac{-k}{\mu} \frac{\mathrm{d}(p + \rho gz)}{\mathrm{d}l} \ , \tag{2.8}$$

In three dimensions, this equation can be written as

$$\mathbf{u} = -\frac{k}{\mu}\mathbf{grad}\,(p \pm \rho gz), \tag{2.9}$$

The plus sign is used when z is pointing vertically upward. The minus sign is used when z is pointing vertically downward. Typical values for the hydraulic conductivity are shown in Table 2.2 below.

To convert from the hydraulic conductivity in [m/s] to permeability in Darcy multiply by 1.04×10^5.

2.2.2 EXERCISE, EXCEL NAMING

About naming variables in EXCEL.

Plot the downward Darcy velocity in a vertical column of water that is subject to only gravity forces as a function of the hydraulic conductivity. The pressure at both the inlet and the outlet is atmospheric (Figure 2.3). Indicate in the plot the various lithologies given in Table 2.2. Use an EXCEL spreadsheet for the calculations and use the procedure in Figure 2.3 to name the relevant parameters (name indicated in the top left corner)[1]

2.2.3 EMPIRICAL RELATIONS FOR PERMEABILITY (CARMAN–KOZENY EQUATION)

The general form of all these relations is $k =$ shape factor $\times$ porosity factor $(f(\varphi)) \times$ square of grain size diameter. The relation that is most often used is the Carman-Kozeny relation [27]. It is

---

[1] New versions of EXCEL may have slightly different syntax.

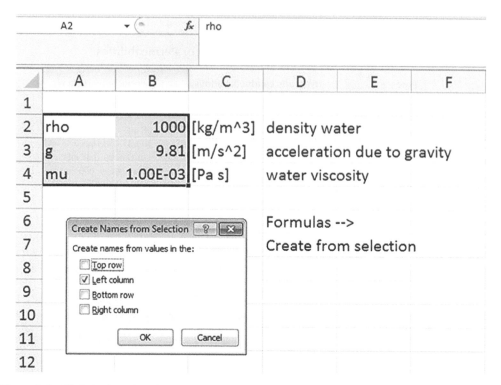

Figure 2.3 Click the formulas tab. First, highlight the cells containing the values at the right and the names at the left. In the formulas tab, choose "create from selection" and indicate that you want to create the names from the value in the left column. After clicking "OK", cell B2 is named "rho", B3 is named "g" and B3 is named "mu". The names of the cells are now indicated in the top left row, where now the name of cell A2 is indicated. Instead of indicating cells by row and column numbers, we can now write for instance $= \rho * g / \mu$.

based on "some kind of derivation", which involves flow in tubes. Poiseuille law for the average velocity v in a cylindrical tube of radius R is described by

$$v = \frac{R^2}{8\mu} \frac{\phi_o - \phi_L}{L} = \frac{\phi_o - \phi_L}{2\mu L} R_h^2. \qquad (2.10)$$

where the hydraulic radius $R_h = \frac{\pi R^2}{2\pi R} = R/2 =$ cross-section/ wetted perimeter. The hydraulic radius is equal to the cross-section available to flow divided by the wetted perimeter as follows: volume available to flow/total wetted surface $=$ volume available to flow /volume bed/(wetted surface/volume bed) $= \frac{\varphi}{a}$. Here, a is defined as the (wetted surface/volume bed) We obtain

$$R_h = \frac{\varphi}{a}. \qquad (2.11)$$

In a porous medium with spherical grains of radius R, the Volume bed=(grains + void) of one unit cell $= (4\pi/3R^3)/(1 - \varphi) =$ volume grains/(volume grains/volume bed). The surface in one unit cell $= 4\pi R^2$. It thus follows that

$$a = \frac{4\pi R^2 (1 - \varphi)}{4\pi / 3 R^3} = \frac{6(1 - \varphi)}{D_p}. \qquad (2.12)$$

The Darcy velocity is given as the pore velocity times the porosity

$$u = v\varphi = \varphi \frac{\phi_o - \phi_L}{2\mu L} R_h^2 = \frac{\phi_o - \phi_L}{L} \frac{\varphi^3}{(1-\varphi)^2} \frac{D_p^2}{72\mu} . \tag{2.13}$$

However, in a porous medium, we have contractions and the fluid follows tortuous paths, and thus we use an empirical correction of 150/72 to account for the tortuosity. Some texts use 180/72. When using 150/72 the permeability k is given by

$$k = \frac{\varphi^3}{(1-\varphi)^2} \frac{D_p^2}{150\mu} . \tag{2.14}$$

2.3 EXAMPLES THAT HAVE AN ANALYTICAL SOLUTION

2.3.1 ONE DIMENSIONAL FLOW IN A TUBE

Physical Model

Consider a vertical core with a diameter of 6 *cm* as shown Figure 2.4.

The permeability of the top fine sand with at $L/2 \leq z \leq L$ is k_2 and of the bottom $0 \leq z \leq L/2$ coarse sand with length $L/2$ is k_1. The total length $L = L_1 + L_2 = 24$ cm and $L_1 = L_2$. On the top of the sand we have a water column of length $L_w = 12$ cm. In this water column, viscous forces outside the porous medium during flow can be neglected. In other words, all friction forces outside the sand can be disregarded. The pressure at the top of the water column is atmospheric to which we assign a value $p_{top} = 0$. At the bottom of the bottom sand, we have also atmospheric pressure to which we again assign a value $p_{top} = 0$. The bottom of the bottom sand is also our datum level $z = 0$. The viscosity of the water $\mu = 10^{-3}$ Pa s and constant. The density of the water $\rho_w = 1000$ kg/m^3 is also assumed to be constant. The acceleration due to gravity $g = 9.81$ m/s ^{2} The flow rate is denoted as Q.

We assume that Darcy's law of single-phase flow is valid. Furthermore, we assume that the fluids and the porous medium are incompressible. The flow is $1 - D$ in the vertical z−direction. We choose the z−direction in the upward direction.

Model Equations

To obtain the equation in one space dimension, we need to integrate the equation over the cross-section. This is completely analogous to the method with which we obtained the equation in two

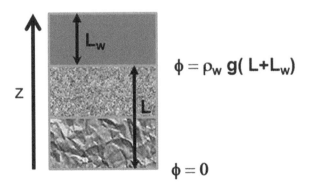

Figure 2.4 flow in layered column.

space dimensions. It is satisfied as long as the permeability and/or porosity are assigned to layers perpendicular to the axial flow direction. The permeability and the porosity are cross-sectional averages over the permeability and the porosity. We omit the averaging bars in the permeability $\overline{k}$ and porosity $\overline{\varphi}$, but we have to understand that the permeability and porosity are cross-sectional averages.

The model equations for a single-phase flow can be derived by substitution of Darcy's Law. We take plus z as pointing vertically upward. We substitute Darcy's law $u = -\frac{k}{\mu}\left(\frac{dp}{dz} + \rho g\right) = -\frac{k}{\mu}\frac{d}{dz}(p + \rho g z) = -\frac{k}{\mu}\frac{d\phi}{dz}$ into the mass balance equation for incompressible flow in $1 - D$, i.e., $\frac{du}{dz} = 0$ and obtain

$$\frac{d}{dz}\left(\frac{k}{\mu}\frac{d\phi}{dz}\right) = 0, \tag{2.15}$$

where we have the boundary condition that at $z = 0$, $p = 0$ and therefore also $\phi = 0$ and at $z = L$ the pressure is $p(z = L) = \rho_w g L_w$ and thus $\phi = \rho_w g L_w + \rho_w g L$.

The model equation for two sands is given in the same way, but we distinguish between two regions, i.e., $0 < z < L_1$ and $L_1 < z < L$. We obtain

$$\begin{aligned}
\frac{d}{dz}\left(\frac{k_1}{\mu}\frac{d\phi}{dz}\right) &= 0 \quad \text{for} \quad 0 < z < L_1, \\
\frac{d}{dz}\left(\frac{k_2}{\mu}\frac{d\phi}{dz}\right) &= 0 \quad \text{for} \quad L_1 < z < L,
\end{aligned} \tag{2.16}$$

where we have the additional boundary conditions that at $z = L_1$ the pressure p is continuous and thus for this case where we have the same densities also the potential ϕ is continuous and the continuity of flux, i.e.,

$$\left(\frac{k_1}{\mu}\frac{d\phi}{dz}\right) = \left(\frac{k_2}{\mu}\frac{d\phi}{dz}\right) \quad \text{at} \quad z = L_1. \tag{2.17}$$

As in each of the domains $0 < z < L_1$ and $L_1 < z < L$, the permeability is different but constant we obtain in both domains

$$\frac{d^2\phi}{dz^2} = 0. \tag{2.18}$$

Analytical Solution

The solution of Eq. (2.18) is linear in z in both domains

$$\begin{aligned}
\phi &= A + Bz \quad \text{for} \quad 0 < z < L_1 \\
\phi &= C + Dz \quad \text{for} \quad L_1 < z < L.
\end{aligned} \tag{2.19}$$

As for $z = 0$ we have that $\phi = 0$ we obtain that $A = 0$. Substitution of the other boundary conditions leads to

$$\begin{aligned}
BL_1 &= C + DL_1, \\
k_1 B &= k_2 D, \\
C + DL &= \rho_w g (L_w + L).
\end{aligned} \tag{2.20}$$

The solution is obtained with Maple, i.e.,

$$D = \rho_w g k_1 \frac{L_w + L}{k_2 L_1 - L_1 k_1 + k_1 L} ,$$

$$B = k_2 \rho_w g \frac{L_w + L}{k_2 L_1 - L_1 k_1 + k_1 L} ,$$

$$C = \frac{\rho_w g (L_w + L) L_1 (k_2 - k_1)}{k_2 L_1 - L_1 k_1 + k_1 L} .$$

$$A = 0. \tag{2.21}$$

2.3.2 EXERCISE, TWO LAYER SAND PACK

About analytical solution of vertical flow in a two-layer sand pack with a water layer on top; illustration of flow against a pressure gradient.[2]

- Substitute the expressions for A, B, C, D into the boundary conditions to show that they are correct,
- use Maple to obtain the solutions (2.21) for B, C, D from Eqs (2.20),
- use EXCEL to make a list of the variables required for the computations,
- plot the potential as a function of the height,
- plot the pressure as a function of the height $\phi = p \pm \rho g z$. What is the correct expression for the potential $\phi = p + \rho g z$ or $\phi = p - \rho g z$, when we define z to be the vertically upward direction?
- can flow "against" the pressure gradient occur? The gradient of which quantity will be a more fundamental driving force for flow?
- how would the equations change if the cylinder is one degree out of its vertical position?

2.3.3 EXERCISE, NUMERICAL MODEL

About a numerical model for a two-layer sand pack with a water layer on top; comparison between numerical and analytical solutions.

- derive the numerical equations we follow [181] (Figure 2.5). We start with equations $\frac{d}{dz}\left(\frac{k}{\mu}\frac{d\phi}{dz}\right) = 0$ and we integrate between z_{i-1} and z_i and use a numerical approximation for the potential gradient, i.e.,

$$\int_{PS}^{PN} \frac{d}{dz}\left(\frac{k}{\mu}\frac{d\phi}{dz}\right) = 0,$$

$$\left(\frac{k}{\mu}\frac{d\phi}{dz}\right)_{PS} = \left(\frac{k}{\mu}\frac{d\phi}{dz}\right)_{PN},$$

$$\left(\frac{k}{\mu}\right)_{PS}\frac{\phi_P - \phi_S}{\Delta z} = \left(\frac{k}{\mu}\right)_{PN}\frac{\phi_N - \phi_S}{\Delta z},$$

$$\phi_P = \frac{\left(\frac{k}{\mu}\right)_{PS}\phi_S + \left(\frac{k}{\mu}\right)_{PN}\phi_N}{\left(\frac{k}{\mu}\right)_{PS} + \left(\frac{k}{\mu}\right)_{PN}}, \tag{2.22}$$

---

[2]The velocity is in the downward direction , whereas the z-coordinate points in the upward direction.

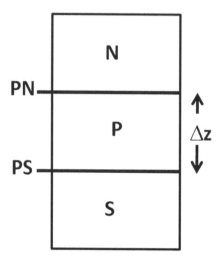

Figure 2.5 Numerical grid. For EXCEL convenience we label the cells going downward with increasing j.

where the central cell P is bounded at the top by cell N and at the bottom by cell S. The subscript PS refers to the interface between cell P and cell S, whereas the subscript PN refers to the interface between cell P and cell N. Furthermore, $\Delta z =$ is the grid cell size. We use an equidistant grid and the harmonic average at the cell boundaries, i.e.,

$$(k)_{PN} = \frac{2}{\left(\frac{1}{k_P} + \frac{1}{k_N}\right)}. \tag{2.23}$$

- implement that $\phi = 0$ at $z = 0$ and $\phi = \rho g(L + L_w)$ at $z = L$.
- what is the advantage and disadvantage for numerical solutions compared to an analytical solution

2.3.4   EXERCISE, EXCEL NUMERICAL 1-D SIMULATION

About the implementation of the 1-D numerical solution with EXCEL.

- Make a list of control variables in the same way as described below for the $2 - D$ flow problem. Use column F and columns to the right of F for this,
- include the definition of the number of grid cells, the number of grid cells in the bottom part N_2 and the grid cell size Δz in the list of variables used for the calculation of the analytical solution (above). You cannot name the variable N_2 as this is ambiguous in EXCEL (why?) and thus use N_b. Also define a variable start, which can assume values zero or one, meaning initialization and running, respectively,
- start at cell $B3$ and list a column of mobility values k_2/μ for cells $i = 0, .., N_2$ attached to the column of mobility values k_1/μ for cells $i = N_2 + 1, .., N$,
- in column C next to the column of mobility values assign the harmonic averages. They are equal to either k_1/μ or k_2/μ except in the cell $i = N_2$,
- put in the boundary values at cell $D3$ and cell $D3 + N$, i.e., when $N = 100$ at cell $D103$,
- put in statement 2.22 in the cells between $D3$ and $D3 + N := if(start = 0, 0, (c3 * d3 + c4 * d5)/(c3 + c4))$ in cell $D4$. Copy all the way down,

- use (file→Options→formula→Calculation options) to put the calculation on manual as shown in Figure 2.28,
- put start equal to one and press "Shift F9" in the "numeric" sheet a few times until the numbers do not change anymore,
- compare analytical and numerical solutions,
- compare the computed results to the experimental results obtained in the laboratory. Express the reading of your pressure device in terms of the total head, the elevation head and the pressure head and also in terms of the potential ϕ.

2.3.5 RADIAL INFLOW EQUATION

Equation (3.16) can be straightforwardly transformed into an equation in radial coordinates. We prefer, however, to do the derivation in the cylindrical symmetrical setting because it is more transparent from the engineering point of view, and it also gives a scheme for the numerical solution of the equations.

The stationary mass balance with respect to the disk-shaped control volume without accumulation reads

$$(\rho q)_{r_w} - (\rho q)_{r_e} = 0, \tag{2.24}$$

$$\frac{(q)_{r_w} - (q)_{r_e}}{(r_e - r_w)} = 0, \tag{2.25}$$

where r_e is the outer radius of the disk and r_w is the inner radius of the disk, and H is its height. Equation (2.25) follows from Eq. (2.24) after division by the nonzero term $(r_e - r_w)$ and the constant density ρ. We use that the layer is horizontal and therefore we can ignore the gravity term in $\phi = p + \rho g z$, i.e., $\frac{d\phi}{dr} = \frac{dp}{dr}$. Furthermore, we assume that we can write the flow averaged over the ζ−direction, by introducing the arithmetically averaged permeability $\bar{k}$ and by using Darcy's law as

$$q = 2\pi \frac{\bar{k} H}{\mu} r \frac{\partial p}{\partial r}. \tag{2.26}$$

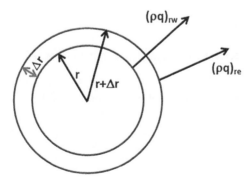

Figure 2.6 Control volume in a radial setting with $r_w \leq r \leq r_e$. The subscripts w and e denote west and east respectively. The third dimension perpendicular to the drawing would show a layer of height H. Therefore ρq is the mass flux through a cylinder of height H.

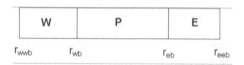

Figure 2.7   Radial grid with central point P and adjacent points $W(est)$ and $E(ast)$. We have indicated the west-boundary of the cell by its radius $r = r_{wb}$ and the east-boundary by $r = r_{eb}$. In addition we have indicated the west-boundary of the west-cell W by $r = r_{wwb}$ and the east-boundary of the cell E by $r = r_{eeb}$.

In the limit that $\Delta r \to 0$ we can get the differential $\left((q)_{r_e} - (q)_{r_w} \right) / (r_e - r_w) \approx \frac{d(q)}{dr}$. The result is

$$\frac{1}{r}\frac{d}{dr}\left(\frac{\bar{k}}{\mu} r \frac{dp}{dr} \right) = 0. \tag{2.27}$$

For the numerical scheme, we use the grid cell division as shown in Figure 2.7.

The explanation of Figure 2.7 is shown in the caption. We like to substitute a numerical approximation of Eq. (2.26) into Eq. (3.17), and we obtain an equation that can be used as a numerical scheme for solving the transient porous medium equation. Therefore, we use

$$(\rho q)_{eb} \approx -\frac{2\pi \rho H}{\mu}\frac{(p_P - p_W)}{\frac{1}{\lambda_W}\ln\frac{r_{wb}}{r_W} + \frac{1}{\lambda_P}\ln\frac{r_P}{r_{wb}}}$$

$$(\rho q)_{wb} \approx -\frac{2\pi \rho H}{\mu}\frac{(p_E - p_P)}{\frac{1}{\lambda_P}\ln\frac{r_{eb}}{r_P} + \frac{1}{\lambda_E}\ln\frac{r_E}{r_{eb}}}.$$

The mass balance equation reads such that $(\rho q)_{eb} = (\rho q)_{wb}$ For slightly compressible liquids we may assume that $\rho_{wb} = \rho_{eb} = \rho_P = \rho$ and find after division of both sides by $2\pi \rho H$

$$\left(\frac{(p_P - p_W)}{\frac{1}{\lambda_W}\ln\frac{r_{wb}}{r_W} + \frac{1}{\lambda_P}\ln\frac{r_P}{r_{wb}}} - \frac{(p_E - p_P)}{\frac{1}{\lambda_P}\ln\frac{r_{eb}}{r_P} + \frac{1}{\lambda_E}\ln\frac{r_E}{r_{eb}}} \right) = 0.$$

From this we can derive the numerical scheme with successive overrelaxation as

$$p_P = \omega \frac{p_W / \left(\frac{1}{\lambda_W}\ln\frac{r_{wb}}{r_W} + \frac{1}{\lambda_P}\ln\frac{r_P}{r_{wb}} \right) + p_E / \left(\frac{1}{\lambda_P}\ln\frac{r_{eb}}{r_P} + \frac{1}{\lambda_E}\ln\frac{r_E}{r_{eb}} \right)}{1 / \left(\frac{1}{\lambda_W}\ln\frac{r_{wb}}{r_W} + \frac{1}{\lambda_P}\ln\frac{r_P}{r_{wb}} + \frac{1}{\lambda_P}\ln\frac{r_{eb}}{r_P} + \frac{1}{\lambda_E}\ln\frac{r_E}{r_{eb}} \right)} + (1 - \omega) p_P, \tag{2.28}$$

where ω is the overrelaxation factor, see [192, 193], which assumes values between $1 < \omega < 2$.

2.3.6   BOUNDARY CONDITIONS FOR RADIAL DIFFUSIVITY EQUATION

We specify one Von Neumann boundary condition at the well, i.e., at $r = r_w$ and one Dirichlet boundary condition at the well. In the problems of our interest, we specify the production flow rate qB_o at the well. Here B_o is the formation volume factor, i.e., the ratio of the volume of oil in the reservoir and in the tank. We use

$$2\pi \frac{k}{\mu}\left(r\frac{dp}{dr} \right)_{r=r_w} = \frac{qB_o}{H}. \tag{2.29}$$

Furthermore, we need to specify the pressure at the well, i.e.,

$$p(r_w) = p_{wf}. \tag{2.30}$$

2.3.7 EXERCISE, RADIAL DIFFUSIVITY EQUATION

About a stationary solution of the radial diffusivity equation and average pressure.

Show that the stationary solution of the radial diffusion equation (2.27) subjected to the boundary conditions (2.29) and (2.30) leads to

$$p(r) - p_{wf} = \frac{q\mu B_o}{2\pi kH} \ln \frac{r}{r_w}, \tag{2.31}$$

where B_o is the formation volume factor that converts stocktank barrels into reservoir barrels. Also show that in a disk of radius R, we obtain for the average pressure

$$\overline{p} - p_{wf} = \frac{q\mu B_o}{2\pi kH} \left(\frac{R^2 \ln \frac{R}{r_w}}{R^2 - r_w^2} - \frac{1}{2} \right) \approx \frac{q\mu B_o}{2\pi kH} \left(\ln \frac{R}{r_w} - \frac{1}{2} \right). \tag{2.32}$$

Hint: take into account that the average of $p(r) - p_{wf}$ is given by

$$\frac{2\pi \int_{r_w}^{R} r \left(p(r) - p_{wf} \right) dr}{\pi (R^2 - r_w^2)}. \tag{2.33}$$

Use Maple to perform the integration.

2.4 MODIFICATIONS OF DARCY'S LAW

It is not always possible to use the simplest form of Darcy's law, viz., as in Eq. (2.9). We will discuss a number of situations that require a different form.

2.4.1 REPRESENTATIVE ELEMENTARY VOLUME

All properties of a porous medium that are relevant for flow are averaged quantities. This means that a quantity, e.g., the porosity or saturation, must be averaged over a sufficiently large volume of the porous medium [18]. The volume must be small with respect to our problem of interest and large enough such that the averaged quantity does not change significantly if we increase the averaging volume by a factor of, say, 2 (Figure 2.8). For porosity (or saturation) the concept can be well understood. The porosity is the void space (space not occupied by rock) divided by the volume of the porous medium. Indeed the porosity in the middle of a pore is 100%. As we are increasing the volume the porosity will become less because part of the porous medium space is now occupied by grains. Initially, the calculated porosity will drop below the average, before it increases again as the averaging volume start to cover the volume outside the first layer of grains. It starts to wiggle around the average until a more or less constant value is obtained. The smallest volume that gives the acceptable average is the "Representative Elementary Volume (REV)". If such an REV cannot be found, it may be necessary to derive another model equation for transport in such a porous medium. In practice, Darcy's law is often still used, but the user must be aware of this aspect.

There are many geological situations in which the proper definition of a representative elementary volume is not possible. For instance, in fluvial deposits, so-called cross-bed sets [151] form (Figure 2.9). Typical X-bed sets are river deposits and have dimensions of $L \times W \times H$ of (2–20) m $\times$ (0.5–5) m $\times$ (0.14–0.4) m. These are shown in the left part of Figure 2.9. The cross-beds consist of stacks of coarse foreset lamine with a thickness of 4–14 mm, fine foreset laminae with

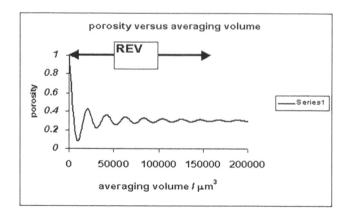

Figure 2.8 Definition of representative elementary volume.

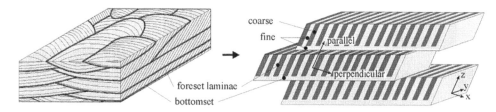

Figure 2.9 On the right a 3D representation of vertically stacked layers in cross-beds with distinctive foreset and bottomset facies. On the left the schematized model. The cross-bed sets are characterized by box like structures with a specific length, height and width. The foreset and bottom set laminae with this box-like structure have characteristic dimensions. [151].

a thickness of 2–10 mm and bottom sets with a thickness of 10–100 mm. Typical permeabilities of the coarse foreset lamina is 200 mD and for the fine foreset lamina it is 50 mD. The typical structure occurs due to the intricacies of fluvial grain transport, but that is outside the scope of a text on fluid flow transport.

For the pressure and the potential, small-scale variations are not to be expected as small fluctuations of pressure on a small scale would lead to large gradients and thus to large flows that would smoothen the initial pressure fluctuations. On the other hand, there can be large fluctuations in the pressure or potential gradient and averaging over an REV [18] is necessary for deriving macroscopic core scale relations.

For a quantity like a permeability determination of a REV is less clear. This becomes already obvious from the fact that on a macroscopic (core scale level) there are many averages such as the arithmetic, the geometric average and the harmonic average. This means that as opposed to the porosity, the permeability is structure-dependent.

For fluid properties, it is in general not considered necessary to consider the averaging volume in porous media. However, if the pore throats become smaller than the mean free path of the molecules these property values can no longer be described by their bulk values.

A typical example is the occurrence of slip. At low gas pressures, the mean free path of the gas molecules, i.e., the length of a path that a molecule can travel without colliding to another molecule becomes of the order or larger than the pore radius. Klinkenberg (1941) [136]

Table 2.3
Permeabilities versus Pressure

Pressure [psi]	Permeability [1] [mDarcy]	Permeability [2] [mDarcy]
10	66.0	41.0
20	56.0	33.0
30	52.0	31.0
40	50.5	29.0
50	49.5	28.5
60	49.0	28.0
80	48.5	27.5
100	48.0	27.0

proposed

$$k_g = k_l(1 + \frac{4c\lambda}{r}) = k_l(1 + \frac{b}{p}) , \qquad (2.34)$$

where k_g is the gas permeability and k_l is the liquid permeability and λ is the mean free path, c is proportionality factor of order one and r is the radius of the tube [m]. We call b the slip factor.
The mean free path can be calculated from

$$\lambda = \frac{1}{\sqrt{2}\pi N d^2} , \qquad (2.35)$$

where d is here and only here the diameter of the molecule and N is the number of molecules per m^3. The diameter of the molecule can be related to the parameter b in the Van der Waals formula $(p + a^2/V)(V - n\tilde{b}) = nRT$ with $\tilde{b} = 2/3\pi N_{av}d^3$ where N_{av} is the Avogadro number. These considerations are important for the interpretation of mini-permeameter measurements [245] of permeability on out-crops.

2.4.2 EXERCISE, SLIP FACTOR

About plotting the permeability versus the inverse pressure to obtain the liquid permeability and the slip factor.

- Plot (Insert → scatter → scatter with straight lines and markers) the permeability values in Table 2.3 versus the inverse pressure and determine the slip factors. Use formulas → define name to name the columns for pressure "psi", and the columns for the permeability "perm1" and "perm2", respectively. From that make a column with "1/psi",
- use the "add trend line option" by clicking the line with the right mouse button to determine the slip factors (Figure 2.10),
- put in axes titles like in the plot shown in Figure 2.11.

2.4.3 SPACE DEPENDENT DENSITY

In case of space-dependent density, we need to replace Eq. (2.9). Such location-dependent densities occur for example for varying salt concentration.)

$$u_x = -\frac{k}{\mu}\frac{\partial p}{\partial x}$$

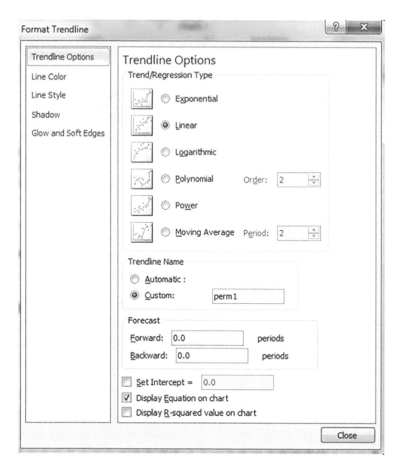

Figure 2.10 Right clicking on the plotted line allows to insert a trendline. Use the options as displayed.

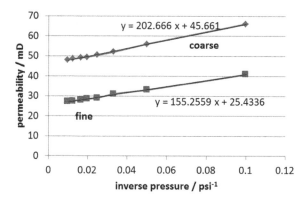

Figure 2.11 Plot of permeability versus the inverse pressure. If one clicks on the figure and uses WIN-DOWS 10, three menu choices are added in the top right corner, viz., "Design, Layout and Format". These options can be used to insert axis titles etc.

$$u_y = -\frac{k}{\mu}\frac{\partial p}{\partial y}$$

$$u_z = -\frac{k}{\mu}\left(\frac{\partial p}{\partial z}+\rho g\right). \tag{2.36}$$

Indeed the use of a space-dependent density would cause in Eq. (2.9) to obtain a term $gz\partial\rho/\partial z$ term, which is not correct. This correctly derived Darcy's law is due to Lorentz [156].

It is possible to understand why Eq. (2.36) is the correct formulation for spatially dependent densities by considering the Navier Stokes equation. A precise connection between the Navier-Stokes equation and Darcy's law requires some upscaling procedure like averaging [12, 13, 18, 37, 252]. Unfortunately, such a rigorous derivation is far beyond the scope of this monograph. Here and in the paragraph below, we show this relation admittedly in an intuitive manner rather than in an exact derivation. There is an extensive literature on how to do this correctly of which some are mentioned above. Our aim here is to improve the intuitive understanding of Darcy's law. For all of this, we require the notion of representative elementary volume (REV). The idea is that we need to average quantities such as porosity, saturation, pressure, etc. over a volume. The volume of the REV must be small with respect to the volume of our problem of interest and large enough such that the averaged quantity does not change significantly if we increase the averaging volume by a factor of, say, 2 (Figure 2.8).

Navier Stokes is the continuum form of Newton's second law $F = ma = m\frac{dv}{dt}$. It expresses that inertia forces (ma) are equal to the force brought about by pressure variations (F_{pres}), gravity force (F_{grav}) and viscous force F_{visc}. All integrations below are over the fluid part of the medium.

$$m\frac{d\mathbf{v}}{dt} = -\oint_S p dA - m\mathbf{g} + \mathbf{F_{visc}} \tag{2.37}$$

where the pressure p is directed in the outward normal direction with respect to the surface A. Note that we assume that z is pointing in the upward direction. The viscous force F_{visc} is the average viscous force on the fluid element. The references [13, 252] show that our approach leads to the correct result. The handling of the pressure and gravity term will show that the driving force is proportional to the sum of an average pressure gradient term $< \nabla p >$ and $\rho g\mathbf{e_z}$, which is only for a constant density equal to $< \nabla(p + \rho gz) >$. We apply

$$\oint_S p d\mathbf{A} = \int_V \mathbf{grad} p dV. \tag{2.38}$$

This can be easily validated in cubic geometry by substitution of $dV = dxdydz$ with $\mathbf{grad}\, p = \mathbf{e_x}\partial p/\partial x + \mathbf{e_y}\partial p/\partial y + \mathbf{e_z}\partial p/\partial z$ and integration of the terms in Eq. (2.38) versus x, y, z, respectively. When we write the mass as a volume integral over the density ρ, we obtain

$$\int_V \rho\frac{d\mathbf{v}}{dt}dV = -\int_V \mathbf{grad} p dV + \int_V \rho\mathbf{g}dV - \int_V \frac{\mu\mathbf{u}}{k}dV, \tag{2.39}$$

where we have considered that the viscous forces in the porous medium are proportional to the Darcy velocity u (some average velocity) and proportional to the viscosity. Note that we assume that z is pointing in the downward direction. The proportionality constant is written in the denominator and will be equated to the permeability k. Eq. (2.39) holds for a representative elementary volume (Figure 2.8). We assume that inertia forces represented by the left term of Eq. (2.39) are negligible. We recover Darcy's equation when we ignore correlations between the Darcy velocity and a local permeability. As the equation is valid for any volume beyond the REV, we can omit the integrals in Eq. (2.39). In that case, we obtain

$$\mathbf{u} = -\frac{k}{\mu}\left(\mathbf{grad} p + \rho\, g\,\mathbf{e_z}\right), \tag{2.40}$$

where we have defined the positive z direction in the same direction as the direction of the acceleration due to gravity, leading to change in sign. If ρ is constant, then this equation is equivalent to

$$\mathbf{u} = -\frac{k}{\mu}\mathbf{grad}\phi \,, \tag{2.41}$$

where $\phi = p + \rho gz$ is the potential. Hydrologists like to write

$$\mathbf{u} = -\frac{k\rho g}{\mu}\left(\mathbf{grad}\frac{P}{\rho g} + \mathbf{e_z}\right) = -K\,\mathbf{grad}\mathrm{h} \,, \tag{2.42}$$

which is only correct for constant density.

2.4.4 WHY IS THE FLOW RESISTANCE PROPORTIONAL TO THE SHEAR VISCOSITY?

With hindsight, we surprisingly find that Darcy's law only depends on shear flow resistance. This is correct because all velocity gradients lead to pressure losses proportional to the shear viscosity. It is therefore important to note that also elongational viscosity [28], which describes the resistance to a velocity gradient in the flow direction is for Newtonian fluids also proportional to the shear viscosity. Figure 2.1 shows the highly complex flow fields that can be expressed in such a simple law as Darcy's law. Assuming in this figure a constant interstitial at the south (bottom boundary) of 1 [m/s] (velocity can be arbitrarily chosen as much smaller because creeping flow is linear). The interstitial velocity is $v = u/\varphi$, correcting for the fact that we disregard flow inside the grains. The true average velocity is still higher, as the fluid particles do not follow a straight line, but a tortuous path with tortuosity . In order to derive Darcy's law, we must average over volumes (Representative Elementary Volume), much larger than the volume shown in Figure 2.1.

2.4.5 FORCHHEIMER EQUATION MUST BE USED FOR HIGH VALUES OF THE REYNOLDS NUMBER

When the Reynolds number

$$Re = \frac{\rho v d_p}{\mu} \tag{2.43}$$

becomes much larger than one, inertia forces in the Navier Stokes equation can no longer be disregarded. In the Reynolds number ρ is the density, $v = u/\varphi$ is the Darcy velocity divided by the porosity φ, d_p is the average grain diameter and μ is the viscosity. Practically, this becomes of concern when $Re > 10$. In this case, we replace Darcy's law by the empirical Forchheimer's law [94, 96]. Forchheimer's law is an extension of Darcy's law as it considers the inertia term. The equation is stated empirically and reads

$$\mathbf{grad}(\mathrm{p} + \rho\mathrm{gz}) = -\left(\frac{\mu\mathbf{u}}{k} + \beta\rho\mathbf{u}|\mathbf{u}|\right) \,, \tag{2.44}$$

where we introduce the so-called inertia factor $\beta\,[\mathrm{m}^{-1}]$. This is all there is to know, but we gain some insight if we try to use the same procedure as above, but now including the inertia term.

We start with the equation

$$\int_V \rho\frac{d\mathbf{v}}{dt}dV = -\int_V \mathbf{grad}\mathrm{p}dV - \int_V \rho\mathbf{g}dV - \int_V \frac{\mu\mathbf{u}}{k}dV \,. \tag{2.45}$$

Indeed in Newton's law we mean with $F = ma$ with $a = \frac{dv}{dt}$ that a is the acceleration of the moving object. In continuum mechanics, the acceleration is at a particular position not as in Newton's law moving along with the object. We use that $\mathbf{v} = \mathbf{v}(x, y, z.t)$ or

$$d\mathbf{v} = \frac{\partial \mathbf{v}}{\partial x}dx + \frac{\partial \mathbf{v}}{\partial y}dy + \frac{\partial \mathbf{v}}{\partial z}dz + \frac{\partial \mathbf{v}}{\partial t}dt , \tag{2.46}$$

and take the derivative toward t. We obtain

$$\frac{d\mathbf{v}}{dt} = \frac{\partial \mathbf{v}}{\partial x}\frac{dx}{dt} + \frac{\partial \mathbf{v}}{\partial y}\frac{dy}{dt} + \frac{\partial \mathbf{v}}{\partial z}\frac{dz}{dt} + \frac{\partial \mathbf{v}}{\partial t} . \tag{2.47}$$

In shorthand notation

$$\rho\frac{d\mathbf{v}}{dt} = \rho\frac{\partial \mathbf{v}}{\partial t} + \rho\mathbf{v}.\mathbf{grad}\mathbf{v} . \tag{2.48}$$

We use the vector relation

$$\frac{1}{2}\mathbf{grad}v^2 = \mathbf{v} \times \mathbf{curl}\ \mathbf{v} + \mathbf{v}.\mathbf{grad}\mathbf{v} , \tag{2.49}$$

and obtain

$$\int_V \rho(\frac{\partial \mathbf{v}}{\partial t} + \frac{1}{2}\mathbf{grad}v^2 - \mathbf{v} \times \mathbf{curl}\ \mathbf{v})dV = -\int_V \mathbf{grad}p\,dV - \int_V \rho g\,dV - \int_V \frac{\mu\mathbf{u}}{k}dV. \tag{2.50}$$

As we are dealing with stationary flows, we have that the $\rho\frac{\partial \mathbf{v}}{\partial t} = 0$. As closure relation, we assume Forchheimer's Law, i.e., that

$$\int_V -\rho\mathbf{v} \times \mathbf{curl}\ \mathbf{v}dV = \int_V \beta\rho\mathbf{u}|\mathbf{u}|dV, \tag{2.51}$$

whereas before we assume that $\mathbf{u}$ is the averaged velocity over a representative elementary volume. The coefficient β is called the inertia factor. We note here that the introduction of the Forchheimer term is entirely empirical. We, therefore, obtain the following equation if the density is constant

$$\int_V \mathbf{grad}(p + \rho gz + \frac{1}{2}\rho v^2)dV = -\int_V \left(\frac{\mu\mathbf{u}}{k} + \beta\rho\mathbf{u}|\mathbf{u}|\right) dV . \tag{2.52}$$

We only need the assumption of constant density in dealing with the "kinetic energy term" $\frac{1}{2}\rho v^2$. Again this equation is valid for any volume larger than the REV. As we did already assume that all variables on the right of the equation are averaged quantities over the representative elementary volume we obtain

$$\mathbf{grad}(p + \rho gz + \frac{1}{2}\rho v^2) = -\left(\frac{\mu\mathbf{u}}{k} + \beta\rho\mathbf{u}|\mathbf{u}|\right) . \tag{2.53}$$

In the absence of friction and inertia, the RHS of Eq. (2.53) is zero and the mechanical energy $p + \rho gz + \frac{1}{2}\rho v^2$ is constant along a flow line. There are two aspects of physical significance, which follow Eq. (2.53). First, the form of Eq. (2.51) suggests that the inertia term can indeed give rise to a square dependence on the velocity. The second interesting aspect is that a term $\frac{1}{2}\rho v^2$ should be added to the pressure and gravity term to define the mechanical energy.

For constant density, we can divide by ρg and write the equation as it is used by hydrologists

$$\mathbf{grad}(\frac{p}{\rho g} + z + \frac{1}{2}\frac{v^2}{g}) = -\left(\frac{\mu\mathbf{u}}{k\rho g} + \frac{\beta}{g}\mathbf{u}|\mathbf{u}|\right). \tag{2.54}$$

The term $(\frac{p}{\rho g})$ is called the pressure head, the term z is called the elevation head and the term $(\frac{1}{2}\frac{v^2}{g})$ is called the velocity head (see Table I).

Table 2.4

Darcy Velocity versus Pressure Gradient

Test 1		Test 2		Test 3		Test 4		Test 5	
v (cm/s)	pres grad (kPa /cm)	v (cm/s)	pres grad (kPa /cm)	v (cm/s)	pres grad (kPa /cm)	v (cm/s)	pres grad (kPa /cm)	v (cm/s)	pres grad (kPa /cm)
0.547	2.779	0.44	2.779	0.429	2.367	2.51	0.1	2.51	0.512
0.972	8.027	0.972	9.1592	1.248	14.922	5.28	0.2	5.11	1.15
1.37	16.054	1.3	16.466	1.458	18.936	15.14	1.1	10.14	1.59
2.517	50.015	1.694	27.186	1.716	29.433	27.23	1.2	33.12	3.89
4.719	118.555	1.869	33.1378	3.064	81.918	32.11	1.4	44.28	5.49
4.938	182.67	1.97	38.489	4.496	172.996	53.87	2.5	58.91	8.85
4.96	129.667	2.399	65.246	4.522	165.586	69.28	4	66.37	11.52
		2.78	75.332	4.575	90.357	81.53	5.6	81.26	15.41
		4.421	173.922			94.59	6.8	95.34	17.23
		4.925	189.668			106.29	8.3	103.58	21.81 .

2.4.6   EXERCISE, INERTIA FACTOR

About finding the permeability and inertia factor from a plot of the pressure gradient versus the mass flux ρu.

- Plot the pressure gradient versus the density × the Darcy velocity for one of the examples given in Table 2.4. What is the viscosity (10°C) used for these groundwater examples in the Netherlands?
- insert a trendline allowing for a quadratic equation,
- calculate the permeability and the inertia factor.

2.4.7   ADAPTATION OF CARMAN-KOZENY FOR HIGHER FLOW RATES

For higher flow rates, Eq. (2.14) must be amended as follows

$$\frac{\phi_o - \phi_L}{L} = \frac{(1-\varphi)^2}{\varphi^3}\frac{150\mu u}{D_p^2} + \frac{1.75\rho u^2}{D_p}\frac{1-\varphi}{\varphi^3} . \tag{2.55}$$

The first term on the RHS of Eq. (2.55) is the Blake-Kozeny (Carman-Kozeny) part and the second term is the Burke Plummer part [18]. It shows for which Reynolds number inertia terms become important

$$\frac{\phi_o - \phi_L}{L} = \frac{(1-\varphi)^2}{\varphi^3}\frac{150\mu u}{D_p^2}\left(1+\frac{1.75}{150}\frac{\rho u D_p}{\mu}(1-\varphi)\right)$$

$$:= \frac{(1-\varphi)^2}{\varphi^3}\frac{150\mu u}{D_p^2}\left(1+\frac{1.75}{150}Re(1-\varphi)\right) . \tag{2.56}$$

Therefore, at Re = 10, the inertia correction to the Darcy flow starts to become relevant, i.e., of the order of 10%.

2.4.8 EXERCISE, CARMAN KOZENY

About extension of the Carman Kozeny equation to Re $>>$ 1.

Use Eq. (2.56) to obtain an expression for the permeability and the inertia factor in terms of the porosity and the particle diameter. The correlation between permeability and inertia used by Petroleum Engineers is [64]

$$\beta[ft^{-1}] = \frac{2.73e10}{k[mD]^{1.1045}} \, , \qquad (2.57)$$

where indeed k is in mD and β is in $[ft^{-1}]$ [96].

2.4.9 ANISOTROPIC PERMEABILITIES

The permeability can be different in different directions, i.e., we have a permeability k_x in the x-direction, a permeability k_y in the y-direction and a permeability k_z in the z-direction. Therefore, also the mobilities $\lambda = k/\mu$ i.e., the permeability divided by the viscosity can be different in different directions. The mobility can be considered a tensor ref. ([93], Chapter 31) and

$$\begin{pmatrix} u_x \\ u_y \\ u_z \end{pmatrix} = - \begin{pmatrix} \lambda_{xx} & 0 & 0 \\ 0 & \lambda_{yy} & 0 \\ 0 & 0 & \lambda_{zz} \end{pmatrix} \begin{pmatrix} \frac{\partial \phi}{\partial x} \\ \frac{\partial \phi}{\partial y} \\ \frac{\partial \phi}{\partial z} \end{pmatrix} . \qquad (2.58)$$

However, the mobility (permeability) tensor is not necessarily diagonal in the frame of reference. This can be shown as follows (Figure 2.12). Consider $2 - D$ flow. We assume that the x- and y- components of the pressure gradient are equal. The gradients are defined in the horizontal plane, and therefore, we ignore the gravity terms and thus replace ϕ by p. The mobilities are not equal. The application of Eq. (2.58) shows that the $x-$component u_x and the $y-$component u_y are not equal. Consequently, the Darcy velocity vector has a different direction than the potential

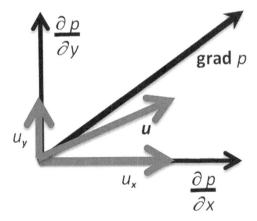

Figure 2.12 For the pressure gradient **grad** p has equal components in the x-direction and y-direction. The figure shows the absolute values of the pressure gradient. Gravity terms are disregarded. The permeability in the y-direction is smaller than in the x-direction. Therefore the Darcy velocity u_y in the y-direction is smaller than the Darcy velocity u_x in the x-direction. Therefore the Darcy velocity **u** has not the same direction as **grad** p.

gradient. In general, we have

$$
\begin{pmatrix} u_x \\ u_y \\ u_z \end{pmatrix} = - \begin{pmatrix} \lambda_{xx} & \lambda_{xy} & \lambda_{xz} \\ \lambda_{yx} & \lambda_{yy} & \lambda_{yz} \\ \lambda_{zx} & \lambda_{zy} & \lambda_{zz} \end{pmatrix} \begin{pmatrix} \frac{\partial \phi}{\partial x} \\ \frac{\partial \phi}{\partial y} \\ \frac{\partial \phi}{\partial z} \end{pmatrix}. \tag{2.59}
$$

Consider only two directions, i.e., the layer direction $//$ and the X-dip direction $\perp$. The direction perpendicular to these two directions is denoted as the y direction. A vector $\nabla \mathbf{P}$ with an arbitrary direction rotated through the y-axis by an angle of α has components $\nabla P_{//} = |\nabla \mathbf{P}| \cos \alpha$ and $\nabla P_{\perp} = |\nabla \mathbf{P}| \sin \alpha$. After rotation of this vector through ϑ degrees its components are $|\nabla \mathbf{P}| \cos(\alpha + \vartheta)$ and $|\nabla \mathbf{P}| \sin(\alpha + \vartheta)$. We use the geometry relations $\cos(\alpha + \vartheta) = \cos(\alpha)\cos(\vartheta) - \sin(\alpha)\sin(\vartheta)$ and $\sin(\alpha + \vartheta) = \cos(\alpha)\sin(\vartheta) + \sin(\alpha)\cos(\vartheta)$. They are easily derived by observing that $\exp(i\alpha\vartheta) = \exp(i\alpha)\exp(i\vartheta)$ and using that $\exp ix = \cos x + i \sin x$. Using this, we can derive for the rotation operator $\mathbf{R}$

$$
\mathbf{R} = \begin{pmatrix} \cos \vartheta & -\sin \vartheta \\ \sin \vartheta & \cos \vartheta \end{pmatrix}, \tag{2.60}
$$

and the inverse rotation matrix is obtained by rotating through $-\vartheta$, which means that we we write $\sin \vartheta \to -\sin \vartheta$ in Eq. (2.60).

Ignoring gravity terms we can write Darcy's law as $\mathbf{u} = \mathbf{k} \cdot \nabla \mathbf{p}/\mu$. If we apply the symmetry operator to this equation we obtain $\mathbf{R} \cdot \mathbf{u} = 1/\mu \quad \mathbf{R} \cdot \mathbf{k} \cdot \mathbf{R}^{-1} \cdot \mathbf{R} \cdot \nabla \mathrm{p}$. Therefore we obtain for $\mathbf{k}$

$$
\mathbf{k} = \begin{pmatrix} \cos \vartheta & -\sin \vartheta \\ \sin \vartheta & \cos \vartheta \end{pmatrix} \begin{pmatrix} k_{//} & 0 \\ 0 & k_{\perp} \end{pmatrix} \begin{pmatrix} \cos \vartheta & \sin \vartheta \\ -\sin \vartheta & \cos \vartheta \end{pmatrix}. \tag{2.61}
$$

After matrix multiplication, we find

$$
\mathbf{k} = \begin{bmatrix} \cos^2 \vartheta \, k_{//} + \sin^2 \vartheta \, k_{\perp} & \cos \vartheta \sin \vartheta \, (k_{//} - k_{\perp}) \\ \cos \vartheta \sin \vartheta \, (k_{//} - k_{\perp}) & \cos^2 \vartheta \, k_{\perp} + \sin^2 \vartheta \, k_{//} \end{bmatrix}. \tag{2.62}
$$

2.4.10 EXERCISE, MATRIX MULTIPLICATION

About using matrix multiplication to convert the diagonalized permeability tensor to a tensor with arbitrary orientation in space; matrix multiplcation with EXCEL.

Use matrix multiplication in EXCEL to validate Eq. (2.62). Take $k_{//} = 10$ and $k_{\perp} = 1$ and choose an angle. Use the EXCEL help function (F1) to obtain the procedure how to use "MMULT".

2.4.11 SUBSTITUTION OF DARCY'S LAW IN THE MASS BALANCE EQUATION

Substitution of Darcy's law (Eq. (2.40)) into Eq. (2.5) leads to

$$
\mathbf{div} \cdot (\rho \mathbf{u}) = \mathbf{div} \cdot \left(\frac{-k\rho}{\mu} (\mathbf{grad} \mathrm{p} + \rho g \mathbf{e_z}) \right) = 0. \tag{2.63}
$$

If ρ is constant we use $\phi = p + \rho gz$ to simplify to

$$
\mathbf{div} \cdot \left(\frac{k}{\mu} \mathbf{grad} \, (\mathrm{p} + \rho g z) \right) = \mathbf{div} \cdot \left(\frac{k}{\mu} \mathbf{grad} \phi \right) = 0. \tag{2.64}
$$

In hydrology, we would use the notation

$$
\mathbf{div} \cdot \mathbf{u} = \mathbf{div} \cdot \left(\frac{-k\rho g}{\mu} (\mathbf{grad}(\frac{\mathrm{p}}{\rho g} + z)) \right) = -\mathbf{div} \cdot (\mathrm{K} \, \mathbf{grad} \mathrm{h}) = 0. \tag{2.65}
$$

2.5   STATISTICAL METHODS TO GENERATE HETEROGENEOUS POROUS MEDIA

2.5.1   THE IMPORTANCE OF HETEROGENEITY

Owing to the ways reservoirs are deposited and due to the diagenesis thereafter, all reservoirs are heterogeneous. This does not mean that nothing can be said about a reservoir without considering heterogeneities. The application of mass conservation laws on a model, which considers the reservoir at a single average pressure (material balance) is a powerful method to analyze reservoir behavior independent of the heterogeneity structure. Water influx into this single pressure oil reservoir is predicted from models [237] that assume a cylindrical homogeneous reservoir structure. Well pressure analysis tests are largely interpreted in terms of a single average reservoir permeability. Early water breakthrough in the Schoonebeek reservoir was predicted with a model that considered a single homogeneous layer [73]. Many other features observed in reservoirs, e.g., water upconing and unstable displacement can be qualitatively explained without considering heterogeneities. Other features need an adequate heterogeneity model to describe the essence of the flow problem.

Reservoir simulators are powerful predictive tools even when they use highly schematic reservoir heterogeneity models and in spite of capillary pressure functions and relative permeability curves that are measured on non-representative samples. It is widely accepted that a biased choice may select samples without small-scale heterogeneities or at least samples that do not fall apart [188]. In view of the above, it is understandable that for reservoir engineers even if they noted that all reservoirs were heterogeneous [80], the primary concern was to understand fluid flow behavior. By the adage do not try to run before you can walk fluid flow behavior in homogeneous reservoirs allows to understand the basic physical features. There are many examples where such an approach in itself is highly fruitful.

The increased interest in flow through heterogeneous reservoirs coincided with the advent of high-speed computers. There are also many examples where flow behavior does not need to be considered. The interconnectedness of sand lenses can be often assessed from a geological study combined with a geostatistical analysis. If the sand lenses do not intersect, we do not need a simulator to prove that there is no flow between them. On the other side of the spectrum, there are many cases where the reservoir structure is of secondary importance as shown above. In other cases where one needs quick results, it is logical to use the simulator to incorporate heterogeneity effects.

We will show methods to generate or schematic heterogeneous reservoir models. Examples will be given to analyze the flow behavior in these reservoirs.

2.5.2   GENERATION OF RANDOM NUMBERS DISTRIBUTED ACCORDING TO A GIVEN DISTRIBUTION FUNCTION

It is useful to be able to generate a heterogeneous permeability field. The methods used in the literature can be classified as follows:

1. Stochastic Random Fields [46]
2. marked Point Processes [101]
3. Markov Chains [229].

There is extensive literature on each of the methods. The purpose of statistical methods is that they give a result inclusive of an error estimate [119]. If there is any information, do not throw it away but try to incorporate it in the models. If one knows that there is an impermeable shale lens at a depth of 840 m with a constant thickness of 20 m put it in the reservoir model, and do not blame the statistics if one doesn't.

The problem with all of the statistical models, as with all models for flow in porous media, is that it uses parameter functions that need to be estimated before the model can be applied. These parameter functions such as the variogram, the shale length distributions and the transitional probabilities in the Markov chains are always estimated with too little data. There is, however, nothing better that we can do and as long as we remain aware of these limitations we can use the statistical models to our advantage.

Stochastic random fields [72] are almost always stationary Gaussian fields that are characterized by the mean, the standard deviation and a variogram, related to the correlation structure. Weak stationarity means that only the variogram and standard deviation are invariant with respect to translation. Strong stationarity means that also the mean is invariant to translation. The theory allows that the fields are constrained by observed data such as hard data obtained from observations in wells and soft data, such as, e.g., obtained from geological experience, which sets low and high limits. Such fields are called conditioned. It is also possible to construct fields with attributes (sand or clay) called Indicator Kriging. The method is largely used to generate permeability and porosity fields. We have to assume that some function of the permeability is normally distributed. Then the mean, the standard deviation and the variogram need to be estimated. Another disadvantage of the method is that the connection between geology and stochastic random fields is rather abstract. Geological data are difficult to incorporate.

Marked point processes are used to distribute objects with given properties in space. A marked point process is a point process attached to or marked by random processes that define type, shape and size of the random objects. Two aspects need to be distinguished: (a) characterization of a single object by a set of markers and (b) the point process that distributes the individual objects in space. The point process is almost always a Poisson process, i.e., a point in space is found by drawing three uniformly distributed random numbers that are normalized to the field size. Normalized means dividing by the block dimensions such that a block $L \times H \times W$ of given dimensions covers the whole reservoir. The object for instance a shale lens is characterized by its shape (e.g., an ellipsoid) and a statistical distribution of its axes. The advantage of the method is that the relation to geology is rather direct. The disadvantage is that simultaneous conditioning to more than one well is practically impossible.

Markov chains [69, 88, 89, 229] can be used to assign a distribution of lithologies (clays, sand and silt) in space, based on transitional probabilities. The method has both a direct relation to geology and can be easily conditioned to the measurements in the field. The rigorous method in more than $1 - D$ is computationally expensive. Approximate methods show promising results but are still in a development phase.

2.5.3 LOG-NORMAL DISTRIBUTIONS AND THE DYKSTRA-PARSON'S COEFFICIENT

It has been found [128] that permeabilities in the field are approximately log-normally distributed. Jensen et al. use the notion $p-$normally distributed, which can be anything between normally ($p = 1$) and log- normally ($p = 0$) distributed. For any other value than $p = 0$ or $p = 1$ analytical expressions of averages become very complex and one has to have a good reason for getting in the trouble of dealing with fractional values of p. Log-normally distributed means nothing else that the logarithm ($\ln k$) of the permeability is normally distributed [243]. In formula

$$P(\ln k)\mathrm{d}\ln k = \frac{1}{\sqrt{2\pi s^2}} \exp -\frac{(\ln k - \mu)^2}{2s^2} \mathrm{d}\ln k, \qquad (2.66)$$

$$P(k)dk = \frac{1}{\sqrt{2\pi s^2}} \frac{1}{k} \exp - \frac{(\ln k - \mu)^2}{2s^2} dk,$$

which can be derived from the fundamental transformation law of probabilities, i.e.,

$$|p(y)dy| = |p(x)dx|$$

$$p(y) = p(x)|\frac{dx}{dy}|. \tag{2.67}$$

In Eq. (2.66) we use μ to denote the average of the logarithm of the permeability and s to denote the standard deviation of the logarithm of the permeability. One of the nice features of the log-normal distribution is that the arithmetically $(E(k))$, geometrically $(\exp(E(\ln k))$ and harmonically $\left(\frac{1}{E(\frac{1}{k})}\right)$ averaged permeability can be simply expressed in terms of μ and s. The derivations are somewhat lengthy and will be omitted here, but the result is

$$E(k) = \int_0^\infty kP(k)dk = \exp\left(\mu + \frac{1}{2}s^2\right),$$

$$\exp(E(\ln k)) = \exp \int_0^\infty \ln k P(k)dk = \exp \mu,$$

$$\frac{1}{E(\frac{1}{k})} = \frac{1}{\int_0^\infty \frac{1}{k}P(k)dk} = \exp\left(\mu - \frac{1}{2}s^2\right). \tag{2.68}$$

Petroleum Engineers like to use the Dykstra-Parson's coefficient (V_{DP}) [141, 143] as a measure of heterogeneity. It is related to the standard deviation of the logarithm of the permeability (Figure 2.13).

$$V_{DP} = 1 - \exp(-s). \tag{2.69}$$

2.5.4 EXERCISE, LOGNORMAL DISTRIBUTION FUNCTIONS

About the use of log-normal distribution as a model for heterogeneous permeabilities.

Plot the distribution $P(k)$ when the permeability k is log-normally distributed and the Dykstra-Parson's coefficient $V_{DP} = 0.7$ and the average permeability is one Darcy $(0.987e{-}12$ m$^2)$ Hint: use Eq. (2.69) to calculate s and Eq. (2.68-a) to calculate the average of the logarithm of the permeability μ. It is convenient to use the function "*NORMDIST*" in EXCEL. Use also EXCEL to obtain the cumulative distribution function $F(k)$. How would you use "*NORMDIST*(x,mean,standard deviation,true (=1))" "to obtain the cumulative distribution function? Hint: $F(k) = F(\ln k)$. How would you use "LOGNORMDIST(x,mean,standard deviation) "to obtain the cumulative distribution function?

$$F(k) = \int_0^k \frac{P(\ln u)}{u} du = \int_0^k P(u)du ,$$

$$F(\ln k) = \int_{-\infty}^{\ln k} P(\ln u)d\ln u .$$

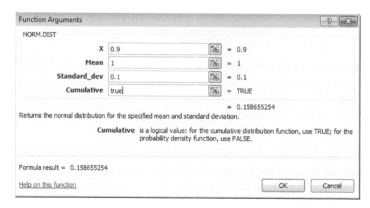

Figure 2.13 The function in EXCEL to calculate the probability density function and the cumulative distribution function.

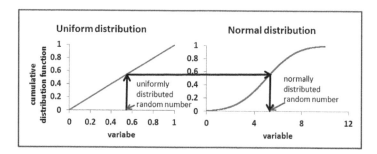

Figure 2.14 Method to generate a set of normally distributed random numbers with an average of 5 and a standard deviation of 1. Start to generate a uniformly distributed random number and follow the arrows upward to the right and downward to find the normally distributed random number.

2.5.5 GENERATION OF A RANDOM FIELD

We want to be able to generate random fields for flow modeling exercises. The following procedure makes it possible to generate a set of data that are distributed according to an arbitrary distribution function (Figure 2.14). We start with a uniformly distributed random number X between zero and one, i.e., $p(x) = 1$ and wish to obtain a random number distributed according to an arbitrary distribution function $p(y)$. We use Eq. (2.67) and obtain

$$\frac{dx}{dy} = p(y) . \tag{2.70}$$

This equation can be integrated to relate the uniformly distributed random number to the cumulative distribution function of [194]

$$x = \int_{-\infty}^{y} p(y')dy' = F(y) ,$$

$$y(x) = F^{-1}(x) .$$

The inverse transformation is shown in Figure 2.14.

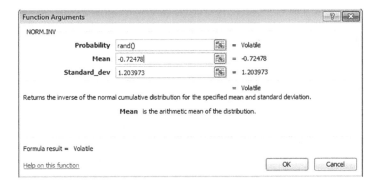

Figure 2.15   Using the function "norminv" to obtain a normal distribution for $\ln(k)$.

2.5.6   EXERCISE, LOG-NORMAL PERMEABILITY FIELD

About generating permeability k values drawn from a log-normal distribution function, i.e., $\ln(k)$ is normally distributed.

Generate 100 permeability data points $\ln k$ that are normally distributed, have an average of one Darcy and a Dykstra-Parson's coefficient $V_{DP} = 0.7$. Hint: use Eq. (2.69) to calculate s and Eq. (2.68a) to calculate the average of the logarithm of the permeability μ. It is convenient to use the function "*NORMINV*" in EXCEL. How do you use the result to obtain 100 data points that are log-normally distributed (Figure 2.15)?

2.5.7   EXERCISE, AVERAGE PERMEABILITY FIELD

About calculation of the average permeability in a lognormally distributed 20×20 random field.
   Consider Figures 2.16–2.20.

- Explain how the average permeability from a set of data in a periodic unit cell can be found.
- explain in detail, given the set of potentials and conductivities between cells, how the average was found.
- compare to the arithmetic average, the harmonic average and the geometric average. Which of the averages gives the smallest value, the largest value and the intermediate value, and the best value?

2.6   UPSCALING OF DARCY'S LAW IN HETEROGENEOUS MEDIA

2.6.1   ARITHMETIC, GEOMETRIC AND HARMONIC AVERAGES

From geological models or geostatistical models, we obtain a much more detailed reservoir heterogeneity structure than we can use in our numerical computations. Consequently, we need some kind of averaging procedure. To understand the problem that averaging is not trivial, it is important to understand that the average permeability is not only a function of the permeability distribution but also of the arrangement of the "permeabilities" in space. As an extreme example of this aspect, we calculate the average permeability for flow perpendicular and parallel to a layered reservoir.

The layered structure with N layers with a total length L and a height H is shown in Figure 2.21. The horizontal direction is the $x-$direction and the vertical direction the

	A	B	C	D	E	F	G	H	I	J	K	L
1	VDP	0.8										
2	s	1.609438										
3	mu	-1.29515	mu+1/2s^2=0									
4	visc	1										
5	dx	0.05			EXP(NORMINV(RAND(),mu,s)) in non-greenall cells							
6	dy	0.05										
7	start	0										
8			1	2	3	4	5	6	7	8	9	
9			0.19758	0.518741	0.005862	9.320868	0.317583	0.163753	0.252985	0.333103	0.131635	0.814
10	1	0.13386	0.004167	0.205623	0.310914	0.018233	0.666433	0.165476	0.068486	1.969627	0.039757	8.642
11	2	1.473821	0.555356	0.250297	0.01352	0.156394	0.126495	0.387631	0.319205	3.109761	0.09861	0.118
12	3	18.29099	0.156385	0.192816	0.091752	0.27062	0.459809	0.137955	0.343025	0.363043	0.09443	0.026
13	4	0.28919	0.072706	0.112386	0.150024	0.643063	0.335731	0.0823	0.894392	31.51067	0.098222	0.037
14	5	0.321735	0.046463	0.359995	0.21501	0.430641	0.462406	0.466792	0.011713	0.0177	0.703478	0.375
15	6	0.005536	1.7884	2.056506	7.903325	0.525861	0.331241	0.243807	0.757865	0.063459	0.651024	0.020
16	7	0.107909	0.13704	1.321829	0.159464	0.354203	0.073841	1.167932	0.577595	0.323167	0.148279	0.156
17	8	23.2577	0.024216	1.609723	0.032815	0.397942	0.710885	0.146572	0.004154	1.322525	0.281563	0.055
18	9	14.52748	0.352831	0.148296	0.184522	0.709518	0.190406	0.088034	0.222568	0.056206	0.130298	0.057
19	10	0.092623	0.381421	0.023423	0.178166	0.405833	0.204927	0.865441	0.132124	0.039391	0.196475	0.007
20	11	0.068075	0.998403	0.286241	1.070367	0.055882	0.01644	0.323148	6.675706	0.374614	0.187905	6.123
21	12	0.752565	0.613247	0.393191	0.24868	0.143459	0.175097	9.106759	0.312949	0.288652	2.738148	0.0
22	13	0.090312	0.281722	0.332753	0.694185	0.330822	0.070042	0.775344	0.03897	0.303094	0.238014	0.187
23	14	0.442069	0.96627	2.509573	0.121519	0.262081	0.47829	0.199065	0.018031	0.041089	0.128132	0.795
24	15	0.099806	0.042757	0.17029	0.064969	0.990644	0.286866	0.727565	4.830234	0.226755	0.072568	0.256
25	16	3.192668	0.113539	0.660429	1.222516	1.630139	0.167784	2.202147	0.414979	0.261662	5.45808	0.934
26	17	0.038614	0.030464	0.021316	0.444166	2.413898	0.031947	2.307768	0.095583	0.244968	0.290952	0.08

Figure 2.16 Input permeabilities drawn from log-normally distributed random numbers, i.e., $exp(norminv(rand(),mu,s))$, where rand() is a uniformly distributed random number. We make a data field with a Dykstra-Parsons' coefficient of $V_{DP} = 0.8$. In this case, the standard deviation for the log-normal permeability field is s = -ln(1-V_{DP}P). For a lognormal field, the average permeability E(k) is $E(k) = exp(\mu + 1/2s^2)$. We generate a permeability field with an expected average permeability of $E(k) = 1$ Darcy. This means that $\mu = -1/2s^2$. The variable "start" is introduced to restart the program.

	A	B	C	D	E
7	start 0				
8			1	2	3
9			=C29	=D29	=E29
10	1	=V10	=SQRT(inputpermIC10*inputpermIC11)	=SQRT(inputpermID10*inputpermID11)	=SQRT(inputpermIE10*inputpermIE11)
11	2	=V11	=SQRT(inputpermIC11*inputpermIC12)	=SQRT(inputpermID11*inputpermID12)	=SQRT(inputpermIE11*inputpermIE12)
12	3	=V12	=SQRT(inputpermIC12*inputpermIC13)	=SQRT(inputpermID12*inputpermID13)	=SQRT(inputpermIE12*inputpermIE13)
13	4	=V13	=SQRT(inputpermIC13*inputpermIC14)	=SQRT(inputpermID13*inputpermID14)	=SQRT(inputpermIE13*inputpermIE14)
14	5	=V14	=SQRT(inputpermIC14*inputpermIC15)	=SQRT(inputpermID14*inputpermID15)	=SQRT(inputpermIE14*inputpermIE15)
15	6	=V15	=SQRT(inputpermIC15*inputpermIC16)	=SQRT(inputpermID15*inputpermID16)	=SQRT(inputpermIE15*inputpermIE16)
16	7	=V16	=SQRT(inputpermIC16*inputpermIC17)	=SQRT(inputpermID16*inputpermID17)	=SQRT(inputpermIE16*inputpermIE17)
17	8	=V17	=SQRT(inputpermIC17*inputpermIC18)	=SQRT(inputpermID17*inputpermID18)	=SQRT(inputpermIE17*inputpermIE18)
18	9	=V18	=SQRT(inputpermIC18*inputpermIC19)	=SQRT(inputpermID18*inputpermID19)	=SQRT(inputpermIE18*inputpermIE19)
19	10	=V19	=SQRT(inputpermIC19*inputpermIC20)	=SQRT(inputpermID19*inputpermID20)	=SQRT(inputpermIE19*inputpermIE20)
20	11	=V20	=SQRT(inputpermIC20*inputpermIC21)	=SQRT(inputpermID20*inputpermID21)	=SQRT(inputpermIE20*inputpermIE21)
21	12	=V21	=SQRT(inputpermIC21*inputpermIC22)	=SQRT(inputpermID21*inputpermID22)	=SQRT(inputpermIE21*inputpermIE22)
22	13	=V22	=SQRT(inputpermIC22*inputpermIC23)	=SQRT(inputpermID22*inputpermID23)	=SQRT(inputpermIE22*inputpermIE23)
23	14	=V23	=SQRT(inputpermIC23*inputpermIC24)	=SQRT(inputpermID23*inputpermID24)	=SQRT(inputpermIE23*inputpermIE24)
24	15	=V24	=SQRT(inputpermIC24*inputpermIC25)	=SQRT(inputpermID24*inputpermID25)	=SQRT(inputpermIE24*inputpermIE25)
25	16	=V25	=SQRT(inputpermIC25*inputpermIC26)	=SQRT(inputpermID25*inputpermID26)	=SQRT(inputpermIE25*inputpermIE26)
26	17	=V26	=SQRT(inputpermIC26*inputpermIC27)	=SQRT(inputpermID26*inputpermID27)	=SQRT(inputpermIE26*inputpermIE27)
27	18	=V27	=SQRT(inputpermIC27*inputpermIC28)	=SQRT(inputpermID27*inputpermID28)	=SQRT(inputpermIE27*inputpermIE28)
28	19	=V28	=SQRT(inputpermIC28*inputpermIC29)	=SQRT(inputpermID28*inputpermID29)	=SQRT(inputpermIE28*inputpermIE29)
29	20	=V29	=SQRT(inputpermIC29*inputpermIC30)	=SQRT(inputpermID29*inputpermID30)	=SQRT(inputpermIE29*inputpermIE30)
30			=C10	=D10	=E10

Figure 2.17 Mobility in the south direction between the central cell P and the cell south of it. Geometric average between the P column and the S south column gives in 2-D the best estimate of this value. The flow leaving via the south boundary of the cell N is equal to the flow entering through the north boundary cell P.

	A	B	C	D
7	start 0			
8			1	2	3
9			=C29	=D29	=E29
10	1	=SQRT(inputperm!B11*inputperm!C11)	=SQRT(inputperm!C11*Inputperm!D11)	=SQRT(inputperm!D11*inputperm!E11)	=SQRT(inp
11	2	=SQRT(inputperm!B12*inputperm!C12)	=SQRT(inputperm!C12*inputperm!D12)	=SQRT(inputperm!D12*inputperm!E12)	=SQRT(inp
12	3	=SQRT(inputperm!B13*inputperm!C13)	=SQRT(inputperm!C13*inputperm!D13)	=SQRT(inputperm!D13*inputperm!E13)	=SQRT(inp
13	4	=SQRT(inputperm!B14*inputperm!C14)	=SQRT(inputperm!C14*inputperm!D14)	=SQRT(inputperm!D14*inputperm!E14)	=SQRT(inp
14	5	=SQRT(inputperm!B15*inputperm!C15)	=SQRT(inputperm!C15*inputperm!D15)	=SQRT(inputperm!D15*inputperm!E15)	=SQRT(inp
15	6	=SQRT(inputperm!B16*inputperm!C16)	=SQRT(inputperm!C16*inputperm!D16)	=SQRT(inputperm!D16*inputperm!E16)	=SQRT(inp
16	7	=SQRT(inputperm!B17*inputperm!C17)	=SQRT(inputperm!C17*inputperm!D17)	=SQRT(inputperm!D17*inputperm!E17)	=SQRT(inp
17	8	=SQRT(inputperm!B18*inputperm!C18)	=SQRT(inputperm!C18*inputperm!D18)	=SQRT(inputperm!D18*inputperm!E18)	=SQRT(inp
18	9	=SQRT(inputperm!B19*inputperm!C19)	=SQRT(inputperm!C19*inputperm!D19)	=SQRT(inputperm!D19*inputperm!E19)	=SQRT(inp
19	10	=SQRT(inputperm!B20*inputperm!C20)	=SQRT(inputperm!C20*inputperm!D20)	=SQRT(inputperm!D20*inputperm!E20)	=SQRT(inp
20	11	=SQRT(inputperm!B21*inputperm!C21)	=SQRT(inputperm!C21*inputperm!D21)	=SQRT(inputperm!D21*inputperm!E21)	=SQRT(inp
21	12	=SQRT(inputperm!B22*inputperm!C22)	=SQRT(inputperm!C22*inputperm!D22)	=SQRT(inputperm!D22*inputperm!E22)	=SQRT(inp
22	13	=SQRT(inputperm!B23*inputperm!C23)	=SQRT(inputperm!C23*inputperm!D23)	=SQRT(inputperm!D23*inputperm!E23)	=SQRT(inp
23	14	=SQRT(inputperm!B24*inputperm!C24)	=SQRT(inputperm!C24*inputperm!D24)	=SQRT(inputperm!D24*inputperm!E24)	=SQRT(inp
24	15	=SQRT(inputperm!B25*inputperm!C25)	=SQRT(inputperm!C25*inputperm!D25)	=SQRT(inputperm!D25*inputperm!E25)	=SQRT(inp
25	16	=SQRT(inputperm!B26*inputperm!C26)	=SQRT(inputperm!C26*inputperm!D26)	=SQRT(inputperm!D26*inputperm!E26)	=SQRT(inp
26	17	=SQRT(inputperm!B27*inputperm!C27)	=SQRT(inputperm!C27*inputperm!D27)	=SQRT(inputperm!D27*inputperm!E27)	=SQRT(inp
27	18	=SQRT(inputperm!B28*inputperm!C28)	=SQRT(inputperm!C28*inputperm!D28)	=SQRT(inputperm!D28*inputperm!E28)	=SQRT(inp
28	19	=SQRT(inputperm!B29*inputperm!C29)	=SQRT(inputperm!C29*inputperm!D29)	=SQRT(inputperm!D29*inputperm!E29)	=SQRT(inp
29	20	=SQRT(inputperm!B30*inputperm!C30)	=SQRT(inputperm!C30*inputperm!D30)	=SQRT(inputperm!D30*inputperm!E30)	=SQRT(inp
30			=C10	=D10	=E10

Figure 2.18 Mobility in the east direction between the central cell P and the cell east of it. Geometric average between the P column and the E east column gives in 2-D the best estimate of this value. The flow leaving through the east boundary of the cell W is equal to the flow entering through the west boundary cell P.

	A	B	C	
1	VDP	0.8		
2	s	=-LN(1-VDP)		
3	mu	=-0.5*s*s	mu+1/2s^2=0	
4	visc	1		
5	dx	=1/20		EXP(
6	dy	0.05		
7	start	1	=IF(start=0,0,1)	
8			1	2
9			=C29	=D29
10	1	=V10+1	=IF(start=0,0,(lpe!C10*D10+lpe!B10*B10+lps!C10*C11+lps!C9*C9)/((lpe!C10+lpe!B10+lps!C10+lps!C9)))	=IF(s
11	2	=V11+1	=IF(start=0,0,(lpe!C11*D11+lpe!B11*B11+lps!C11*C12+lps!C10*C10)/((lpe!C11+lpe!B11+lps!C11+lps!C10)))	=IF(s
12	3	=V12+1	=IF(start=0,0,(lpe!C12*D12+lpe!B12*B12+lps!C12*C13+lps!C11*C11)/((lpe!C12+lpe!B12+lps!C12+lps!C11)))	=IF(s
13	4	=V13+1	=IF(start=0,0,(lpe!C13*D13+lpe!B13*B13+lps!C13*C14+lps!C12*C12)/((lpe!C13+lpe!B13+lps!C13+lps!C12)))	=IF(s
14	5	=V14+1	=IF(start=0,0,(lpe!C14*D14+lpe!B14*B14+lps!C14*C15+lps!C13*C13)/((lpe!C14+lpe!B14+lps!C14+lps!C13)))	=IF(s
15	6	=V15+1	=IF(start=0,0,(lpe!C15*D15+lpe!B15*B15+lps!C15*C16+lps!C14*C14)/((lpe!C15+lpe!B15+lps!C15+lps!C14)))	=IF(s
16	7	=V16+1	=IF(start=0,0,(lpe!C16*D16+lpe!B16*B16+lps!C16*C17+lps!C15*C15)/((lpe!C16+lpe!B16+lps!C16+lps!C15)))	=IF(s
17	8	=V17+1	=IF(start=0,0,(lpe!C17*D17+lpe!B17*B17+lps!C17*C18+lps!C16*C16)/((lpe!C17+lpe!B17+lps!C17+lps!C16)))	=IF(s
18	9	=V18+1	=IF(start=0,0,(lpe!C18*D18+lpe!B18*B18+lps!C18*C19+lps!C17*C17)/((lpe!C18+lpe!B18+lps!C18+lps!C17)))	=IF(s
19	10	=V19+1	=IF(start=0,0,(lpe!C19*D19+lpe!B19*B19+lps!C19*C20+lps!C18*C18)/((lpe!C19+lpe!B19+lps!C19+lps!C18)))	=IF(s
20	11	=V20+1	=IF(start=0,0,(lpe!C20*D20+lpe!B20*B20+lps!C20*C21+lps!C19*C19)/((lpe!C20+lpe!B20+lps!C20+lps!C19)))	=IF(s
21	12	=V21+1	=IF(start=0,0,(lpe!C21*D21+lpe!B21*B21+lps!C21*C22+lps!C20*C20)/((lpe!C21+lpe!B21+lps!C21+lps!C20)))	=IF(s
22	13	=V22+1	=IF(start=0,0,(lpe!C22*D22+lpe!B22*B22+lps!C22*C23+lps!C21*C21)/((lpe!C22+lpe!B22+lps!C22+lps!C21)))	=IF(s
23	14	=V23+1	=IF(start=0,0,(lpe!C23*D23+lpe!B23*B23+lps!C23*C24+lps!C22*C22)/((lpe!C23+lpe!B23+lps!C23+lps!C22)))	=IF(s
24	15	=V24+1	=IF(start=0,0,(lpe!C24*D24+lpe!B24*B24+lps!C24*C25+lps!C23*C23)/((lpe!C24+lpe!B24+lps!C24+lps!C23)))	=IF(s
25	16	=V25+1	=IF(start=0,0,(lpe!C25*D25+lpe!B25*B25+lps!C25*C26+lps!C24*C24)/((lpe!C25+lpe!B25+lps!C25+lps!C24)))	=IF(s
26	17	=V26+1	=IF(start=0,0,(lpe!C26*D26+lpe!B26*B26+lps!C26*C27+lps!C25*C25)/((lpe!C26+lpe!B26+lps!C26+lps!C25)))	=IF(s

Figure 2.19 The potential of the central cell labeled P is a weighted average of the surrounding cells, i.e., $\Phi_P = (\lambda_{PE}\Phi_E + \lambda_{PW}\Phi_W + \lambda_{PS}\Phi_S + \lambda_{PN}\Phi_N)/(\lambda_{PE} + \lambda_{PW} + \lambda_{PS} + \lambda_{PN})$. After iterating, all central cells Φ_P satisfy this equation, thus finding the complete potential field.

	A	B	C	D	E	F
1	VDP	0.8				
2	s	=-LN(1-VDP)				
3	mu	=-0.5*s*s	mu+1/2s^2=0			
4	visc	1				
5	dx	=1/20	EXP(NORMINV(RAND(),mu,s))			
6	dy	0.05				
7	start	1	=IF(start=0,0,1)			
8			1	2	3	4
9			=C29	=D29	=E29	=F29
10	1	=V10+1	=Ipe!C10*(calc!C10-calc!D10)	=Ipe!D10*(calc!D10-calc!E10)	=Ipe!E10*(calc!E10-calc!F10)	=Ipe!F10*(calc!F10-calc!G10)
11	2	=V11+1	=Ipe!C11*(calc!C11-calc!D11)	=Ipe!D11*(calc!D11-calc!E11)	=Ipe!E11*(calc!E11-calc!F11)	=Ipe!F11*(calc!F11-calc!G11)
12	3	=V12+1	=Ipe!C12*(calc!C12-calc!D12)	=Ipe!D12*(calc!D12-calc!E12)	=Ipe!E12*(calc!E12-calc!F12)	=Ipe!F12*(calc!F12-calc!G12)
13	4	=V13+1	=Ipe!C13*(calc!C13-calc!D13)	=Ipe!D13*(calc!D13-calc!E13)	=Ipe!E13*(calc!E13-calc!F13)	=Ipe!F13*(calc!F13-calc!G13)
14	5	=V14+1	=Ipe!C14*(calc!C14-calc!D14)	=Ipe!D14*(calc!D14-calc!E14)	=Ipe!E14*(calc!E14-calc!F14)	=Ipe!F14*(calc!F14-calc!G14)
15	6	=V15+1	=Ipe!C15*(calc!C15-calc!D15)	=Ipe!D15*(calc!D15-calc!E15)	=Ipe!E15*(calc!E15-calc!F15)	=Ipe!F15*(calc!F15-calc!G15)
16	7	=V16+1	=Ipe!C16*(calc!C16-calc!D16)	=Ipe!D16*(calc!D16-calc!E16)	=Ipe!E16*(calc!E16-calc!F16)	=Ipe!F16*(calc!F16-calc!G16)
17	8	=V17+1	=Ipe!C17*(calc!C17-calc!D17)	=Ipe!D17*(calc!D17-calc!E17)	=Ipe!E17*(calc!E17-calc!F17)	=Ipe!F17*(calc!F17-calc!G17)
18	9	=V18+1	=Ipe!C18*(calc!C18-calc!D18)	=Ipe!D18*(calc!D18-calc!E18)	=Ipe!E18*(calc!E18-calc!F18)	=Ipe!F18*(calc!F18-calc!G18)
19	10	=V19+1	=Ipe!C19*(calc!C19-calc!D19)	=Ipe!D19*(calc!D19-calc!E19)	=Ipe!E19*(calc!E19-calc!F19)	=Ipe!F19*(calc!F19-calc!G19)
20	11	=V20+1	=Ipe!C20*(calc!C20-calc!D20)	=Ipe!D20*(calc!D20-calc!E20)	=Ipe!E20*(calc!E20-calc!F20)	=Ipe!F20*(calc!F20-calc!G20)
21	12	=V21+1	=Ipe!C21*(calc!C21-calc!D21)	=Ipe!D21*(calc!D21-calc!E21)	=Ipe!E21*(calc!E21-calc!F21)	=Ipe!F21*(calc!F21-calc!G21)
22	13	=V22+1	=Ipe!C22*(calc!C22-calc!D22)	=Ipe!D22*(calc!D22-calc!E22)	=Ipe!E22*(calc!E22-calc!F22)	=Ipe!F22*(calc!F22-calc!G22)
23	14	=V23+1	=Ipe!C23*(calc!C23-calc!D23)	=Ipe!D23*(calc!D23-calc!E23)	=Ipe!E23*(calc!E23-calc!F23)	=Ipe!F23*(calc!F23-calc!G23)
24	15	=V24+1	=Ipe!C24*(calc!C24-calc!D24)	=Ipe!D24*(calc!D24-calc!E24)	=Ipe!E24*(calc!E24-calc!F24)	=Ipe!F24*(calc!F24-calc!G24)
25	16	=V25+1	=Ipe!C25*(calc!C25-calc!D25)	=Ipe!D25*(calc!D25-calc!E25)	=Ipe!E25*(calc!E25-calc!F25)	=Ipe!F25*(calc!F25-calc!G25)
26	17	=V26+1	=Ipe!C26*(calc!C26-calc!D26)	=Ipe!D26*(calc!D26-calc!E26)	=Ipe!E26*(calc!E26-calc!F26)	=Ipe!F26*(calc!F26-calc!G26)
27	18	=V27+1	=Ipe!C27*(calc!C27-calc!D27)	=Ipe!D27*(calc!D27-calc!E27)	=Ipe!E27*(calc!E27-calc!F27)	=Ipe!F27*(calc!F27-calc!G27)
28	19	=V28+1	=Ipe!C28*(calc!C28-calc!D28)	=Ipe!D28*(calc!D28-calc!E28)	=Ipe!E28*(calc!E28-calc!F28)	=Ipe!F28*(calc!F28-calc!G28)
29	20	=V29+1	=Ipe!C29*(calc!C29-calc!D29)	=Ipe!D29*(calc!D29-calc!E29)	=Ipe!E29*(calc!E29-calc!F29)	=Ipe!F29*(calc!F29-calc!G29)
30			=C10	=D10	=E10	=F10
31		sum	=SUM(C10:C29)	=SUM(D10:D29)	=SUM(E10:E29)	=SUM(F10:F29)
32		arithmetic	=AVERAGE(inputperm)			
33		geometric	=GEOMEAN(inputperm)			
34		harm	=HARMEAN(inputperm)			

Figure 2.20 Columns 1 calculates the flow between the left cell (C) and the right cell (D) of column one. Adding all elements in a single column calculates the flow rate from left to right. If done correctly this added values will be the same after summing the elements in the second, third etc. column. As we have chosen the viscosity equal to one, the total potential difference equal to one, the size of the domain equal to one, the summed flow rates satisfy $Q = -K/\mu(=1)\Delta\Phi/\Delta x = K$ because $\Delta\Phi/\Delta x$ is the potential gradient over the entire domain.

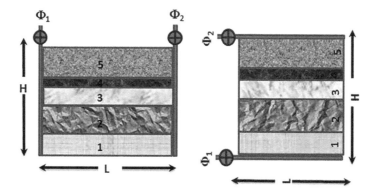

Figure 2.21 Flow through a stack of horizontal layers with characteristic thickness h_i and permeability k_i with $i = 1,...,5$. In both cases the flow is from Φ_1 to Φ_2, i.e., left to right or bottom to top.

$y-$direction. Each of the $i = 1,...,N$ has a height h_i, a porosity φ_i and a permeability k_i. We want to express the permeability of the structure in terms of a single permeability tensor in Darcy's law

$$\mathbf{u} = -\frac{\mathbf{k}}{\mu} \cdot (\mathbf{grad}\, p + \rho g \mathbf{e}_z) \, , \tag{2.71}$$

where k [m^2] denotes the permeability tensor, μ [Pa s] denotes the viscosity, ρ [kg/m^3] the density, g [m/s^2] the acceleration due to gravity. Furthermore, p [Pa] is the pressure and $\mathbf{e}_z$ is the unit vector in the vertical upward direction. The Darcy velocity $\mathbf{u}$ also called the specific discharge is a volume flux $[m^3/m^2/s]$. As the density is considered constant, we can define a potential $\phi = p + \rho g z$ such that

$$\mathbf{u} = -\frac{\mathbf{k}}{\mu} \cdot \mathbf{grad}\ \phi\ . \tag{2.72}$$

First, we consider horizontal flow (Figure 2.21a). We apply a potential $\phi_1 = p_1 + \rho g z_1$ on the left and $\phi_2 = p_2 + \rho g z_2$ on the right. The total discharge Q_x can be written as

$$Q_x = \sum_{i=1}^{N} u_{ix} h_i = \sum_{i=1}^{N} \frac{k_i h_i}{\mu} \frac{\phi_1 - \phi_2}{L} = \frac{k_h H}{\mu} \frac{\phi_1 - \phi_2}{L}\ , \tag{2.73}$$

and it follows for the effective permeability k_h in horizontal direction

$$k_h = \sum_{i=1}^{N} \frac{k_i h_i}{H}, \quad k_h = \frac{1}{H} \int_0^H k dx\ . \tag{2.74}$$

In other words the permeability k_h is the "arithmetic" average of the permeabilities in each layer. Second, we consider horizontal flow as in Figure 2.21b. We apply a potential $\phi_1 = p_1 + \rho g z_1$ on the left and $\phi_2 = p_2 + \rho g z_2$ on the right. The potential gradient $\frac{d\phi}{dx} = -\frac{\mu u_x}{k_i}$. We thus obtain

$$\int_0^H \frac{d\phi}{dx} dx = -\mu u_x \int_0^H \frac{1}{k(x)} dx = -\frac{\mu u_x}{k_v} H$$

$$\phi(H) - \phi(0) = -\mu u_x \sum_{i=1}^{N} \frac{h_i}{k_i} = -\mu u_x \frac{H}{k_v}\ , \tag{2.75}$$

and it follows that for the effective permeability k_v

$$k_v = \frac{1}{\sum_{i=1}^{N} \frac{h_i}{k_i H}} \quad k_v = \frac{H}{\int_0^H \frac{1}{k} dx}\ . \tag{2.76}$$

In other words the permeability k_v is the harmonic average of the permeabilities in each layer.

We do want to mention the geometric average. It states that the logarithm of the effective permeability is the arithmetic mean of the logarithm of the constituting permeabilities. It is applied when an area is divided into a number of blocks of equal size. If the permeabilities of the different blocks are uncorrelated, the geometric average is appropriate (see, however, the last exercise) in 2D

$$k_{geo} = \prod_{i=1}^{n} \sqrt[n]{k_i}\ . \tag{2.77}$$

Below we discuss more sophisticated methods [134] to find average permeabilities.

2.6.2   THE AVERAGED PROBLEM IN TWO SPACE DIMENSIONS

We consider a block $\Omega_3 \in R^3$ with length L, width W and with height H. The length coincides with the x-direction and the width with the y-direction. The x-direction and y-direction are direction in the dip and strike direction, constituting the so-called layer directions. The third direction ζ is perpendicular to a plane embedded in the layer and is also called the X-dip (cross-dip) direction. We like to reserve here the z-direction as the vertical upward direction. This z-axis can make any angle with the x-axis and the y-axis without loss in generality. We like to integrate equation over the ζ direction between the bottom of the layer and the top of the layer. We assume that the layer is bounded from above by an impermeable cap-rock and from below by the impermeable base rock. We integrate equation

$$\frac{\partial}{\partial x}\left(\frac{k}{\mu}\frac{\partial \phi}{\partial x}\right) + \frac{\partial}{\partial y}\left(\frac{k}{\mu}\frac{\partial \phi}{\partial y}\right) + \frac{\partial}{\partial \zeta}\left(\frac{k}{\mu}\frac{\partial \phi}{\partial \zeta}\right) = 0, \qquad (2.78)$$

where we have replaced z by ζ to indicate that the X-dip direction is not necessarily vertical. The assumption that is required to get a useful result if we integrate over the coordinate ζ, is that $\phi(x,y,\zeta) \rightarrow \phi_o(x,y)$. In other words we assume that the potential does not depend on ζ. This assumption appears to be reasonable (see [143], p. 206) if

$$\sqrt{\frac{k_\zeta}{k_{//}}}\frac{L}{H} >> 1, \qquad (2.79)$$

the permeability ratio corrected for the aspect ratio L/H is much bigger than one. If the permeability is heterogeneous, one may replace L by the correlation length. If such a condition is not satisfied in a numerical calculation, it means that we need to subdivide the reservoir into more grid blocks in the ζ−direction. In the case that the potential does not depend on ζ, we find a straightforward equation after integration versus ζ. The height of the layer $H = H(x,y)$, i.e., is a function of the coordinates in the dip and strike direction. As the upper boundary for integration, we drop this dependence in the notation.

$$\int_0^H \left(\frac{\partial}{\partial x}\left(\frac{k}{\mu}\frac{\partial \phi}{\partial x}\right) + \frac{\partial}{\partial y}\left(\frac{k}{\mu}\frac{\partial \phi}{\partial y}\right) + \frac{\partial}{\partial \zeta}\left(\frac{k}{\mu}\frac{\partial \phi}{\partial \zeta}\right)\right) d\zeta = 0. \qquad (2.80)$$

Note that the last term in Eq. (2.78) drops out upon integration. The porosity φ and the permeability k are the only quantities that depend on ζ. Moreover, we use the following abbreviations $\bar{k}H = \int_0^H k d\zeta$ and $\bar{\varphi}H = \int_0^H \varphi d\zeta$. We obtain

$$\frac{\partial}{\partial x}\left(\frac{\bar{k}}{\mu}\frac{\partial \phi}{\partial x}\right) + \frac{\partial}{\partial y}\left(\frac{\bar{k}}{\mu}\frac{\partial \phi}{\partial y}\right) = 0. \qquad (2.81)$$

This equation looks exactly the same as Eq. (2.78) except that the dependence on ζ has dropped out and that the permeability and porosity are X-dip averaged quantities. This means that in many text books Eq. (2.81) is presented as being as good as the corresponding $3 - D$ equation. We now know that Eq. (2.81) is only valid when condition (2.79) is satisfied

2.6.3   EFFECTIVE MEDIUM APPROXIMATION

The effective medium approximation was designed for calculating average resistances in resistor networks. It considers one resistor drawn from a distribution of resistors embedded in a "sea" of average resistances. The equation tells us that the average contribution to the excess voltage (potential) from all the resistances drawn from the distribution must be zero. For permeabilities with a probability density distribution $h(k)$ the result reads (see Appendix 2.C),

$$\int_0^\infty h(k)\frac{k - k_{eff}}{k + (\gamma^{-1} - 1)k_{eff}}dk = 0, \qquad (2.82)$$

where k_{eff} is the average permeability where we are after. Being designed for networks it contains a geometric factor γ, which we take $\frac{3}{10}$ for our purposes. This geometric factor γ is in $2D$ equal to the inverse percolation threshold. All one needs is a integration routine to compute the integral for a given estimated value of k_{eff} and a "goal seek" routine to find the value of k_{eff} from Eq. (2.82). An idea of a derivation of Eq. (2.82) is given in appendix 2.C.

2.6.4   PITFALL: A CORRECTLY AVERAGED PERMEABILITY CAN STILL LEAD TO ERRONEOUS PRODUCTION FORECASTS

It is very simple to show that a correctly averaged permeability can lead to erroneous production forecasts.

We suppose a two-layered (Figure 2.22a) structure originally filled with oil. Layer 1 has a permeability of one Darcy, and layer 2 has a permeability of two Darcy. The oil is displaced with a solvent, which differs as to its physical properties from the oil only by its color. We disregard

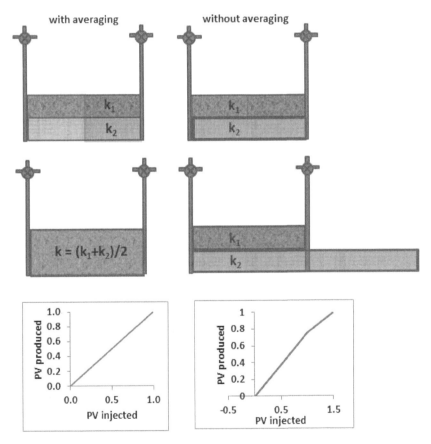

Figure 2.22   A perfect averaging procedure for the permeability can still lead to an incorrect production forecast. In this example, the oil and water have the same physical properties and can only be distinguished by their color. Here oil is represented by the reddish color and water by the blue color. Left the production history with an averaged permeability; one pore volume of water injected leads to one porevolume (PV) of oil produced. Right represents the right plots with $k_2 = 2k_1$; now one PV of oil is only produced after injection of 1.5 PV of water.

mixing effects. We assume that the porosities in both layers are the same. From Darcy's law, it follows that the Darcy velocity in layer 2 is twice the Darcy velocity in layer 1, i.e., $u_2 = 2u_1$. After injection of $\frac{3}{4}$ pore volume breakthrough of solvent occurs. All oil has been recovered only after injection of $\frac{3}{2}$ pore volume. This is shown in figure 2.22b. When we replace the two-layered structure by a homogeneous reservoir of with the average permeability of $\frac{3}{2}$ Darcy we obtain the right potential distribution. However, breakthrough will occur after injection of one pore volume, simultaneously with the moment that all oil has been recovered. This simple example shows how easy it is to misinterpret results obtained with a simulator. This may be felt as unfair after all the trouble which was put in to obtain the effective permeability [107]. We conclude that an effective permeability hopefully leads to the right potential distribution but may fail to predict the right recovery curve (see however [117, 139]).

2.7   NUMERICAL UPSCALING

2.7.1   FINITE VOLUME METHOD IN $2 - D$; THE PRESSURE FORMULATION

In the numerical calculations in Ω_2, i.e., in two-dimensions, we use the grid cell shown in Figure 2.23. The central cell is labeled $P(oint)$. It contains four boundaries, i.e., the north, south, west and east boundary, which are abbreviated as nb, sb, sb, eb respectively. The flow [m³/m/s] over these boundaries is denoted by $q_{ny}, q_{py}, q_{wx}, q_{px}$. The first index denotes the cell label (north, point and west) and the second label the direction. There are two flows labeled for each cell, i.e., the one passing over the east boundary in the $x-$direction and the one passing over the south boundary in the $y-$direction. This labelling is arbitrary, but we like to keep this notation throughout this booklet, as it avoids many errors. The corners of the central grid cell are labeled nw, ne, se, sw indicating northwest, etc. Not all of these indicators are used for the numerical potential flow calculation, but they will pop up in some form in our examples

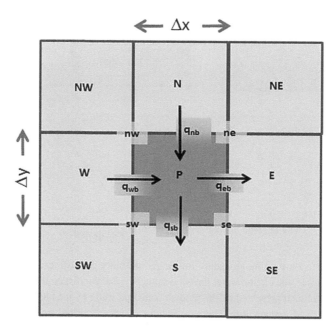

Figure 2.23   The grid cells used in 2-D numerical calculations.

later and we like to keep a uniform nomenclature. The adjacent cells to the central cell are $NW, N, NE, E, SE, S, SW, W$ corresponding to wind directions in obvious notation. The grid cell length in the x-direction is Δx and in the y-direction Δy. In the examples here, all grid cells have the same size.

The conservation law applied for the grid cell reads

$$q_{wx} - q_{px} + q_{ny} - q_{py} - q_{prod} + q_{inj} = 0. \tag{2.83}$$

We approximate the flows using Darcy's law as follows

$$q_{wx} = -\frac{k_{WP} H \Delta y}{\mu} \frac{\phi_P - \phi_W}{\Delta x} := -\lambda_{WP} (\phi_P - \phi_W),$$

$$q_{px} = -\frac{k_{PE} H \Delta y}{\mu} \frac{\phi_E - \phi_P}{\Delta x} := -\lambda_{PE} (\phi_E - \phi_P),$$

$$q_{ny} = -\frac{k_{NP} H \Delta x}{\mu} \frac{\phi_P - \phi_N}{\Delta y} := -\lambda_{NP} (\phi_P - \phi_N),$$

$$q_{py} = -\frac{k_{PS} H \Delta x}{\mu} \frac{\phi_S - \phi_P}{\Delta y} := -\lambda_{PS} (\phi_S - \phi_P). \tag{2.84}$$

Here we have used k_{WP} to denote the geometric average permeability between the permeability of the west cell and the central cell and equivalently for the other combinations, i.e.,

$$k_{WP} = \sqrt{k_W k_P}, \ k_{PE} = \sqrt{k_E k_P},$$

$$k_{NP} = \sqrt{k_N k_P}, \ k_{PS} = \sqrt{k_S k_P}.$$

where

$$k_\beta = \frac{1}{H} \int_0^H (k(y))_\beta \, dy \quad \text{with} \beta = P, W, E, S, N, \tag{2.85}$$

i.e., the arithmetically averaged permeability over the height. Naively one may argue that instead of a geometric average permeability a harmonic average permeability must be used. As shown in [190], only 1-D simulations the harmonic average should be used, but for 2-D simulations, the geometric average is preferred, whereas for 3-D simulations, a third-order powerlaw [144, 190] gives better results.

We note here that the cross-dip (perpendicular to the layer) permeability is in practice a factor of 10–100 lower than the areal permeabilities, i.e., in the $x-$direction and $y-$direction. Such an averaging would we correct if (see Lake, "Enhanced Oil Recovery") we would have vertical equilibrium, i.e., if the potential difference over the grid cell in the areal direction is much bigger than the potential gradient in the $z-$direction. If the following dimensionless inequality is satisfied, then such an arithmetical averaging procedure can be defended. i.e.,

$$\sqrt{\frac{k_z}{k_{//}}} \frac{L}{H} \gg 1. \tag{2.86}$$

If the permeability is heterogeneous, one may replace L by the correlation length. If such a condition is not satisfied in a numerical calculation, it means that we need to subdivide the reservoir into more grid blocks in the $\zeta-$direction. In a $3-D$ simulation, one can also use this criterium as a way to determine the number of grid cells in the $\zeta-$direction, by using $H = \Delta \zeta$ in the equation above. In general, the grid size lengths are smaller than the correlation lengths.

The ratio of harmonically averaged permeabilities and the viscosity are the mobilities. Here we use the symbols λ_{WP} etc. for a product of quantities also involving grid cell dimensions, e.g., $\lambda_{WP} = \frac{k_{WP} H \Delta y}{\mu \Delta x}$ etc. (see Eq. (2.84)), in the interest of brief notation.

We obtain

$$-\lambda_{WP} \left(\phi_P - \phi_W \right) + \lambda_{PE} \left(\phi_E - \phi_P \right) - \lambda_{NP} \left(\phi_P - \phi_N \right) + \lambda_{PS} \left(\phi_S - \phi_P \right) - q_{prod} + q_{inj} = 0. \quad (2.87)$$

Note that q_{prod}, q_{inj} as used here mean the average. We distinguish between the source terms where the flow is given and source terms that result from a given well flow potential at the well (see [64]), i.e.,

$$q_{prod} = q_p + PI \left(\phi_P - \phi_{wf} \right), \quad (2.88)$$

$$q_{inj} = q_i - II \left(\left(\phi_P - \phi_{wf} \right). \quad (2.89)$$

Pieceman [184] proposed the following equation for the productivity (PI) /injection (II) indices (see also [185]

$$PI = \frac{2\pi k H}{\mu B_o \left(\ln \left(\frac{0.2\sqrt{\Delta x \Delta y}}{r_w} \right) + S \right)}, \quad (2.90)$$

$$II = \frac{2\pi k H}{\mu B_o \left(\ln \left(\frac{0.2\sqrt{\Delta x \Delta y}}{r_w} \right) + S \right)}, \quad (2.91)$$

where H is the height of the layer and k is the arithmetically averaged permeability over the height (as in Eq. (2.85)). Furthermore, we have that A is the area considered (in a numerical calculation the grid surface ($\Delta x \Delta y$)). Finally S is the skin factor, which takes into account the well impairment, i.e., a reduction of the flow rate due to an increased resistance due to a damaged formation around the well (particle clogging) (Figure 2.24).

In this case, there are two convenient forms of Eq. (2.87). One of them is to explicitly write ϕ_P in terms of the rest, i.e.,

$$\phi_P = \frac{\lambda_{WP} \phi_W + \lambda_{PE} \phi_E + \lambda_{NP} \phi_N + \lambda_{PS} \phi_S - q_p + q_i + PI \phi_{wf} + II \phi_{wf}}{\lambda_{WP} + \lambda_{PE} + \lambda_{NP} + \lambda_{PS} + PI + II}. \quad (2.92)$$

The other method uses almost the same equation, i.e., (see Numerical Recipes p. 868),

$$\phi_P^{new} = \omega \frac{\lambda_{WP} \phi_W + \lambda_{PE} \phi_E + \lambda_{NP} \phi_N + \lambda_{PS} \phi_S - q_p + q_i + PI \phi_{wf} + II \phi_{wf}}{\lambda_{WP} + \lambda_{PE} + \lambda_{NP} + \lambda_{PS} + PI + II} + (1 - \omega) \phi_P^{old}, \quad (2.93)$$

where the overrelaxation parameter (a typical value would be $\omega = 1.5$) can be adjusted for the best results. Note that for $\omega = 1$ we just have Eq. (2.92).

2.7.2   THE FINITE AREA METHOD; THE STREAM FUNCTION FORMULATION

The two most common terms for transport phenomena are expressed either in terms of the divergence of the mass flux is equal to some source term or that the curl of a gradient of a vector quantity is equal to zero. The divergence term can be expressed in terms of the mass flux out of a given volume, thus giving rise to the finite volume method described above. The terms with the curl give rise to the circulation over an enclosed area is zero or equal to some "source" term. In this case, we have to consider the area and analogously to the finite volume method we call the method to derive numerical schemes the finite area method. An application of the finite area

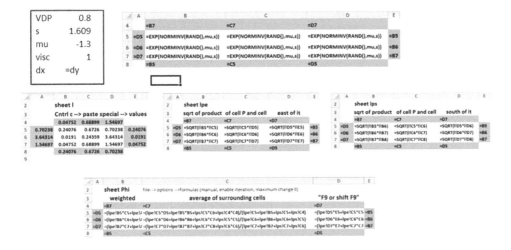

Figure 2.24   Procedure to carry out a numerical calculation of a heterogeneous permeability field.

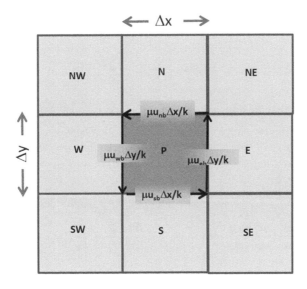

Figure 2.25   For the streamfunction formulation we use that $\oint_S \frac{\mu\vec{u}}{k} \cdot d\mathbf{s} = 0$.

method is when we try to derive a numerical scheme for the stream function. We rewrite Darcy's law

$$\frac{\mu\mathbf{u}}{k} = (\mathbf{grad}\,(p + \rho gz))\;\; or\;\; \oint_S \frac{\mu\vec{u}}{k} \cdot ds = 0. \tag{2.94}$$

The circular integral can be numerically approximated (see Eq. (2.25))

$$\oint_S \frac{\mu\mathbf{u}}{k} \cdot d\mathbf{s} \quad \approx -\frac{\Delta y}{2}\left(\frac{\mu}{k_P} + \frac{\mu}{k_E}\right)u_{y,eb} - \frac{\Delta x}{2}\left(\frac{\mu}{k_S} + \frac{\mu}{k_P}\right)u_{x,sb}, \tag{2.95}$$

$$+\frac{\Delta y}{2}\left(\frac{\mu}{k_W} + \frac{\mu}{k_P}\right)u_{y,wb} + \frac{\Delta x}{2}\left(\frac{\mu}{k_P} + \frac{\mu}{k_N}\right)u_{x,nb} = 0. \tag{2.96}$$

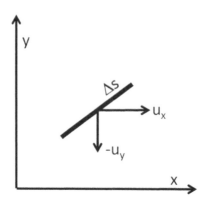

Figure 2.26   Flow through a surface Δs.

This can be combined with the mass conservation law **div** $u = 0$ (see Eq. (2.5)), with the density $\rho =$ constant), which implies in two dimensions that we can define a streamfunction ψ that satisfies

$$u_x = \frac{\partial \psi}{\partial y} \qquad\qquad u_y = -\frac{\partial \psi}{\partial x}$$

$$\mathbf{div} \cdot \mathbf{u} \;=\; \frac{\partial u_x}{\partial x} + \frac{\partial u_y}{\partial y} = \frac{\partial^2 \psi}{\partial x \partial y} - \frac{\partial^2 \psi}{\partial y \partial x} = 0 \,. \tag{2.97}$$

Figure 2.26 shows that the flow q through a surface Δs is equal to $dq = u_x dy - u_y dx = \frac{\partial \psi}{\partial y} dy + \frac{\partial \psi}{\partial x} dx = d\psi$. This has a very simple interpretation, i.e., that between one curve of constant value of $\psi = \psi_1$ and one curve of constant value of $\psi = \psi_2$ the flow $q = \psi_2 - \psi_1$.

We substitute $u_{y,PE} = u_{y,eb} = -(\psi_E - \psi_P)/\Delta x$ and $u_{x,PN} = u_{x,nb} = (\psi_N - \psi_P)/\Delta y$ to obtain

$$\frac{\Delta y}{2\Delta x}\Big(\frac{\mu}{k_P} + \frac{\mu}{k_E}\Big)(\psi_E - \psi_P) - \frac{\Delta x}{2\Delta y}\Big(\frac{\mu}{k_S} + \frac{\mu}{k_P}\Big)(\psi_P - \psi_S),$$

$$-\frac{\Delta y}{2\Delta x}\Big(\frac{\mu}{k_W} + \frac{\mu}{k_P}\Big)(\psi_P - \psi_W) + \frac{\Delta x}{2\Delta y}\Big(\frac{\mu}{k_P} + \frac{\mu}{k_N}\Big)(\psi_N - \psi_P) = 0,$$

and obtain the following scheme in terms of the stream function ψ

$$\psi_P = \frac{\xi_{y,S}\,\psi_S + \xi_{y,P}\,\psi_N + \xi_{x,P}\,\psi_E + \xi_{x,W}\,\psi_W}{\xi_{y,S} + \xi_{y,P} + \xi_{x,P} + \xi_{x,W}}, \tag{2.98}$$

where $\xi_{x,P} = \frac{\Delta y}{2\Delta x}\Big(\frac{\mu}{k_P} + \frac{\mu}{k_E}\Big)$, $\xi_{y,P} = \frac{\Delta x}{2\Delta y}\Big(\frac{\mu}{k_P} + \frac{\mu}{k_N}\Big)$, $\xi_{y,P} = \frac{\Delta x}{2\Delta y}\Big(\frac{\mu}{k_S} + \frac{\mu}{k_P}\Big)$, $\xi_{x,W} = \frac{\Delta y}{2\Delta x}\Big(\frac{\mu}{k_W} + \frac{\mu}{k_P}\Big)$ and thus contain both the arithmetic average reverse mobility, i.e., μ/k and the grid size ratio's. Wells cannot be modeled when we have a streamfunction. The values between streamlines (contours that have the same streamfunction values) give the amount of flow between them. Streamlines are perpendicular to a constant potential surface. When we have such a constant potential surface for instance at the left (Figure 2.25), then the gradient of the stream function perpendicular to the constant potential surface will be zero, i.e., $\psi_P = \psi_W$ and we obtain

$$\psi_P = \frac{\xi_{y,S}\,\psi_S + \xi_{x,P}\,\psi_E + \xi_{y,P}\,\psi_N}{\xi_{y,S} + \xi_{x,P} + \xi_{y,P}}. \tag{2.99}$$

2.7.3   FINITE ELEMENT METHOD (AFTER F. VERMOLEN)

The general opinion is that for a field with piecewise constant permeabilities, the finite element method is superior. Here we like to give a formulation (in 2-D) that can also easily implemented in EXCEL. Clearly there are packages, which do this work for you like SEPRAN and FEM-lab, and COMSOL. Consider the equation

$$\nabla \cdot (\mathcal{K} \nabla \Phi) = 0, \tag{2.100}$$

where $\mathcal{K} = \mathcal{K}(x, y)$ denotes the mobility (permeability/viscosity) field. We integrate the equation over the whole domain Ω with boundary $\delta\Omega$ and after multiplying by a test function v. The test function is of compact support, i.e., only non zero in a small domain. Moreover v is differentiable. We obtain

$$\int_{\Omega} \nabla \cdot (\mathcal{K} \nabla \Phi) v dA, \tag{2.101}$$

where $A = dxdy$. We apply Green's theorem to rewrite Eq. (2.101) as

$$\int_{\Omega} \nabla \cdot (\mathcal{K} \nabla \Phi) v dA = \int_{\Omega} \nabla \cdot (\mathcal{K} \nabla \Phi v) dA - \int_{\Omega} \mathcal{K} \nabla \Phi \cdot \nabla v dA \tag{2.102}$$

and we apply Gauss to obtain

$$\int_{\Omega} \nabla \cdot (\mathcal{K} \nabla \Phi) v dA = \int_{\delta\Omega} \mathcal{K} \frac{\partial \Phi}{\partial n} v ds - \int_{\Omega} \mathcal{K} \nabla \Phi \cdot \nabla v dA = 0. \tag{2.103}$$

This is called the weak formulation. We choose $v = 0$ whenever Φ is given and we assume that we can write

$$\Phi = \sum_j \phi_j v_j. \tag{2.104}$$

Substitution in the second term after the equal sign leads to

$$\int_{\Omega} \mathcal{K} \nabla \Phi \cdot \nabla v dA = \int_{\Omega} \mathcal{K} \phi_j \nabla v_j \cdot \nabla v dA. \tag{2.105}$$

Subsequently we obtain a linear system in terms of the coefficients ϕ_j. Let us assume that our elements are squares with the node points p, n, ne etc. (Figure 2.27).

We use a bilinear basis function, with support $2\Delta x \times 2\Delta x$. In other words, we use square grids. The basis function is of compact support and only nonzero in a domain Ω. It can be represented as for any grid cell as

$$v_{ij} = \left(1 - \frac{|x - x_i|}{\Delta x}\right)\left(1 - \frac{|y - y_j|}{\Delta x}\right) \text{ in } \Omega = \left\{x, y| \; |x - x_i| \le \Delta x, |y - y_j| \le \Delta x\right\}. \tag{2.106}$$

The point p is at the origin, i.e., $x_i = 0, y_i = 0$. The derivatives of the basis functions can be written as

$$\frac{\partial v_{ij}}{\partial x} = -\frac{sign(x - x_i)}{\Delta x}\left(1 - \frac{|y - y_j|}{\Delta x}\right),$$

$$\frac{\partial v_{ij}}{\partial y} = -\frac{sign(y - y_j)}{\Delta x}\left(1 - \frac{|x - x_i|}{\Delta x}\right).$$

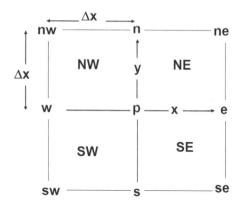

Figure 2.27 Point centered grid for finite element calculations.

Figure 2.28 The setting in the calculation tab for doing iterative computations. This tab is found starting at "Files-Options-Formulas". Use the paragraph on "Calculation Options". Set the "workbook calculation to manual" and switch off the "calculation before save". Enable iterative calculation and choose the maximum number of iterations, e.g., 100. Set the maximum change equal to zero. Press "OK".

We use that the central cell p is interacting with all other cells α i.e $\alpha = (p, n = e, ne, nw = se, w = s, sw)$

$$\sum_{j=1}^{N} \int_{-L/2}^{L/2} \int_{-L/2}^{L/2} \mathcal{K}\phi_j \nabla v_j \cdot \nabla v_p dA = \sum_{\alpha} \int_0^{\Delta x} \int_0^{\Delta x} \phi_\alpha \mathcal{K}_{NE} \nabla v_\alpha \cdot \nabla v_p dA$$

$$+ \sum_{\alpha} \int_{-\Delta x}^{0} \int_0^{\Delta x} \phi_\alpha \mathcal{K}_{NW} \nabla v_\alpha \cdot \nabla v_p dA + \sum_{\alpha} \int_0^{\Delta x} \int_{-\Delta x}^{0} \phi_\alpha \mathcal{K}_{SE} \nabla v_\alpha \cdot \nabla v_p dA$$

$$+ \sum_{\alpha} \int_{-\Delta x}^{0} \int_{-\Delta x}^{0} \phi_\alpha \mathcal{K}_{SW} \nabla v_\alpha \cdot \nabla v_p dA = 0,$$

where the sum runs over all grid cells N over the entire domain $L \times L$. However, only the base functions $(p, n = e, ne, nw = se, w = s, sw)$ contribute and therefore the integration is limited to the four squares shown in Figure 2.27. We use MAPLE to evaluate the integrals

$$\sum_{\alpha} \int_0^{\Delta x} \int_0^{\Delta x} \phi_\alpha \mathcal{K}_{NE} \nabla v_\alpha \cdot \nabla v_p dA = \mathcal{K}_{NE} \left(-\frac{1}{3}\phi_{ne} - \frac{1}{6}\phi_n + \frac{2}{3}\phi_p - \frac{1}{6}\phi_e \right)$$

$$\sum_{\alpha} \int_{-\Delta x}^{0} \int_{0}^{\Delta x} \phi_{\alpha} \mathcal{K}_{NW} \nabla v_{\alpha} \cdot \nabla v_p dA = \mathcal{K}_{NW} \left(-\frac{1}{3}\phi_{nw} - \frac{1}{6}\phi_n + \frac{2}{3}\phi_p - \frac{1}{6}\phi_w \right)$$

$$\sum_{\alpha} \int_{0}^{\Delta x} \int_{-\Delta x}^{0} \phi_{\alpha} \mathcal{K}_{SE} \nabla v_{\alpha} \cdot \nabla v_p dA = \mathcal{K}_{SE} \left(-\frac{1}{3}\phi_{se} - \frac{1}{6}\phi_s + \frac{2}{3}\phi_p - \frac{1}{6}\phi_e \right)$$

$$\sum_{\alpha} \int_{-\Delta x}^{0} \int_{-\Delta x}^{0} \phi_{\alpha} \mathcal{K}_{SW} \nabla \mathbf{v}_{\alpha} \cdot \nabla \mathbf{v}_p dA = \mathcal{K}_{SW} \left(-\frac{1}{3}\phi_{sw} - \frac{1}{6}\phi_s + \frac{2}{3}\phi_p - \frac{1}{6}\phi_w \right).$$

Consequently our algorithm to obtain the potential field is as follows

$$\phi_p = \mathcal{K}_{NE} \left(-\frac{1}{3}\phi_{ne} - \frac{1}{6}\phi_n + \frac{2}{3}\phi_p - \frac{1}{6}\phi_e \right) + \mathcal{K}_{NW} \left(-\frac{1}{3}\phi_{nw} - \frac{1}{6}\phi_n + \frac{2}{3}\phi_p - \frac{1}{6}\phi_w \right)$$

$$+ \mathcal{K}_{SE} \left(-\frac{1}{3}\phi_{se} - \frac{1}{6}\phi_s + \frac{2}{3}\phi_p - \frac{1}{6}\phi_e \right) + \mathcal{K}_{SW} \left(-\frac{1}{3}\phi_{sw} - \frac{1}{6}\phi_s + \frac{2}{3}\phi_p - \frac{1}{6}\phi_w \right).$$

$$\phi_p = \frac{ \begin{aligned} &2\left(\mathcal{K}_{NE}\phi_{ne} + \mathcal{K}_{SE}\phi_{se} + \mathcal{K}_{SW}\phi_{sw} + \mathcal{K}_{NW}\phi_{nw}\right) \\ &+ \left(\mathcal{K}_{NE}+\mathcal{K}_{NW}\right)\phi_n + \left(\mathcal{K}_{SW}+\mathcal{K}_{SE}\right)\phi_s + \left(\mathcal{K}_{NW}+\mathcal{K}_{SW}\right)\phi_w + \left(\mathcal{K}_{NE}+\mathcal{K}_{SE}\right)\phi_e \end{aligned} }{4\left(\mathcal{K}_{NE}+\mathcal{K}_{SE}+\mathcal{K}_{NW}+\mathcal{K}_{SW}\right)}.$$

2.7.4 FLOW CALCULATION

Our interest is to find the flow from the p cell to the n cell, i.e.,

$$Q = -\int_{0}^{\Delta y} \mathcal{K}_{NW} \frac{\partial \Phi}{\partial x} = -\sum_{j} \int_{0}^{\Delta y} \mathcal{K}_{NW}\phi_j \frac{\partial v_j}{\partial x} \tag{2.107}$$

In the domain of interest, we can write with Eq. (2.107). We like to use the abbreviation a for Δx

$$\frac{Q}{\mathcal{K}_{NW}} = -\int_{0}^{a} \left(\phi_p \frac{\partial v_p}{\partial x} + \phi_n \frac{\partial v_n}{\partial x} + \phi_{nw} \frac{\partial v_{nw}}{\partial x} + \phi_w \frac{\partial v_w}{\partial x} \right) dy$$

$$= -\int_{0}^{a} \left(\frac{\phi_p}{a}\left(1-\frac{y}{a}\right) + \frac{\phi_n}{a}\left(1-\frac{a-y}{a}\right) - \frac{\phi_{nw}}{a}\left(1-\frac{a-y}{a}\right) - \frac{\phi_w}{a}\left(1-\frac{y}{a}\right) \right) dy$$

$$= -\int_{0}^{a} \left(\frac{\phi_p}{a}\left(1-\frac{y}{a}\right) + \frac{\phi_n}{a}\frac{y}{a} - \frac{\phi_{nw}}{a}\frac{y}{a} - \frac{\phi_w}{a}\left(1-\frac{y}{a}\right) \right) dy$$

$$= \frac{1}{2}\left(\phi_w - \phi_p - \phi_n + \phi_{nw}\right).$$

In the same way, it is possible to compute

$$\frac{Q}{\mathcal{K}_{NE}} = -\int_{0}^{a} \left(\phi_p \frac{\partial v_p}{\partial x} + \phi_n \frac{\partial v_n}{\partial x} + \phi_{ne} \frac{\partial v_{ne}}{\partial x} + \phi_e \frac{\partial v_e}{\partial x} \right) dy$$

$$= \frac{1}{2}\left(\phi_n - \phi_e + \phi_p - \phi_{en}\right).$$

2.A   FINITE VOLUME METHOD IN EXCEL

In this case, we choose to use eight sheets (Figure 2.29). The first sheet contains the input data and the control variables to run the worksheet. The sheet *perm* contains the permeability-height product distribution in the layer. The sheets *labpe* and *labps* are required to obtain the grid size length corrected geometrically averaged mobilities, i.e., λ_{PE} and λ_{PS} in the $x-$direction and $y-$direction, respectively. The sheet *phiwf* contains the well flow pressures in the cells containing production or injection wells. The sheet *pindex* contains the productivity or injectivity index in the grid cells containing wells. The sheet *wells* contains flow rates in the cells containing a well for which the rate is specified. The sheet *flow* performs the actual Gauss-Seidel iteration with or without over-relaxation. We have to keep in mind that EXCEL likes to calculate the sheets in an alphabetical order of the sheet names. The calculations in flow are iterative and involve circular references, i.e., the calculation in a cell needs the value of the cell itself to do the calculation. This is possible by setting the calculation procedure in the appropriate mode. On the menu, go to *Tools-Options* and choose the calculation tab. Use the setting as shown in Figure 2.28

Initialize the sheet by assigning zero to the start value in the *data* sheet (this is the cell to the right of the word "start") and press *F*9 a few times. Then go to the sheet *flow* and press *Shift F*9. The maximum number of iterations is 32767. Of course one can do more iterations if one writes a macro, but this we leave to the interested expert in spreadsheet programming.

2.A.1   THE DATA SHEET

We use control variables for running the sheet and actual input data (Figure 2.30). The control variables running the sheet are *start*, which is given a value zero when the sheet is to be initialized. One likes to press "*F*9", i.e., calculate the worksheet to put all the values at their initial values. Press "*F*9" a few times to make sure that this is actually done. Put the value *start* = 1 for running the sheet. *dx* is the length of the grid cell in the x-direction and *dy* is the length of the grid cell in the y-direction. Furthermore, we define *omega* as the over-relaxation factor, which enhances the convergence rate of the numerical calculation performed in sheet "*flow*" (see Press page 866 ff).

The input data include the viscosity of the fluid μ (mu) in [Pas] and the height of the reservoir H. We have added a number of variables to facilitate the use of boundary conditions in the sheet "*flow*". For instance, we use *phibound*1 to express the value of ϕ at a boundary of the domain. We note that it is convenient to have the boundary coinciding with the center of the grid cell. In the same way, we use *phibound*2. In every grid cell we can have a source term. It is not important where in the grid cell the flow occurs, i.e., whether it is in the middle due to a well or across the boundary. In our formulation, we use an average flow rate over the height of the reservoir, i.e., *flow*1, *flow*2 are expressed in [m^3/s], i.e., volume rate of flow. Sometimes one likes that the boundary of a grid cell on which the flow is given (or flow is zero) coincides with the boundary of adjacent grid cells in which the flow potential is given. This means that such a boundary grid cell has half the size. It is not always that one will be able to specify the flow rate in a grid cell. One may know the well pressure, and the radius of the well. Therefore, we give *phiwf*1, *phiwf*2

Figure 2.29   The name of the sheets required to perform the computations.

	A	B	C	D
potentialflow:1				
1	start	1	[-]	if start=0, initialize, otherwise run
2				
3	dx	1	[m]	grid cell length x-direction
4	dy	1	[m]	grid cell length y-direction
5	omega	1.55	[-]	overrelaxation factor
6				
7	mu	0.001	[Pa s]	dynamic viscosity
8	height	5	[m]	height reservoir
9				
10	phibound1	1000000	[Pa]	pressure on boundary 1
11	phibound2	2000000	[Pa]	pressure on boundary 2
12	flow1	0.000018	[m^3/s]	flow rate 1
13	flow2	0.000018	[m^3/s]	flow rate 2
14	phiwf1	2000000	[Pa]	well flow potential 1
15	phiwf2	1000000	[Pa]	well flow potential 2
16				
17	rprodwell	0.1	[m]	radius production well
18	rinjwell	0.1	[m]	radius injection well
19	CA	25	[-]	shape factor
20	skinprod	4	[-]	skin factor for production well
21	skininj	7	[-]	skin factor for injection well
22	pindex	=2*PI()*height/(mu*(0.5*LI	[-]	productivity index/perm
23	iindex	=2*PI()*height/(mu*(0.5*LI	[-]	injectivity index/perm
24				
25	VDP	0.7	[-]	Dykstra Parson's coefficient
26	kav	=0.000000000001*height	[m^2]	average permeability height product
27	sdev	=-LN(1-VDP)	[-]	standard deviation for log perm
28	muav	=LN(kav)-0.5*sdev*sdev	[-]	average ln(k/ko), ko=1 [m^2]
29				

Figure 2.30 Input and control parameters.

as flow potentials [Pa] in the well. In addition, the well radii, i.e., $rprodwell, rinjwell$ [m] are given as radii of a production and an injection well. A well is usually partly clogged and there is extra resistance in the neighborhood of the well. This is expressed in terms of a skin effect, with parameters $skinprod, skininj$ for a production well and an injection well, respectively. Such a skin effect uses that we may assume to have a quasi-steady state situation near the well. Problems with injection water are well known and large injection skins can occur. The flow rates of a production well and an injection well are given by Eqs (2.88), (2.89). These equations are too lengthy to be displayed in Figure 2.30 and are therefore displayed below.

$$PI = 2 * PI() * height / (mu * (0.5 * LN(4 * dx * dy / $$
$$(1.781 * CA * rprodwell * rprodwell)) + skinprod)),$$

$$II = 2 * PI() * height / (mu * (0.5 * LN(4 * dx * dy / $$
$$(1.781 * CA * rinjwell * rinjwell)) + skininj)).$$

The Dykstra-Parsons coefficient V_{DP} describes the heterogeneity of the reservoir. For a log-normal field, the Dykstra-Parsons coefficient is related to the standard deviation s of the logarithm of the permeability $V_{DP} = 1 - \exp(-s)$. The average permeability height product is given by k_{av} and is expressed in units $[m^2]$. For those of us used to Darcy as a unit for permeability, we have $1D = 0.987 \times 10^{-12}[m^2]$. We use $s = -\ln(1 - VDP)$. The average of the log of the permeability μ is obtained from [44, 46], i.e.,

$$\mu = \ln(kav) - 0.5 * sdev * sdev \tag{2.108}$$

The standard deviation and the average of the log of the permeability can be used to calculate random uncorrelated log-normally distributed permeability field. The relations mentioned above can be found in many standard textbooks of statistics. One may also apply existing software to generate correlated random fields [46, 72].

2.A.2   THE SHEET FOR CALCULATION OF THE X-DIP AVERAGED PERMEABILITY

In our example, we assume that the X-dip averaged permeability is log-normally distributed without any correlation between adjacent cells. The entire procedure is explained in (Eq. 2.109) and Figures 2.31 and 2.32. Indeed, we will have the statement in all the cells in $(b5:d7)$

$$= LOGINV(RAND(), muav, sdev), \tag{2.109}$$

which generates an uncorrelated log-normally distributed random field with the average of the logarithm of the permeability given by μ and a standard deviation of s (see [192], Chapter VII, pp. 287–288). If one likes to use more cells, e.g., use cells $(b5..AY40)$, and add the periodic equations appropriately; one has to do so in every sheet. Note that if the whole worksheet is computed all permeability values will change. If one likes to avoid it one needs to use *copy* the whole array *, paste special* on cell $C3$ choosing the appropriate values. It is also possible to leave the statements as above and press *shift F9* in the sheet *flow* to perform only calculations in the sheet *flow*.

2.A.3   THE HARMONICALLY AVERAGED GRID SIZE CORRECTED MOBILITY IN THE x-DIRECTION

Often it is claimed that for the flow in the east direction (and south direction see below) we need the harmonically averaged permeability of the cell P and the cell east, i.e., E. However, as explained in [190], this is incorrect. In 2-D calculations, it is preferred to use the geometric average. It is convenient to incorporate Δy to convert from Darcy velocity **u** to flow rate **q**, to divide by Δx, which is used to calculate the potential gradient, and to incorporate the viscosity μ to obtain the mobility.

	A	B	C	D	E
4		=B7	=C7	=D7	
5	=D5	=dy*SQRT(!!B5*!!C5)/(mu*dx)	=dy*SQRT(!!C5*!!D5)/(mu*dx)	=dy*SQRT(!!D5*!!E5)/(mu*dx)	=B5
6	=D6	=dy*SQRT(!!B6*!!C6)/(mu*dx)	=dy*SQRT(!!C6*!!D6)/(mu*dx)	=dy*SQRT(!!D6*!!E6)/(mu*dx)	=B6
7	=D7	=dy*SQRT(!!B7*!!C7)/(mu*dx)	=dy*SQRT(!!C7*!!D7)/(mu*dx)	=dy*SQRT(!!D7*!!E7)/(mu*dx)	=B7
8		=B5	=C5	=D5	

Figure 2.31   The geometrically averaged grid size corrected mobility in the x-direction between the central P and the cell to the right E. Sheet I contains the permeabilities.

	A	B	C	D	E
4		=B7	=C7	=D7	
5	=D5	=dx*SQRT(!!B5*!!B6)/(mu*dy)	=dx*SQRT(!!C5*!!C6)/(mu*dy)	=dx*SQRT(!!D5*!!D6)/(mu*dy)	=B5
6	=D6	=dx*SQRT(!!B6*!!B7)/(mu*dy)	=dx*SQRT(!!C6*!!C7)/(mu*dy)	=dx*SQRT(!!D6*!!D7)/(mu*dy)	=B6
7	=D7	=dx*SQRT(!!B7*!!B8)/(mu*dy)	=dx*SQRT(!!C7*!!C8)/(mu*dy)	=dx*SQRT(!!D7*!!D8)/(mu*dy)	=B7
8		=B5	=C5	=D5	

Figure 2.32 Geometrically averaged mobilities in the y-direction between the central cell P and the cell south, i.e., S.

	A	B	C	D	E
2		**Sheet wells**			
3		**insert well potential**			
4		=B7	=C7	=D7	
5	=D5	0	0	0	=B5
6	=D6	0	=phiwf1	0	=B6
7	=D7	0	0	0	=B7
8		=B5	=C5	=D5	

Figure 2.33 Put the well flow potentials into the cells that contain wells for which the well flow potential is given. We can put any value in the other cells (as they have productivity indexes equal to zero), but we prefer to use zero for convenience. Instead of $phiwf1$ one can also put in the appropriate numerical value.

2.A.4 THE GEOMETRICALLY AVERAGED GRID SIZE CORRECTED MOBILITY IN THE y-DIRECTION BETWEEN THE CENTRAL P AND THE CELL S.

For the flow in the south direction, we also use the geometrically averaged permeability of the cell P and the cell south, i.e., S. It is convenient to incorporate Δx to convert from Darcy velocity $\mathbf{u}$ to flow rate $\mathbf{q}$, to divide by Δy, which is used to calculate the potential gradient, and to incorporate the viscosity μ to obtain the mobility.

2.A.5 THE SHEET FOR THE WELL FLOW POTENTIAL

In some of the grid cells in the field, there is a well for which the flow potential is given. In this sheet $phiwf$ (Figure 2.33), we give the flow potentials in the grid cells that contain a well for which the flow potential is given. The other cells are considered to have a zero productivity (PI) or injectivity index(II). One can put any number here in the sheet $phiwf$ in a cell with zero PI or II. For convenience, we use the value zero (Figure 2.34).

2.A.6 THE SHEET FOR PRODUCTIVITY/INJECTIVITY INDEXES

The sheet $pindex$ for the productivity and injectivity indexes has the same outlay as the sheet $phiwf$ shown in Figure 2.33 shown above. Now it is important to put the value zero in all cells that do not contain a well for which the flowing potential is given. The two cells that do not contain zero should be at exactly the same location as the cells containing $phiwf1$ and $phiwf2$ in Figure 2.33, but only we are putting now the values of the productivity indexes or injectivity indexes at these locations

Note that the parameters $pindex$ and $iindex$ in the data sheet define the productivity index except for the permeability, which is different in every grid cell. One may prefer to put hard numeric

	A	B	C	D	E
2		**Sheet pindex**			
3		**insert well potential**			
4		=B7	=C7	=D7	
5	=D5	0	0	0	=B5
6	=D6	0	=pindex*!!C6	0	=B6
7	=D7	=iindex*!!B7	0	0	=B7
8		=B5	=C5	=D5	

Figure 2.34 Productivity and injectivity indexes put in the grid cells containing a well for which the flow potential is given.

data in this field or to define other/additional parameters to define the productivity/injectivity indexes.

2.A.7 THE SHEET FOR THE WELLS

One may like to include wells for which the flow rate is given. This sheet has the same outlay as the sheet *phiwf* shown in Figure 2.33 shown above. Only now we put in *flow*1 and *flow*2 in the cells where a well is located for which the flow rate is specified. Note that for injection *flow*1, *flow*2 is positive and for production it is negative. In general, we expect that the wells for which the flow rate is specified are not in the same grid cell as a well for which the flow potential is specified.

2.A.8 THE SHEET FOR FLOW CALCULATIONS

Furthermore, the flow sheet has the general outlay as shown in Figures 2.33 and 2.24. A cell not at the boundary is given by Eq. (2.93), and it reads in spreadsheet notation in cell $D4$

$$= IF(start = 0, 0, omega * (labpe!C4 * C4 + labpe!D4 * E4 + labps!D3 * D3$$
$$+ labps!D4 * D5 + wells!D4 + pindex!D4 * phiwf!D4)$$
$$/(labpe!C4 + labpe!D4 + labps!D3 + labps!D4 + pindex!D4)$$
$$+ (1 - omega) * D4).$$

If one of the boundaries of the cell (e.g., the west boundary of cell $C4$) is impermeable, we lose the term which contains a reference to a cell which is outside our domain. Indeed, if we copy cell $D4$ to cell $C4$, we obtain

$$= IF(start = 0, 0, omega * (labpe!B4 * B4 + labpe!C4 * D4 + labps!C3 * C3$$
$$+ labps!C4 * C5 + wells!C4 + pindex!C4 * phiwf!C4)$$
$$/(labpe!B4 + labpe!C4 + labps!C3 + labps!C4 + pindex!C4)$$
$$+ (1 - omega) * C4). \tag{2.110}$$

Clearly cell $B4$ is outside the domain. We can prove that if we omit these terms containing $B4$ we arrive at the correct expression, which is

$$
\begin{aligned}
= IF(start = 0,0, & omega*(labpe!C4*D4 + labps!C3*C3 \\
& + labps!C4*C5 + wells!C4 + pindex!C4*phiwf!C4) \\
& /(labpe!C4 + labps!C3 + labps!C4 + pindex!C4) \\
& + (1 - omega)*C4).
\end{aligned}
$$

The same procedure can be applied if we like to know expressions for no flow boundaries at the north, south or east side of the domain Ω_2. Note that the no flow boundary coincides with the edge of the cell, i.e., those indicated in Figure 2.23 as eb, nb, wb, sb respectively.

If we like to indicate that the potential of a cell in the middle of the cell (at point P) has a given value we can just put that in the cell. However, now we do not indicate the edge of the cell. If the flow potential at one of the edges is given, e.g., at the west edge of cell $C4$ is equal to $phibound2$, we use the following statement

$$
\begin{aligned}
= IF(start = 0,0, & omega*(2*perm!C4*dy/dx/mu*phibound2 + labpe!C4*D4 \\
& + labps!C3*C3 + labps!C4*C5 + wells!C4 + pindex!C4*phiwf!C4) \\
& /(2*perm!C4*dy/dx/mu + labpe!C4 + labps!C3 + labps!C4 + pindex!C4) \\
& + (1 - omega)*C4).
\end{aligned}
$$

In other words, we use the fact that the flow potential gradient from the edge of the cell to the center of the cell runs over a distance $\Delta x/2$. In addition, the permeability-height product between the edge of the cell and the center is considered constant and equal to the permeability-height product of the cell under consideration. This approximation is only first-order accurate. Again the procedure is to copy statement (2.110) to the boundary and amend the terms which contain references to cells outside the domain Ω_2. As a second example, we have copied the statement (2.110) for cell $D4$ to cell $D3$ and we obtain

$$
\begin{aligned}
= IF(start = 0,0, & omega*(labpe!C3*C3 + labpe!D3*E3 + labps!D2*D2 \\
& + labps!D3*D4 + wells!D3 + pindex!D3*phiwf!D3) \\
& /(labpe!C3 + labpe!D3 + labps!D2 + labps!D3 + pindex!D3) \\
& + (1 - omega)*D3).
\end{aligned}
$$

Clearly, all references to cell $D2$ are outside the domain. Therefore, we have to modify this statement. We want to implement the boundary condition that on the north edge of cell $D3$ the potential has a value $phibound1$. We use the fact that the flow potential gradient from the edge of the cell to the center of the cell runs over a distance $b/2$. Also, the permeability-height product between the edge of the cell and the center is considered constant and equal to the permeability-height product of the cell under consideration. Therefore, we obtain

$$
\begin{aligned}
= IF(start = 0,0, & omega*(labpe!C3*C3 + labpe!D3*E3 \\
& + 2*perm!D3*dx/dy/mu*phibound1 \\
& + labps!D3*D4 + wells!D3 + pindex!D3*phiwf!D3) \\
& /(labpe!C3 + labpe!D3 + 2*perm!D3*dx/dy/mu + labps!D3 + pindex!D3) \\
& + (1 - omega)*D3).
\end{aligned}
$$

Finally, one may have the boundary condition of no flow not on the edges but in the center of the cell. In this case, one will perform the balance equation on half of the grid cell and so if flow to the north is zero in the grid cell (Figure 2.23), we observe that q_{wx} and q_{px} will assume half of the values as flow is only through $b/2$ instead of b. Furthermore $q_{ny} = 0$ and q_{py} remains unchanged. Assume again that we want to have a no flow boundary to the north for cell $D3$, but now though the middle of the cell. Then we obtain

$$= IF(start = 0, 0, omega * (\frac{1}{2}labpe!C3 * C3 + \frac{1}{2}labpe!D3 * E3$$
$$+ labps!D3 * D4 + wells!D3 + pindex!D3 * phiwf!D3)$$
$$/(\frac{1}{2}labpe!C3 + \frac{1}{2}labpe!D3 + labps!D3 + pindex!D3) + (1 - omega) * D3).$$

If a given flow comes into the grid cell, either through the wells or the edges, one just indicates this amount in the sheet "*wells*".

After one has given the statements of all the cells, both in interior domain and at its boundaries one proceeds as follows. Go to the data sheet and put the variable *start* (right to the name start) equal to zero. Press *F9* a few times. Determine the number of iterations one wants (one can always exit by pressing the escape button). Set the calculation tab as in Figure 2.28 and put in the number of iterations one wants. Put the value of *start* equal to one. Then go to the sheet "*flow*" and press *shift F9* (one can repeat this as often as one wants).

2.B   FINITE ELEMENT CALCULATIONS

Performing a finite element calculation in EXCEL can just as easily be done as the finite volume method discussed above. However, we note that finite element uses a point-centered grid. Therefore, we need to specify the permeabilities with respect to point P, as the northeast (NE), southeast (SE), the northwest (NW) and the southwest (SW) parts. Therefore, it is useful to use two sheets. One sheet contains the NE permeabilities with respect to cell P and the other the southeast (SE) permeabilities. However, in this case, these values are not independent and in fact could have been found from the (NE) sheet. The southeast sheet has all permeabilities of the northeast sheet, but now shifted one cell to the north. Be aware that the edge of the problem represents the midpoint of a cell. Use a drawing to assign the (NE, SE) permeabilities. Finite element calculations are more accurate (because we used higher order) than the finite volume calculations.

2.C   SKETCH OF PROOF OF THE EFFECTIVE MEDIUM
APPROXIMATION FORMULA

The proof of the effective medium approximation is not simple and requires understanding of resistances in electric circuits. The proofs I found in Kirkpatrick [135] and in [216] page 296 are more or less understandable. To solve the problem use is made of Thevenin's theorem, which states (Figure 2.35) that the effect of a different conductivity in a single bond in the network can be seen as the added conductance G'_{AB}. This added conductance can be calculated by inserting a current i_0 at A, which originates from infinity and inserting a current i_0 at B, which flows away to infinity. The superposition of these two currents cancel out at infinity and lead to a closed circuit.

Consider a network made up of equal conductances, g_m, connecting nearest neighbors on a cubic mesh. As the criterium to fix g_m we require that the extra voltages induced, i.e., the local fields, when individual conductances g_{ij} replace g_m, in this medium average to zero. Consider Figure 2.35, where the conductance $g_o = g_{AB}$ is surrounded by effective conductances g_m.

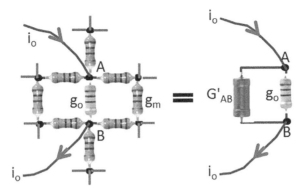

Figure 2.35 Constructions used in calculating the average voltage across one conductance g_o surrounded by a uniform medium.

To the uniform (only conductances g_m) field solution, in which the voltage increase by a constant amount V_m, per row, we add the effects of a fictitious current, i_o, introduced at A and extracted at B. Since the uniform solution fails to satisfy current conservation at A and B, the magnitude of i_o is chosen to correct for this, i.e.,

$$i_o = V_m (g_m - g_o). \tag{2.111}$$

In the equivalent network the voltage between A and B is given by

$$V_o = \frac{i_o}{g_o + G'_{AB}}. \tag{2.112}$$

We first obtain the conductance G_{AB} in the uniform effective medium, i.e., for the situation that g_o in the equation above is replaced by the average conductance g_m. In this case, we obtain $G_{AB} = G'_{AB} + g_m$. Let us write the current distribution in the left part of Figure 2.35, with $g_o = g_m$, i.e., in the uniform network as the sum of two contributions, a current i_o introduced at A and extracted at very large distances in all directions and a current i_o introduced at infinity and extracted at B. In each case, the current flowing through each of the z equivalent bonds is $\frac{i_o}{z}$, so that a total current of $\frac{2i_o}{z}$ flows through the AB bond. Therefore, we find that

$$V_o = \frac{i_o}{g_m + G'_{AB}} = \frac{i_o}{G_{AB}} = \frac{2i_o}{z g_m} \tag{2.113}$$

and $G_{AB} = \frac{z}{2} g_m$ and $G'_{AB} = \left(\frac{z}{2} - 1\right) g_m$. We substitute this into Eqs. (2.111) and (2.112) and obtain

$$V_o = \frac{i_o}{g_o + G'_{AB}} = V_m \frac{g_m - g_o}{g_o + G'_{AB}} = V_m \frac{g_m - g_o}{g_o + \left(\frac{z}{2} - 1\right) g_m}. \tag{2.114}$$

The bonds are distributed according to $h(g)dg$. We require that the average contribution of all conductances drawn from this distribution (such as g_o) vanish" or

$$\int_0^\infty h(g) \frac{g_m - g}{g + (\frac{z}{2} - 1)g} dg = 0. \tag{2.115}$$

In two dimensions $\frac{2}{z}$ is the percolation threshold, but we like to replace $\frac{z}{2} = \frac{1}{\mathcal{P}_c}$ for all cases, i.e., we obtain

$$\int\limits_0^\infty h(g)\frac{g_m - g}{g + (\frac{1}{\mathcal{P}_c} - 1)g}dg = 0. \qquad (2.116)$$

2.D HOMOGENIZATION

Homogenization for incompressible one-phase flow leads to a result that is also intuitively obvious. It tells us that we need to run a simulation on a small cubical part of the reservoir with the equation

$$\frac{\partial}{\partial x}\left(\frac{k}{\mu}\frac{\partial\phi}{\partial x}\right) + \frac{\partial}{\partial y}\left(\frac{k}{\mu}\frac{\partial\phi}{\partial y}\right) + \frac{\partial}{\partial\zeta}\left(\frac{k}{\mu}\frac{\partial\phi}{\partial\zeta}\right) = 0, \qquad (2.117)$$

which is subjected to a global potential gradient in one of the main x, y or ζ directions. The potential is periodic in the other main directions, i.e., the directions in which the potential gradient is not applied. The integrated flow, e.g., through plane intersecting the center of the cube perpendicular to the main flow direction determines the permeability tensor. Here below we give an outline of the homogenization procedure.

Homogenization is a powerful method of upscaling. It considers the problem at two scales, i.e., the small scale and the large scale. All scaling factors, inclusive the value of dimensionless numbers are expressed in terms of the scaling ratio ε between the small scale and the large scale. In addition, we consider a periodic lattice in three-dimensional space which consists of unit cells with a permeability distribution schematically represented by the octahedra in Figure 2.36, which by periodical continuation builds an entire lattice. The length of the entire lattice is L and the length of a unit cell is l, with $\frac{l}{L} = \varepsilon << 1$. A two-dimensional picture of such a lattice is shown in Figure 2.36. The entire lattice is supposed to be a model of a large field and the unit cell is considered to be a representative of a grid volume in a simulation. We assume that the flow

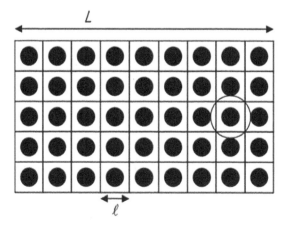

Figure 2.36 A periodic lattice. Each cell represents a permeability distribution at a small scale. The cell is periodically continued and builds the entire lattice. The length of the entire lattice is L and the length of a single unit cell is l, where $\varepsilon = \frac{l}{L} << 1$.

in the unit cell is governed by Darcy's equation in a heterogeneous medium, i.e.,

$$\mathbf{div}\left(\frac{k}{\mu}\mathbf{grad}(\phi)\right) = 0, \tag{2.118}$$

where $\mu\,[Pas]$ is the viscosity, $\phi\,[Pa] = p + \rho gz$ is the potential, where p is the pressure, $\rho\,[kg/m^3]$ is the density, $g\,[m/s^2]$ is the acceleration due to gravity and $z\,[m]$ the coordinate in the vertical direction. We obtain

$$\mathbf{u} = -\frac{k}{\mu}\mathbf{grad}\phi, \tag{2.119}$$

$$\mathbf{div}\ \mathbf{u} = 0. \tag{2.120}$$

We assume that we can find an approximate solution $\phi^\varepsilon(\mathbf{r}_b, \mathbf{r}_s), \mathbf{u}^\varepsilon(\mathbf{r}_b, \mathbf{r}_s)$ of Eq. (2.119) and (2.120), where $\mathbf{r}_b$ refers to the large (b(ig)) scale with length L, and $\mathbf{r}_s$ refers to the local (s(mall)) scale in the unit cell with length ℓ. Here below we drop the superscript ε. The solutions are written in terms of a series in terms of ε (see also below), and we like to equate the terms of the same order of magnitude. For this reason, it is convenient to write the equations in a dimensionless form.

We will do this by inspection, i.e., we will write the dependent and independent variables X as a product of a reference quantity X_R and a dimensionless quantity X_D with an added off-set X_{off} (which we can all equate to zero for this specific case). We note that the differentiation is both with respect to $\mathbf{r}_b$ and with respect to $\mathbf{r}_s$ and usually assume that the (differentiation) contributions at the small scale are $1/\varepsilon$ times the contribution at the large scale.

For instance, if we differentiate a quantity c, we write

$$\frac{\partial c}{\partial x} = \frac{\partial c}{\partial x_b} + \frac{\partial c}{\partial x_s}, \tag{2.121}$$

i.e., a sum of a small-scale and a large-scale derivation (Figure 2.37) and assume that $\partial/\partial x_s$ is one order of magnitude larger than $\partial/\partial x_b$., Consequently $\frac{\partial c}{\partial x_b} \sim \varepsilon \frac{\partial c}{\partial x_s}$.

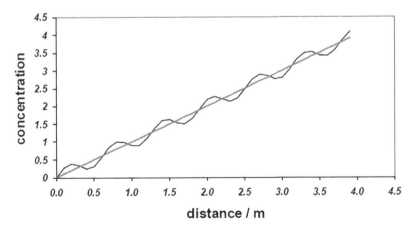

Figure 2.37 Global and local variation of the concentration in a porous medium. We can split the derivative in a term that describes the global differentiation and a term that describes the local differentiation. In homogenization, it is usually assumed that the global contribution is one order of magnitude smaller respect to ε than the local contribution.

Application of these concepts to Eq. (2.119) and Eq. (2.120) and non-dimensionalizing the equations (for $\mathbf{x}_b$ we use the reference length L and for $\mathbf{x}_s$ we use the reference length ℓ) leads to

$$u_R \mathbf{u}_D = -\frac{k\phi_R}{\mu L}\mathbf{grad}_{bD}\phi_D - \frac{k\phi_R}{\mu\ell}\mathbf{grad}_{sD}\phi_D, \tag{2.122}$$

$$\frac{u_R}{L}\mathbf{div}_{bD}\mathbf{u}_D + \frac{u_R}{\ell}\mathbf{div}_{sD}\mathbf{u}_D = 0. \tag{2.123}$$

Here $\mathbf{grad}_{bD}, \mathbf{div}_{bD}$ is determined by dimensionless differentiation with respect to the large-scale coordinate $\mathbf{r}_b$ and $\mathbf{grad}_{sD}, \ \mathbf{div}_{sD}$ is determined by dimensionless differentiation with respect to the small-scale coordinate $\mathbf{x}_s$. After non-dimensionalization $\mathbf{grad}_{bD}\phi_D \sim \varepsilon \ \mathbf{grad}_{sD}\phi_D$ and $\mathbf{div}_{bD}\mathbf{u}_D \sim \varepsilon \ \mathbf{div}_{sD}\mathbf{u}_D$. We can choose the reference potential difference ϕ_R as the potential difference applied to the entire lattice.

We now drop the subscript D to write the dimensionless equations and use $\varepsilon = \frac{\ell}{L}$. Moreover, we use the reference velocity $u_R = \frac{k_{ref}\phi_R}{\mu L}$, where k_{ref} is some averaged permeability. We obtain for the dimensionless velocity $\mathbf{u}$,

$$\mathbf{u} = -\frac{k}{k_{ref}}\mathbf{grad}_b\phi - \frac{k}{\varepsilon k_{ref}}\mathbf{grad}_s\phi \qquad \text{and} \tag{2.124}$$

$$\mathbf{div}_b\mathbf{u} + \frac{1}{\varepsilon}\mathbf{div}_s\mathbf{u} = 0. \tag{2.125}$$

We write the potential and velocity in terms of a series in ε

$$\phi = \phi_0 + \varepsilon\phi_1 + \varepsilon^2\phi_2 + \dots \text{ and} \tag{2.126}$$

$$\mathbf{u} = \mathbf{u}_0 + \varepsilon\mathbf{u} + \varepsilon^2\mathbf{u}_2 + \dots \qquad , \tag{2.127}$$

where the coefficients ϕ_o, ϕ_1, ϕ_2 and also the components of $\mathbf{u}_0, \mathbf{u}_1, \mathbf{u}_2$ are of the same order of magnitude with respect to ε. Note that we do not have to use the periodicity of the lattice to come up with these equations, i.e., we just derive the equations for a problem living on two different scales. It must be mentioned that $\ell \to 0$, and therefore $\varepsilon \to 0$ has no meaning from the physical point of view. Taking $\varepsilon \to 0$ is used in the mathematical analysis only.

We substitute Eqs. (2.126) and (2.127) into Eqs. (2.124) and (2.125), and we find to the lowest order of ε

$$\frac{1}{\varepsilon}\mathbf{div}_s\mathbf{u}_0 = 0, \tag{2.128}$$

$$\mathbf{div}_b u_0 + \mathbf{div}_s u_1 = 0, \tag{2.129}$$

$$-\frac{k}{k_{ref}\varepsilon}\mathbf{grad}_s\phi_0 = 0, \tag{2.130}$$

$$\mathbf{u}_o = -\frac{k}{k_{ref}}\mathbf{grad}_b\phi_0 - \frac{k}{k_{ref}}\mathbf{grad}_s\phi_1. \tag{2.131}$$

Indeed $|\mathbf{div}_s\mathbf{u}_0/\varepsilon| >> |\mathbf{div}_b u_0|$ and $\mathbf{div}_b u_0$ and $\mathbf{div}_s u_1$ are both of the same order of magnitude. Moreover, $\left|\frac{k}{k_{ref}}\mathbf{grad}_b\phi_0\right| << \left|\frac{k}{\varepsilon k_{ref}}\mathbf{grad}_s\phi_0\right|$ and again $\left|\frac{k}{k_{ref}}\mathbf{grad}_b\phi_0\right|$ and $\left|\frac{k}{k_{ref}}\mathbf{grad}_s\phi_1\right|$ are both of the same order of magnitude. In essence, this is all there is to it, but we have some tedious algebra to do before we get to the final result. It follows from $\frac{k}{k_{ref}\varepsilon}\mathbf{grad}_s \ \phi_o = 0$ that $\phi_o = \phi_o(\mathbf{r}_b)$, i.e., it is only dependent on the macroscopic scale $\mathbf{r}_b$. Note that $\mathbf{grad}_b \ \phi_o$ is the globally appied potential gradient. If this gradient were zero, there would be no flow at all and $\phi_1 = 0$. As the flow would be proportional to the applied gradient also the components of $\mathbf{grad}_s\, \phi_1$ would be proportional to the components of $\mathbf{grad}_b\, \phi_o$. The most general form to describe this is

$$\mathbf{grad}_s\phi_1 = \mathbf{w} \cdot \mathbf{grad}_b\phi_o := \mathbf{grad}_s\mathbf{v} \cdot \mathbf{grad}_b\phi_0. \qquad (2.132)$$

The second equality follows from the fact that the $\mathbf{curl}_s$ of each of the three row vectors representing $\mathbf{w}$ is zero, and therefore each of the row vectors can be represented by the gradient of a vector $\mathbf{v}$. In matrix notation, we can write

$$\begin{pmatrix} \frac{\partial \phi_1}{\partial x_s} \\ \frac{\partial \phi_1}{\partial y_s} \\ \frac{\partial \phi_1}{\partial z_s} \end{pmatrix} = \begin{pmatrix} w_{xx} & w_{xy} & w_{xz} \\ w_{yx} & w_{yy} & w_{yz} \\ w_{zx} & w_{zy} & w_{zz} \end{pmatrix} \begin{pmatrix} \frac{\partial \phi_o}{\partial x_b} \\ \frac{\partial \phi_o}{\partial y_b} \\ \frac{\partial \phi_o}{\partial z_b} \end{pmatrix} = \begin{pmatrix} \mathbf{grad}_s v_x \\ \mathbf{grad}_s v_y \\ \mathbf{grad}_s v_z \end{pmatrix} \cdot \begin{pmatrix} \frac{\partial \phi_o}{\partial x_b} \\ \frac{\partial \phi_o}{\partial y_b} \\ \frac{\partial \phi_o}{\partial z_b} \end{pmatrix}, \qquad (2.133)$$

and substitution into Eq. (2.131) leads to

$$\begin{aligned} \mathbf{u}_o &= -\frac{k}{k_{ref}}\mathbf{grad}_b\phi_0 - \frac{k}{k_{ref}}\mathbf{grad}_s\mathbf{v} \cdot \mathbf{grad}_b\phi_0 \\ &= -\frac{k}{k_{ref}}\left(\mathbf{grad}_s\mathbf{v} + \mathbf{I}\right) \cdot \mathbf{grad}_b\phi_0 = 0, \end{aligned}$$

where $\mathbf{I}$ is the unit matrix.

After integration over, the periodic unit cell we obtain

$$< \mathbf{u}_0 > = - < \frac{k}{k_{ref}}\left(\mathbf{grad}_s\mathbf{v} + \mathbf{I}\right) > \cdot \mathbf{grad}_b\phi_0.$$

Clearly, we like to combine with Eq. (2.128), but before we do this it is useful for the understanding to make specific choices for $\mathbf{grad}_b\, \phi_0$. Let is first assume that $\mathbf{grad}_b\, \phi_o = -\mathbf{e}_x$. Then, we obtain

$$\mathbf{u}_o = \frac{k}{k_{ref}}\left(\mathbf{grad}_s v_x + \mathbf{e}_x\right) = \frac{k}{k_{ref}}\mathbf{grad}_s\left(v_x + x\right) := \frac{k}{k_{ref}}\mathbf{grad}_s v_x \quad \text{and}$$

$$\mathbf{div}_s\mathbf{u}_0 = \mathbf{div}_s\left(\frac{k}{k_{ref}}\mathbf{grad}_s v_x\right) = 0. \qquad (2.134)$$

We note that $v_x = v_x + x$; In view of the periodicity of v_x in the periodic unit cell, it follows v_x is periodic in the y, z direction and that in the x-direction $v_x (x_s = 1, y_s, z_s) = v_x (x_s = 0, y_s, z_s) + 1$. We can use Eqs. (2.134) to define an equation with the appropriate boundary conditions to determine $\mathbf{v}$.

The interpretation is that we find the velocity vector $\mathbf{u}_o$ when we subject the periodic unit cell (PUC: usually a cube) to a unit potential gradient in the x-direction. If we average the velocity in the x-direction over the unit cell, we get the average permeability for the velocity in the x-direction for a potential also applied in the x-direction. This is the xx component of the permeability tensor or k_{xx}. In other words, we find for the average value of this component of the permeability tensor

$$\frac{k_{xx}}{k_{ref}} = \frac{1}{|\Omega|}\int_{\Omega} \frac{k}{k_{ref}}\frac{\partial v_x}{\partial x}dV. \qquad (2.135)$$

Note that in view of the non-dimensionalizing we do not deal with the viscosity in this equation. I am aware that I am short-circuiting the derivation here; if I would go back to the fully dimensional description, it will be tedious, so let us leave it at this. As due to flux continuity, the flow through every plane in the PUC perpendicular to $\mathbf{e}_x$ is the same, and we can limit the averaging just by averaging the flow through a single plane. In two dimensions, we just add the u_{ox} to obtain the total flow in a column perpendicular to $\mathbf{e}_x$. In the same way, we can also add the flow in the y direction and obtain other components of the permeability tensor. In such a way, we get

$$\frac{k_{yx}}{k_{ref}} = \frac{1}{|\Omega|} \int_\Omega \frac{k}{k_{ref}} \frac{\partial v_x}{\partial y} dV, \ \frac{k_{zx}}{k_{ref}} = \frac{1}{|\Omega|} \int_\Omega \frac{k}{k_{ref}} \frac{\partial v_x}{\partial z} dV. \tag{2.136}$$

We could also have applied the potential gradient in the y-direction, i.e., $\mathbf{grad}_b\, \phi_o = -\mathbf{e}_y$. In this case we obtain completely analogously that

$$\frac{k_{xy}}{k_{ref}} = \frac{1}{|\Omega|} \int_\Omega \frac{k}{k_{ref}} \frac{\partial v_y}{\partial x} dV, \ \frac{k_{yy}}{k_{ref}} = \frac{1}{|\Omega|} \int_\Omega \frac{k}{k_{ref}} \frac{\partial v_y}{\partial y} dV, \ \frac{k_{zy}}{k_{ref}} = \frac{1}{|\Omega|} \int_\Omega \frac{k}{k_{ref}} \frac{\partial v_y}{\partial z} dV. \tag{2.137}$$

I leave it to the reader to obtain the expressions for k_{xz}, k_{yz} and k_{zz}. In practice the procedure boils down to obtain the potential field in the heterogeneous medium with periodic $((v_x (x_s, y_s + 1, z_s) = v_x (x, y_s, z_s)))$ and semi periodic $(v_x (x_s = 1, y_s, z_s) = v_x (x_s = 0, y_s, z_s) - 1)$ boundary conditions and use the ensuing potential field v_x to obtain the average permeability tensor component k_{xx}. In a statistically homogeneous permeability field the off-diagonal elements of the upscaled $\mathbf{k}$ are theoretically zero, so their nonzero value is due to statistical noise.

Note that the derivation above is not easy; perhaps one of the few examples going beyond extended college mathematics. The derivation is not part of a conventional course in reservoir engineering. It obtains average permeabilities from permeability distributions. One takes a part of the reservoir that is considered representative and calls it a periodic unit cell. An often mentioned, criticism toward homogenization is the apparent central role played by periodic boundary conditions, which are considered physically unrealistic. The question is, however, whether effective properties converge as the size of the periodic unit cell increases independent of the boundary conditions (periodic, Dirichlet, Neumann) imposed on the boundary. Bourgeat and Piatnitski [36] show that, if separation of scale is possible, effective properties in random media converge as the scale of the unit cell increases, independent of the type of boundary conditions. The idea is very analogous to the representative elementary volume (REV) discussed above. The REV volume is a volume that has as average property the property of any other REV. If we would slightly increase the REV, the result would remain the same. A periodic unit cell is part of a larger periodic structure consisting of identical cells. In principle, the REV and PUC are the same, but it is tempting to choose the volume of the PUC smaller considering that some kind of periodicity would allow this. Therefore, a PUC is usually smaller than the REV. In such a PUC, one applies a unit potential gradient in one of the main directions and computes the average flow. One assumes periodic boundary conditions and finds the flow component in the direction of the potential gradient and also in two directions perpendicular to it. The flow is averaged in the three directions, and Darcy's law is applied to the complete cell. Thus, one derives three components of the permeability tensor. However, note that finding significant nonzero off-diagonal permeability components with homogenization in a field that is statistically isotropic indicates lack of convergence. Now one continues to apply the potential gradient in another direction perpendicular to the first. Completely analogously one finds three other components of the permeability tensor. Finally, one applies it to the last direction and one finds the complete averaged permeability tensor. The advantage of homogenization is that it shows exactly which assumptions are required to find the upscaled permeability tensor. The method can be used in almost any case where we have a clear separation of scales.

3 Time Dependent Problems in Porous Media Flow

OBJECTIVE OF THIS CHAPTER

- To understand the relevance of analytical solutions and methods in the interpretation of field data of time-dependent pressure behavior in porous media flow, i.e., the use of a similarity transformation to reduce the partial differential equation to an ordinary differential equation, the use of Laplace transform to find short time and long time limiting expressions, and finally the application of a dimensional analysis to simplify the relevant equations,
- to appreciate that these insights cannot be obtained from bull-headed numerical solutions,
- to derive the relevant equations and boundary conditions in their dimensionless form,
- to use Laplace transformation to find a solution of the radial diffusion equation and compare it to the solution that is obtained with a similarity transformation,
- to compare solutions with numerical Laplace transformation and analytical inversion formulae,
- to be able to derive the average permeability and the skin factor by combining pressure well test pressure data for draw-down and buildup.

3.1 TRANSIENT PRESSURE EQUATION

Consider[1] a porous medium in a domain $\Omega \in R^3$ with coordinates x, y, z. The porosity $\varphi(x, y, z)$ and permeability $k(x, y, z)$ may vary in space and can be functions of x, y, z. We use mass conservation and Darcy's law for single-phase flow in porous media to derive the transient pressure equation. The conceptual ideas are evident from a derivation in one space dimension and a subsequent generalization to three space dimensions.

A control volume consists of a block with length Δx and a cross-section $A = \Delta y \Delta z$. The mass of fluid in the pores is $m = \varphi \rho A \Delta x = \varphi \rho \Delta x \Delta y \Delta z$. Mass conservation states that the rate of change of mass dm/dt in the control volume equals the mass flow $(\rho u \Delta y \Delta z)_{in}$ flowing into the control volume minus the mass flow $(\rho u \Delta y \Delta z)_{out}$ flowing out of the control volume. Also note that u is not the actual velocity in the pores but a volumetric flux, i.e., the volumetric flow into the control volume per unit cross-section.

$$\frac{dm}{dt} = \frac{d\left(\varphi \rho \Delta x \Delta y \Delta z\right)}{dt} = (\rho u \Delta y \Delta z)_{in} - (\rho u \Delta y \Delta z)_{out}. \tag{3.1}$$

Here we use $\rho\,[\mathrm{kg/m^3}]$ for the density. In the examples of our interest, we assume that the density is only a function of the pressure and not of the position coordinates. In other words, the equation is not meant to describe salt water with a salt concentration depending on position. It is also the

---

[1] These notes are an extension of Lecture notes with Prof. Ir. C.P.J.W. van Kruijsdijk (1990).

DOI: 10.1201/9781003168386-3

volumetric flux u or Darcy velocity, which appears in Darcy's law. After division of Eq. (3.1) by $\Delta y \Delta z$ to

$$\frac{d(\rho\varphi)}{dt} = -\frac{(\rho u)_{out} - (\rho u)_{in}}{\Delta x}.$$

This is the equation one uses in numerical calculations. In the limit that $\Delta x \to 0$, we note that the fraction on the right defines differentiation toward x. On the left hand, we have to indicate that the time derivative for $\Delta x \to 0$ is to be taken at x and thus needs to be replaced by the partial derivative. Consequently, we obtain

$$\frac{\partial(\rho\varphi)}{\partial t} = -\frac{\partial(\rho u)}{\partial x}.$$

The derivation in three space dimensions is completely analogous. We have to distinguish the Darcy velocities in the three space dimensions, i.e., $\mathbf{u} = (u_x, u_y, u_z)$.

Mass conservation states that the rate of change of mass dm/dt in the control volume equals the sum of the mass flow through the faces of the cube or

$$\frac{dm}{dt} = \Delta x \Delta y \Delta z \frac{d(\varphi\rho)}{dt} = (\rho u_x \Delta y \Delta z)_{in} - (\rho u_x \Delta y \Delta z)_{out} + (\rho u_y \Delta x \Delta z)_{in}$$
$$- (\rho u_y \Delta x \Delta z)_{out} + (\rho u_z \Delta x \Delta y)_{in} - (\rho u_z \Delta x \Delta y)_{out}. \tag{3.2}$$

Note that the Darcy velocity components (u_x, u_y, u_z) are some convenient average over the face of the cube. In this sense, the equation is an approximation, if one is using the Darcy velocity at the center of the cube face. In the limit of zero grid size length, the equation becomes again exact. Dividing the equation by $\Delta x \Delta y \Delta z$, and taking the limits $\Delta x \to 0, \Delta y \to 0$ and $\Delta z \to 0$ and replacing the total derivative versus time by the partial derivative because now we mean the time derivative at fixed position coordinates we obtain

$$\frac{\partial(\rho\varphi)}{\partial t} = -\frac{\partial(\rho u_x)}{\partial x} - \frac{\partial(\rho u_y)}{\partial y} - \frac{\partial(\rho u_z)}{\partial z}. \tag{3.3}$$

This can also be written in short notation by introducing the divergence for the right side, i.e., $\frac{\partial(\rho\varphi)}{\partial t} = -\mathbf{div}(\rho\mathbf{u})$ or even shorter $\frac{\partial(\rho\varphi)}{\partial t} = -\nabla \cdot (\rho\mathbf{u})$. The difference equation (3.2) can be used in the numerical computations, but we still need to incorporate the equation of motion, e.g., Darcy's law. This we shall do in a next section. The differential equation is used if we are after analytical solutions. Before we can fully specify the equation, we need a law that relates the flow to the pressure. We use Darcy's law, i.e.,

$$\mathbf{u} = -\frac{k}{\mu}\left(\mathbf{e}_x \frac{\partial\phi}{\partial x} + \mathbf{e}_y \frac{\partial\phi}{\partial y} + \mathbf{e}_z \frac{\partial\phi}{\partial z}\right) := -\frac{k}{\mu}\mathbf{grad}\,\phi. \tag{3.4}$$

where $\mathbf{e}_x$ is the unit vector in the x-direction, etc. Note that we use "'phi=ϕ" to denote the potential $\phi = p + \rho g z$ and "'varphi=φ" to denote the porosity. Some people prefer the shorthand notation for the gradient of ϕ, i.e., $\nabla\phi$ instead of $\mathbf{grad}\,\phi$. The potential ϕ depends on the pressure but is "corrected" for the potential energy per unit volume $\rho g z$. The gradient of ϕ is the driving force for the motion of fluids. When we substitute Darcy's equation (3.4) into the mass balance equation (3.3), i.e., substitute $u_x = -\frac{k}{\mu}\frac{\partial\phi}{\partial x}$, etc. we obtain the porous medium equation, i.e.,

$$\frac{\partial(\rho\varphi)}{\partial t} = \frac{\partial}{\partial x}\left(\rho\frac{k}{\mu}\frac{\partial\phi}{\partial x}\right) + \frac{\partial}{\partial y}\left(\rho\frac{k}{\mu}\frac{\partial\phi}{\partial y}\right) + \frac{\partial}{\partial z}\left(\rho\frac{k}{\mu}\frac{\partial\phi}{\partial z}\right). \tag{3.5}$$

It is not immediately obvious what we can do with this equation. First of all, it is nonlinear as the density is pressure-dependent. Moreover, the porosity can also be pressure-dependent. We deal first with the nonlinearity on the right side. Here we apply a so-called convenience approximation in assuming that the density is only weakly dependent on the space coordinates at least for rather incompressible fluids like liquid oil. For gases, we describe a procedure below. For liquids, we assume that the density dependence on the space coordinates, i.e., $\rho = \rho(x, y, z)$ is so weak that we can pull out the density and put it in front of the differentiation. As a result, we obtain after division by ρ

$$\frac{1}{\rho} \frac{\partial(\rho\varphi)}{\partial t} = \frac{\partial}{\partial x}\left(\frac{k}{\mu}\frac{\partial\phi}{\partial x}\right) + \frac{\partial}{\partial y}\left(\frac{k}{\mu}\frac{\partial\phi}{\partial y}\right) + \frac{\partial}{\partial z}\left(\frac{k}{\mu}\frac{\partial\phi}{\partial z}\right). \tag{3.6}$$

However, for the weakly compressible liquids, we cannot ignore that also the porosity depends on the pressure, i.e., the porous medium behaves somewhat like a sponge. If you want to deal with this aspect precisely, you have to pull a bookshelf over your head, and this gets outside the scope of this monograph. I know that at least one of you will become an expert in this matter in his lifetime and that all of you have to have some understanding to know when you have to consult this expert. Therefore, I will shortly clarify this matter. The result is that we can write

$$\frac{1}{\rho}\frac{\partial(\rho\varphi)}{\partial t} = \frac{1}{\rho}\left(\frac{\partial(\rho\varphi)}{\partial P}\right)_T \frac{\partial P}{\partial t} = \varphi c_{eff}\frac{\partial P}{\partial t}, \tag{3.7}$$

where the effective compressibility $c_{eff} = c_o S_o + c_w S_{wc} + c_f$. We denote the oil saturation by S_o and the connate water saturation by S_{wc}. We use c_o to denote the isothermal compressibility of oil and c_w to denote the isothermal compressibility of water. The pore compressibility is denoted by c_f and its value can be obtained from laboratory tests [227, 64], i.e.,

$$c_\alpha = -\frac{1}{V_\alpha}\left(\frac{\partial V_\alpha}{\partial P}\right)_T = \frac{1}{\rho_\alpha}\left(\frac{\partial\rho_\alpha}{\partial P}\right)_T \quad \text{with } \alpha = \text{oil,water} \tag{3.8}$$

$$c_f = \frac{1}{\varphi V_b}\left(\frac{\partial V_b}{\partial P}\right)_T, \tag{3.9}$$

where V_b is the bulk volume of the porous medium, which includes the solids and the fluids in the pores. This definition of the pore compressibility is operational in the sense is that we can devise an experiment to measure it. We use isothermal compressibilities because due to the high heat capacity of the rock temperature effects can be disregarded. One of the unexpected aspects in the derivation is that we have to define the fluid velocity in Darcy's law with respect to the moving grains (see Appendix 3.B). Indeed, during compression, grains move. If we would have disregarded this aspect, we would have underestimated the grain compressibility by a factor of $1 - \varphi$.

In this respect, there is a nice anecdote. Jacob [127] derived the correct equations for the pore compressibility, which were used with great success in groundwater flow. After a decade of use, it was, however, found that Jacob made an error in his differentiation which essentially boiled down to state that $(uv)' = u'v$ instead of the correct $(uv)' = u'v + uv'$. However, this equation with correct differentiation underestimated pore compressibility with a factor of $1 - \varphi$. It was only after it was realized that one had to define Darcy's law with respect to the moving grains that the problem was solved [25]. A method of measuring the pore compressibility that was introduced by Teeuw [227]) is discussed in [64, p. 99]. It becomes clear that this is not an easy measurement, prone to bad reproducibility. Combined with the fact that reservoirs are heterogeneous we can say that it will be difficult to get more than an indication of the value of the pore compressibility in practice. This is different for problems related to groundwater flow. For Petroleum Engineers, the concept of pore compressibility is important because it makes us realize that the effective compressibility is a parameter that needs to be matched within a certain range of possible values.

In summary, for the linearized porous medium equation

$$\varphi \left(c_o S_o + c_w S_{wc} + c_f \right) \frac{\partial P}{\partial t} = \frac{\partial}{\partial x} \left(\frac{k}{\mu} \frac{\partial \phi}{\partial x} \right) + \frac{\partial}{\partial y} \left(\frac{k}{\mu} \frac{\partial \phi}{\partial y} \right) + \frac{\partial}{\partial z} \left(\frac{k}{\mu} \frac{\partial \phi}{\partial z} \right),$$ (3.10)

we have to keep in mind that

1. The density variation as a function of distance is disregarded
2. The pore compressibility is difficult to measure.

As Eq. (3.10) is presented, we can use heterogeneous porosity fields, heterogeneous pore compressibility fields and heterogeneous connate water saturation distributions. For weakly compressible fluids, we usually take the compressibility independent of pressure as otherwise the equation becomes again nonlinear. One may be inclined to think that the oil compressibility is not dependent on location, but of course, there are infamous examples in practice where this is not the case. The same is true of the viscosity. The permeability field is usually very heterogeneous. The problem is not that Eq. (3.10) cannot deal with it but that we usually do not have enough data to give a reliable representation. In many of the examples below, we use a single value for the permeability and interpret such a single value as an averaged permeability.

For gases, the assumption of a pressure-independent compressibility is clearly not correct. For gases, the effect of gravity is usually less relevant, and we can replace the potential in Eq. (3.10) with the pressure p. Subsequently, we apply the so-called Kirchoff integral transform. To apply this idea to the porous medium equation is due to [7, 15, 59, 110],

$$m(p) = \left(\frac{\mu}{\rho k} \right)_{ref} \int\limits_{p_{ref}}^{p} \left(\frac{\rho k}{\mu} \right) dp'.$$ (3.11)

This definition is different from the derivation used in the original article. The integral is normalized by normalizing the integral and consequently m has units pressure. In addition, we have incorporated the permeability in the integrand. This will lead to a simpler derivation. We like to determine the gradient of m, i.e., grad m and the first time derivative. For this, we use Leibnitz' rule

$$\frac{d}{d\lambda} \left(\int\limits_{u(\lambda)}^{v(\lambda)} f(p,\lambda) \, dp \right) = \int\limits_{u(\lambda)}^{v(\lambda)} \frac{\partial f(p,\lambda)}{\partial \lambda} dp + f(v,\lambda) \frac{\partial v}{\partial \lambda} - f(u,\lambda) \frac{\partial u}{\partial \lambda}.$$

The only term that survives in the differentiation of m is the term represented by $f(v,\lambda) \frac{\partial v}{\partial \lambda}$. Clearly, the lower boundary of the integral is a constant, i.e., p_{ref} and differentiating a constant gives zero. However, what about the first term? Do we not need to differentiate $\left(\frac{\rho k}{\mu} \right)$ in the integrand toward p? The answer is no! The integral is over the pressure p', but it does not contain the pressure p explicitly. If you do not believe it, you may become convinced if you differentiate $\int_0^x u^2 \, du = \int_0^x x'^2 dx'$ toward x. You can do this by first carrying out the integral and then doing the differentiation or using Leibnitz' rule. Therefore, we obtain

$$\mathbf{grad}\ m = \left(\frac{\mu}{\rho k} \right)_{ref} \left(\frac{\rho k}{\mu} \right) \mathbf{grad}\ p,$$

$$\frac{\partial m}{\partial t} = \left(\frac{\mu}{\rho k} \right)_{ref} \left(\frac{\rho k}{\mu} \right) \frac{\partial p}{\partial t}.$$

We substitute all of this in our general Eq. (3.3), which will be repeated here with the potential replaced by the pressure

$$\frac{\partial (\rho \varphi)}{\partial t} = \left(\frac{d(\rho \varphi)}{dP} \right)_T \frac{\partial P}{\partial t} = \frac{\partial}{\partial x} \left(\rho \frac{k}{\mu} \frac{\partial p}{\partial x} \right) + \frac{\partial}{\partial y} \left(\rho \frac{k}{\mu} \frac{\partial p}{\partial y} \right) + \frac{\partial}{\partial z} \left(\rho \frac{k}{\mu} \frac{\partial p}{\partial z} \right). \qquad (3.12)$$

In addition, with gases we assume that the reservoir is isothermal as the heat capacity of the sand grains is very large. We obtain after dividing both sides by $\left(\frac{\rho k}{\mu} \right)_{ref}$

$$\left(\frac{\mu}{\rho k} \right) \frac{d(\rho \varphi)}{dP} \frac{\partial m}{\partial t} = \frac{\partial}{\partial x} \left(\frac{\partial m}{\partial x} \right) + \frac{\partial}{\partial y} \left(\frac{\partial m}{\partial y} \right) + \frac{\partial}{\partial z} \left(\frac{\partial m}{\partial z} \right).$$

In the case where we deal with gases the pore compressibility is usually negligible and can be taken outside the differentiation toward p. We use that the isothermal compressibility of the gas is defined as $c_g = \frac{1}{\rho} \left(\frac{d\rho}{dP} \right)_T$. We use the abbreviation for the right side $\Delta m = \frac{\partial}{\partial x} \left(\frac{\partial m}{\partial x} \right) + \frac{\partial}{\partial y} \left(\frac{\partial m}{\partial y} \right) + \frac{\partial}{\partial z} \left(\frac{\partial m}{\partial z} \right)$ and we obtain

$$\left(\frac{\varphi \mu c_g}{k} \right) \frac{\partial m}{\partial t} = \Delta m.$$

The danger is that this makes you very happy. In this case, you like to ignore that $\left(\frac{\varphi \mu c_g}{k} \right)$ is not a weak function of the pressure. For an ideal gas, the compressibility is one over the pressure, i.e., $c_g = 1/p$. For the same ideal gas, the viscosity is independent of pressure. All the same we assume that the transient solutions for the weakly compressible oil is also valid for gas. For small pressure changes, this may not be as bad as it seems, but you need to be aware of the limitation. Indeed for stationary problems, i.e., without the time derivative $\Delta m = 0$ and you can use all the solutions for the pressure equation both for slightly compressible liquids and gases without any limitation.

3.1.1 BOUNDARY CONDITIONS

The porous medium equation concerns second-order derivatives with respect to the space coordinates and a first-order derivative for the time derivative. A first-order derivative with respect to time means that we have to define an initial condition in the whole space domain.

$$\phi(x, y, z, t = 0) = \phi_o(x, y, z).$$

The second derivative versus space coordinates requires a boundary condition on the whole bounding surface Γ of the domain $\Omega \in R^3$ in which the we have the porous medium flow. In general, we have a set of boundaries $\Gamma_1, \Gamma_2, \ldots \Gamma_n$ of which the union is the total bounding surface, i.e., $\Gamma_1 \cup \Gamma_2 \cup \ldots \Gamma_n = \Gamma$. We can have Dirichlet conditions, which state that the potential has certain values on a part boundary $\mathbf{r} \in \Gamma_i$, where $\mathbf{r}$ has components (x, y, z)

$$\phi(\mathbf{r}) = b(\mathbf{r}) \qquad \mathbf{r} \in \Gamma_i.$$

We can also have Neumann conditions, i.e., where we specify the derivative Here we can have that for example

$$\frac{d\phi(\mathbf{r})}{dn} = c(\mathbf{r}) \qquad \mathbf{r} \in \Gamma_j.$$

Finally, you may also have mixed conditions

$$\alpha \phi (\mathbf{r}) + \beta \frac{d\phi (\mathbf{r})}{dn} = d (\mathbf{r}) \qquad \mathbf{r} \in \Gamma_k.$$

The one thing to keep in mind that you have to specify one of these three conditions (Neumann, Dirichlet or mixed) on the entire bounding surface Γ.

3.1.2 THE AVERAGED PROBLEM IN TWO SPACE DIMENSIONS

We consider a block $\Omega_3 \in R^3$ with length L, width W and with height H. The length coincides with the $x-$direction and the width with the $y-$direction. The $x-$direction and $y-$direction are direction in the dip and strike direction, constituting the so-called layer directions. The third direction ζ is perpendicular to a plane embedded in the layer and is also called the X-dip direction. We like to reserve the $z-$direction as the vertical upward direction. This $z-$axis can make any angle with the $x-$axis and the $y-$axis without loss in generality. We like to integrate equation over the ζ direction between the bottom of the layer and the top of the layer. We assume that the layer is bounded from above by an impermeable cap-rock and from below by the impermeable base rock. We integrate equation

$$\frac{1}{\rho} \frac{\partial (\rho \varphi)}{\partial t} = \frac{\partial}{\partial x} \left(\frac{k}{\mu} \frac{\partial \phi}{\partial x} \right) + \frac{\partial}{\partial y} \left(\frac{k}{\mu} \frac{\partial \phi}{\partial y} \right) + \frac{\partial}{\partial \zeta} \left(\frac{k}{\mu} \frac{\partial \phi}{\partial \zeta} \right), \qquad (3.13)$$

where we have replaced z by ζ to indicate that the X-dip direction is not necessarily vertical. Here we use the equation that already contains the assumption that the density has only a weak spatial dependence. The assumption, which is required to get a useful result if we integrate over the coordinate ζ, is that $\phi(x,y,\zeta) = \phi_o(x,y)$. In other words, we assume that the potential does not depend on ζ. This assumption appears to be reasonable [141] if

$$\sqrt{\frac{k_\zeta}{k_{//}}} \frac{L}{H} \gg 1, \qquad (3.14)$$

i.e., if the permeability ratio corrected aspect ratio L/H is much bigger than one [143]. If the permeability is heterogeneous, one may replace L by the correlation length. If such a condition is not satisfied in a numerical calculation, it means that we need to subdivide the reservoir into more grid blocks in the $\zeta-$direction. In the case that the potential does not depend on ζ, find a straightforward equation after integration versus ζ. The height of the layer $H = H(x,y)$, i.e., is a function of the coordinates in the dip and strike direction. As the upper boundary for integration, we drop this dependence in the notation.

$$\int_0^H \frac{1}{\rho} \frac{\partial (\rho \varphi)}{\partial t} d\zeta = \int_0^H \left(\frac{\partial}{\partial x} \left(\frac{k}{\mu} \frac{\partial \phi}{\partial x} \right) + \frac{\partial}{\partial y} \left(\frac{k}{\mu} \frac{\partial \phi}{\partial y} \right) + \frac{\partial}{\partial \zeta} \left(\frac{k}{\mu} \frac{\partial \phi}{\partial \zeta} \right) \right) d\zeta. \qquad (3.15)$$

Note that the last term in Eq. (3.13) drops out upon integration. Moreover, the porosity φ and the permeability k are the only quantities that depend on ζ. We use the following abbreviations $\overline{k}H = \int_0^H k d\zeta$ and $\overline{\varphi}H = \int_0^H \varphi d\zeta$. Consequently, we obtain

$$\frac{1}{\rho} \frac{\partial (\rho \overline{\varphi})}{\partial t} = \frac{\partial}{\partial x} \left(\frac{\overline{k}}{\mu} \frac{\partial \phi}{\partial x} \right) + \frac{\partial}{\partial y} \left(\frac{\overline{k}}{\mu} \frac{\partial \phi}{\partial y} \right). \qquad (3.16)$$

This equation looks exactly the same as Eq. (2.78) except that the dependence on ζ has dropped out and that the permeability and porosity are X-dip averaged quantities. This means that in many text books Eq. (3.16) is presented as being as good as the corresponding 3−D equation. We now know that Eq. (3.16) is only valid when condition (2.79) is satisfied.

3.1.3   THE PROBLEM IN RADIAL SYMMETRY

Eq. (3.16) can be straightforwardly transformed into an equation in radial coordinates. We prefer, however, to do the derivation in the cylindrical symmetrical setting because it is more transparent from the engineering point of view, and it also gives us a scheme for the numerical solution of the equations.

The mass balance with respect to the disk-shaped control volume reads

$$\frac{dm}{dt} = \pi \left(r_e^2 - r_w^2 \right) H \frac{d\left(\varphi\rho\right)}{dt} = \left(\rho q\right)_{r_w} - \left(\rho q\right)_{r_e}, \tag{3.17}$$

where r_e is the outer radius of the disk and r_w is the inner radius of the disk, and H is its height. We use that the layer is horizontal and we can thus ignore the gravity term in $\phi = p + \rho gz$, i.e., $\frac{\partial \phi}{\partial r} = \frac{\partial p}{\partial r}$. Furthermore, we assume that it is possible to write the flow averaged over the ζ−direction, using Darcy's law as

$$\rho q = 2\pi\rho \frac{\bar{k}H}{\mu} r \frac{\partial p}{\partial r}. \tag{3.18}$$

We use that $\left(r_e^2 - r_w^2 \right) = \left(r_e + r_w \right)\left(r_e - r_w \right)$ and that in the limit that $\Delta r \to 0$ we can get the differential $\left((\rho q)_{r_e} - (\rho q)_{r_w} \right) / \left(r_e - r_w \right) \approx \frac{\partial(\rho q)}{\partial r}$. The result is

$$\frac{\partial\left(\varphi\rho\right)}{\partial t} = \frac{1}{r}\frac{\partial}{\partial r}\left(\rho \frac{\bar{k}}{\mu} r \frac{\partial p}{\partial r} \right).$$

In the case that the compressibility is small, i.e., for liquids as opposed to gases we use Eq. (3.7) and use that the density ρ only weakly depends on the space coordinates we obtain $\varphi c_{eff} \frac{\partial P}{\partial t} = \frac{1}{r}\frac{\partial}{\partial r}\left(\frac{\bar{k}}{\mu} r \frac{\partial p}{\partial r} \right)$, which reduces to

$$\frac{\varphi\mu c_{eff}}{\bar{k}} \frac{\partial p}{\partial t} = \frac{1}{r}\frac{\partial}{\partial r}\left(r \frac{\partial p}{\partial r} \right) \tag{3.19}$$

for homogeneous permeability fields.

For the numerical scheme, we use the grid cell division shown in Figure 2.7. The explanation of the figure is shown in the caption. We like to substitute a numerical approximation of Eq. (3.18) into Eq. (3.17) we obtain an equation that can be used as a numerical scheme for solving the transient porous medium equation.

Therefore, we use

$$\left(\rho q\right)_{wb} \approx -\frac{2\pi\rho H}{\mu} \frac{\left(p_P - p_W\right)}{\frac{1}{\lambda_W}\ln\frac{r_{wb}}{r_W} + \frac{1}{\lambda_P}\ln\frac{r_P}{r_{wb}}}$$

$$\left(\rho q\right)_{eb} \approx -\frac{2\pi\rho H}{\mu} \frac{\left(p_E - p_P\right)}{\frac{1}{\lambda_P}\ln\frac{r_{eb}}{r_P} + \frac{1}{\lambda_E}\ln\frac{r_E}{r_{eb}}}$$

For slightly compressible liquids, we may assume that $\rho_{wb} = \rho_{eb} = \rho_P = \rho$, and find after division of both sides by H

$$\pi \left(r_{eb}^2 - r_{wb}^2 \right) H \frac{1}{\rho}\frac{d\left(\varphi\rho\right)}{dt} = -\frac{2\pi H}{\mu}\left(\frac{\left(p_P - p_W\right)}{\frac{1}{\lambda_W}\ln\frac{r_{wb}}{r_W} + \frac{1}{\lambda_P}\ln\frac{r_P}{r_{wb}}} + \frac{\left(p_E - p_P\right)}{\frac{1}{\lambda_P}\ln\frac{r_{eb}}{r_P} + \frac{1}{\lambda_E}\ln\frac{r_E}{r_{eb}}} \right).$$

The left term represents the accumulation term, i.e., $dm/dt/\rho = d\rho V/dt/\rho$, where the division by ρ comes from the flow term on the right side. Indeed, the first term on the right represents the flow entering the grid cell at $r = r_{wb}$, whereas the second term on the right represents the flow leaving the grid cell at $r = r_{eb}$. In cases where the compressibility is small, i.e., when we have liquid phases as opposed to gaseous phases, we like to apply Eq. (3.7) for the left side $\frac{\partial(\rho\varphi)}{\partial t} = \rho\varphi c_{eff}\frac{\partial P}{\partial t}$ and we obtain

$$A\varphi c_{eff}\frac{p_P(t+\Delta t) - p_P(t)}{\Delta t} = -\frac{2\pi}{\mu}\left(\frac{(p_P - p_W)}{\frac{1}{\lambda_W}\ln\frac{r_{wb}}{r_W} + \frac{1}{\lambda_P}\ln\frac{r_P}{r_{wb}}} + \frac{(p_E - p_P)}{\frac{1}{\lambda_P}\ln\frac{r_{eb}}{r_P} + \frac{1}{\lambda_E}\ln\frac{r_E}{r_{eb}}}\right), \quad (3.20)$$

where $A = \pi\left(r_{eb}^2 - r_{wb}^2\right)$ is the area enclosed by the circle with radius r_{eb} and the circle with radius r_{wb}.

There are two basic methods to solve this equation, viz., implicit and explicit. For the explicit term, we assume that all quantities at the right side of the equation are computed at time t. In this case, we can calculate the values of the next time explicitly, i.e., by rewriting Eq. (3.20) as $p_P(t+\Delta t) = rhs(t)$, where $rhs(t)$ denotes all the terms ending at the right side of the equation. The density $(\rho)_{wb}$ is some convenient average between ρ_P and ρ_W. In the same way is $(\rho)_{eb}$ some convenient average between ρ_P and ρ_E. The explicit method is very simple, but it leads to unstable behavior if one chooses large time steps $\Delta t > u/\Delta x$, where u is the local velocity. The other method is called implicit and evaluates all terms on the right at $t+\Delta t$. It erroneously accepts $\Delta t > u/\Delta x$ at the expense of introducing spurious mixing and thus shows unrealistic smoothing behavior. However, the implicit method is still used as the program does not crash due to unstable oscillations.

If we also evaluate the density ρ in the right side terms at time $t+\Delta t$, we need to solve a nonlinear coupled system of equations. It is also possible to evaluate the densities at the right side of the equation at time t. In this case, we obtain an equation of the form

$$\begin{pmatrix} b_1 & c_1 & 0 & 0 & 0\ldots00 & 0 \\ a_2 & b_2 & c_2 & 0 & 0\ldots00 & 0 \\ 0 & a_3 & b_3 & c_3 & 0\ldots00 & 0 \\ & & & & & \\ & & 00\ldots0 & & a_n & b_n \end{pmatrix} \begin{pmatrix} p_1(t+\Delta t) \\ p_2(t+\Delta t) \\ p_3(t+\Delta t) \\ \\ p_n(t+\Delta t) \end{pmatrix} = \begin{pmatrix} r_1 \\ r_2 \\ r_3 \\ \\ r_n \end{pmatrix}.$$

Each row has three or less elements around the diagonal. Such a matrix equation is called a tridiagonal matrix system. It can be solved by a relatively simple algorithm, the so-called Thomas algorithm [192]. If we would have evaluated the densities at time $t+\Delta t$, we would have obtained a similar system of equations as part of the Newton-Raphson solution of a system of equations.

3.1.4 BOUNDARY CONDITIONS FOR RADIAL DIFFUSIVITY EQUATION

For the radial diffusivity equation (3.19), we need to specify one initial condition and two boundary conditions. The initial condition in our problems of interest is usually that the pressure for all radii is equal to an initial pressure, i.e.,

$$p(r, t = 0) = p_i. \quad (3.21)$$

We specify one boundary condition at the well, i.e., at $r = r_w$ and one boundary condition at the boundaries of the domain, i.e., at $r = R$. In the problems of our interest, we specify the production flow rate qB_o at the well. Here B_o is the formation volume factor, i.e., the ratio of the volume of oil in the reservoir and in the tank. We use

$$2\pi r_w \frac{k}{\mu} \left(\frac{\partial p}{\partial r} \right)_{r=r_w} = \frac{qB_o}{H}, \tag{3.22}$$

$$2\pi \frac{k}{\mu} \left(r \frac{\partial p}{\partial r} \right)_{r=r_w} = \frac{qB_o}{H}. \tag{3.23}$$

Furthermore, we assume that there is no flow at the boundary of the well. This implies according to Darcy's law that the pressure gradient at $r = R$ is equal to zero, i.e.,

$$\left(\frac{\partial p}{\partial r} \right)_{r=R} = 0. \tag{3.24}$$

In all of the problems below, we consider the limit that $R \to \infty$.

3.1.5 DIMENSIONAL ANALYSIS FOR THE RADIAL PRESSURE EQUATION; ADAPTED FROM LECTURE NOTES OF LARRY LAKE

In engineering non-dimensionalizing is a standard practice. The following motivations have been given for the application of this procedure.

1. The model equations assume a much simpler form,
2. It shows how the solution depends on the physical parameters. It shows an efficient grouping of these parameters,
3. It shows the order of magnitude of a number of dependent and independent variables; for instance, a characteristic pressure, time, etc.,
4. It is convenient and sometimes essential to formulate numerical models in a dimensionless setting.

There are two ways to develop dimensionless groups from physical laws: dimensional analysis (Buckingham Pi-Theorem) and inspectional analysis. Dimensional analysis relies on there being only a few basic units in the world, and free parameters can only be combined to become dimensionless in manners consistent with this. The method can become technically complicated for many problems of practical relevance, but it is model-free.

The only thing we discuss here is inspectional analysis. Consider equation (3.19) and conditions (3.22), (3.24) and (3.21).

We solve Eqs. 3.19 and boundary conditions (3.21), (3.22) and (3.24) — but we know that the solution will have eight free parameters (see equation (3.19)). It is of importance that the equations are linear in as much as the procedure works best when this is true. The equation is also homogeneous. The best thing is that derivatives are always linear. It does not matter how many equations there are.

Eq. (3.19) and boundary conditions (3.21), (3.22), (3.24) have eight free parameters and without solution we can write the answer as

$$p = p(r,t; \varphi, \mu, c_{eff}, k, r_w, R, p_i, q/H).$$

This notation means that p, the dependent variable, depends on r and t, the independent variables, and on the parameters to the right of the semicolon. The parameters must be constant.

To non-dimensionalize equation (3.19) and boundary conditions (3.21), (3.22) and (3.24) let us define dimensionless variables such that

$$r = r_R r_D + r^*,$$ (3.25)

$$t = t_R t_D + t^*,$$ (3.26)

$$p = p_R p_D + p^*,$$ (3.27)

where the subscript D means dimensionless and $r_R, t_R, p_R, r^*, t^*, p^*$ are scale factors. The scale factors have dimensions but they are arbitrary, at least for now. Note that the transformations (3.25), (3.26) and (3.27) are linear and there will be two for each variable, dependent and independent. There are 6 here.

When these are introduced into Eqs. (3.19) and (3.21), (3.22) and (3.24) we arrive at

$$\frac{1}{r_D} \frac{\partial}{\partial r_D} \left(r_D \frac{\partial p_D}{\partial r_D} \right) = \frac{\phi c_{eff} \mu r_R^2}{k t_R} \frac{\partial p_D}{\partial t_D}$$ (3.28)

for the original partial differential equation and

$$p_D(r_D, t_D = \frac{t - t^*}{t_R}) = \frac{p - p^*}{p_R},$$

$$\left(\frac{\partial p_D}{\partial r_D} \right)_{r_D = \frac{R - r^*}{r_R}} = 0.$$

$$2\pi \frac{k p_R}{\mu} \left(r_D \frac{\partial p_D}{\partial r_D} \right)_{r_D = \frac{r_w - r^*}{r_R}} = \frac{q B_o}{H},$$ (3.29)

for the boundary conditions. Non-dimensionalizing the boundary conditions, including the evaluation points, is as important as non-dimensionalizing the partial differential equation itself. We have done a little rearranging in these equations. Like so much in dimensional analysis, this is somewhat arbitrary. The quantities in brackets are dimensionless because the entire problem statement is dimensionless now.

The scale factors are arbitrary so I can eliminate them by setting the scale factors or the off-sets of the dimensionless groups equal to zero or some physical reference quantity. Which to set to some physical reference quantity and which to set to zero is also arbitrary, but

1. One cannot change the form of the original problem, which means some times that a certain group must be zero or nonzero.
2. One cannot have the scale factors that multiply in Eqs. (3.25), (3.26) and (3.27) be zero also otherwise you are removing a variable from the problem.

This is not as bad as it seems because after a while you see where this is going. I choose to set $\frac{-r^*}{r_R}$ and $\frac{-t^*}{t_R}$ equal to zero in addition to $\frac{p_i - p^*}{p_R}$; this gives

$$r^* = 0,$$

$$t^* = 0.$$

$$p^* = p_i.$$ (3.30)

The choice $p^* = p_i$ is also physically appropriate as p_i is a characteristic reference pressure in our problem. If there is no appropriate characteristic reference quantity as for example for t^*, it is convenient to set the corresponding dimensionless group $(\frac{-t^*}{t_R})$ to zero. Indeed, I can start the production of the reservoir any time I want and it is convenient to call this time $t = 0$. There is nothing against to put t^* equal to the time that has elapsed since the beginning of this century; it is just not convenient. Perhaps you would like to choose $r^* = r_w$ or $r^* = R$. This is, however, again not very convenient. The scale factors X_R must be given a characteristic reference value or one (nonzero). Furthermore, we have a characteristic radius in our problem formulation namely $r_R = r_w$. As said above, the scale factors X_R can never be zero; otherwise, you are removing a variable from the problem. The reference quantity for the pressure p_R we could choose $p_R = \frac{qB_o\mu}{2\pi kH}$ in Eq. (3.22) and obtain:

$$\left(r_D \frac{\partial p_D}{\partial r_D} \right)_{r_D=1} = 1. \tag{3.31}$$

There is, however, no characteristic time in our problem formulation. We are, therefore, free to choose a convenient reference time (see equation (3.28)) as $t_R = \frac{\phi c_{eff}\mu r_w^2}{k}$ (derived from $\frac{\phi c_{eff}\mu r_w^2}{k t_R} = 1$). These completely determine the scale factors, which are no longer arbitrary. We can summarize the dimensionless problem as follows:

$$\frac{1}{r_D} \frac{\partial}{\partial r_D} \left(r_D \frac{\partial p_D}{\partial r_D} \right) = \frac{\partial p_D}{\partial t_D} \tag{3.32}$$

for the original partial differential equation and the associated boundary conditions can be written as

$$\left(r_D \frac{\partial p_D}{\partial r_D} \right)_{r_D=1} = 1, \tag{3.33}$$

$$\left(\frac{\partial p_D}{\partial r_D} \right)_{r_D=r_{eD}} = 0, \tag{3.34}$$

where $r_{eD} = \frac{R}{r_w}$. In the problem considered by us $r_{eD} \to \infty$. or $p_D(r_D \to \infty, t_D) = 0$. The initial condition is

$$p_D(r_D, t_D = 0) = 0. \tag{3.35}$$

The dimensionless solution can be written as

$$p_D = p_D(r_D, t_D; r_{eD}).$$

3.1.6 SOLUTION OF THE RADIAL DIFFUSIVITY EQUATION WITH THE HELP OF LAPLACE TRANSFORMATION

It is always nice if one can find an analytical solution of the partial differential equation (PDE) subjected to its boundary conditions and initial condition. For partial differential equations, this is almost always difficult. You have to develop some skills to do this. This skill can be taught to any engineer with sufficient interest to spoil a number of rainy Sundays to develop it. In the end, you will know where you can find the relevant information in the recommended and other books. For me, references [4, 54, 137] were very useful. The methods used solve the PDE's are (a) the method of characteristics, (b) separation of variables (c) Laplace Transformation and (d) similarity transformations. In this case, we will only illustrate methods (c) and (d).

3.1.7 LAPLACE TRANSFORMATION

The Laplace transform $L[F(t)] = f(s)$ of a function $F(t)$ is defined as

$$f(s) = \int_0^\infty \exp(-st)F(t)dt.$$

It is easy to verify that if $F(t) = c$, i.e., a constant $f(s) = c/s$. It is less easy to verify that $L[\ln(t)] = \frac{-\gamma + \ln s}{s}$, with $\gamma = \exp(0.5772) = 1.781$. The most important thing is that the derivative of $F(t)$ has as its Laplace transform $sf(s) - F(0)$. It is an easy exercise to prove this. If we have a function of two variables, e.g., $p_D(r_D, t_D)$ we can interchange differentiation and integration with respect to r_D and therefore the Laplace transformed Eq. (3.32) leads to

$$\frac{1}{r_D}\frac{\partial}{\partial r_D}\left(r_D\frac{\partial \widetilde{p_D}}{\partial r_D}\right) = s\widetilde{p_D}, \tag{3.36}$$

where we have replaced $\frac{\partial p_D}{\partial t}$ by $s\widetilde{p_D}$ and used the initial condition (3.35). Here $\widetilde{p_D} = L[p_D]$. In this case we find in [4, page 374] the relevant information. It states the solution of the equation

$$z^2\frac{d^2w}{dz^2} + z\frac{dw}{dz} - \left(z^2 - v\right)w = 0. \tag{3.37}$$

We rewrite Eq. (3.36) by writing differentiation of the left side of the equation and subsequently transforming to the independent variable $z = r_D\sqrt{s}$, i.e.,

$$\frac{\partial^2 \widetilde{p_D}}{\partial r_D^2} + \frac{1}{r_D}\frac{\partial \widetilde{p_D}}{\partial r_D} = s\widetilde{p_D},$$

$$s\frac{\partial^2 \widetilde{p_D}}{\partial z^2} + s\frac{1}{z}\frac{\partial \widetilde{p_D}}{\partial z} = s\widetilde{p_D}.$$

Multiplication of the equation by z^2 is indeed Eq. (3.37) with $v = 0$. To see this one needs some experience. First cylindrical problems always one way or the other lead to Bessel functions in the same way as Cartesian problems lead to exponential functions or trigonometric functions. Subsequently, start to look in [4] in the chapter of Bessel functions. Second, play with the differential equation by carrying through the differentiation, such that it appears that Eq. (3.37) is of the right form. Then you need some struggle to get rid of the s in the equation. Abramowitz states that the solution must be a sum of modified Bessel functions that for $v = 0$ are $K_o(z)$ and $I_o(z)$. The modified Bessel function $K_o(z)$ looks somewhat as an exponential $\exp(-z)$, but due to the $1/r$ behavior of the cylindrical problem it becomes infinity at $z = 0$. The modified Bessel function $I_o(z)$ looks somewhat like an exponential $\exp(+z)$, and thus becomes infinity at $z \to \infty$. The general solution is thus $\widetilde{p_D} = AK_o(z) + BI_o(z)$. As the pressure p_D remains bounded for $z \to \infty$, i.e., in a problem where $r_{eD} \to \infty$. Subsequently, we have the solution

$$\widetilde{p_D}(r_D;s) = A(s)K_o\left(r_D\sqrt{s}\right). \tag{3.38}$$

Note that the "constant" A can still be a function of s. The Laplace transformed boundary condition (3.33) reads

$$\left(r_D\frac{\partial \widetilde{p_D}}{\partial r_D}\right)_{r_D=1} = \frac{1}{s}, \tag{3.39}$$

where we have used that $L[1] = \frac{1}{s}$. Subsequently we must differentiate Eq. (3.38) and find the derivative of the modified Bessel function. Again if one has experience in this type of problems reference [4] tells what the result is. If you succeed in doing it yourself, you have been born too late because you could have become famous in the nineteenth century. The result is given on page 376 of reference [4]: $K_o'(z) = -K_1(z)$. The modified Bessel function $K_1(z)$ looks very much like $K_o(z)$, but is not completely the same as we would have had with differentiation of $\exp(-z)$ which leads to $-\exp(-z)$. However, the analogy between the radial case and the Cartesian case must be clear. We substitute (3.37) into (3.39) and obtain

$$\left(\frac{\partial \widetilde{p_D}}{\partial r_D}\right)_{r_D=1} = -A(s)\sqrt{s}\left(K_1\left(r_D\sqrt{s}\right)\right)_{r_D=1} = \frac{1}{s}. \tag{3.40}$$

Therefore

$$\widetilde{p_D}(r_D;s) = -\frac{K_o\left(r_D\sqrt{s}\right)}{s^{\frac{3}{2}}K_1\left(\sqrt{s}\right)}. \tag{3.41}$$

We were unable to find the inverse of the right side of (3.41) in the many tables of inverse Laplace transforms.

Finding a Laplace inversion is very tedious. All the same, we can go through three routes. First, we can use a numerical inverse Laplace transform program. For this, one needs to compute the modified Bessel functions and again fortunately we find how to do this again in reference [4, p. 379]. For diffusion problems, it appears that the Stehfest algorithm is very suitable. It consists of a program of a few lines and one needs to program the Laplace transformed function. For convenience, we have made a few visual basic routines that are coupled to EXCEL (stehfest.xls) to perform the Laplace inversion with the Stehfest algorithm.

The other method uses the analogy between Laplace transforms and Fourier transforms. One likes to substitute $s \to i\omega$. Now one needs to compute the modified Bessel functions of complex arguments. As we shall see, the problem is that our functions are infinity for $\omega \to \infty$. Evert Slob has been sharing with me all his knowledge on this, but I did not succeed to make a robust procedure with the inverse fast Fourier transform.

The third method is to see whether one can find the short time behavior for $s \to \infty$ or the long time behavior for $s \to 0$. We find for the long time behavior (see [4, p. 375])

$$K_o(z) \approx -\left(\ln\left(\frac{1}{2}z\right) + \gamma\right)\left(1 + \frac{z^2}{4}\right) + \frac{z^2}{4},$$

$$K_1(z) \approx \ln\left(\frac{1}{2}z\right)\left(\frac{1}{2}z\right) + \frac{1}{z}\left(1 + \frac{z^2}{4}\right) + \frac{z}{4}\left(2\gamma - 2\right).$$

Note that here $\gamma = 0.5772$. For $K_o(z)$ we only retain $-\left(\ln\left(\frac{1}{2}z\right) + \gamma\right)$ and for $K_1(z)$ we only retain $\frac{1}{z}\left(1 + \frac{z^2}{4}\right)$ We obtain that for the long time behavior

$$\widetilde{p_D}(r_D;s) \approx \frac{\left(\ln\left(\frac{r_D\sqrt{s}}{2}\right) + \gamma\right)}{s} = -\frac{1}{2}\frac{\left(-\ln s + \ln\frac{4}{r_D^2} - 2\gamma\right)}{s}. \tag{3.42}$$

Fortunately we can find the inverse $L^{-1}\left(\frac{1}{s}\ln s\right) = -\gamma - \ln t_D$ [4]. Subsequently we obtain

$$p_D(r_D,t_D) = -\frac{1}{2}\left(\ln\left(\frac{t_D}{r_D^2}\right) - \gamma + 2\ln 2\right) = -\frac{1}{2}\left(\ln\left(\frac{4t_D}{r_D^2}\right) - \gamma\right). \tag{3.43}$$

3.1.8   SELF SIMILAR SOLUTION

Sometimes, it is possible to find some function of the independent variables $\xi = \xi\,(r_D, t_D)$, which allows to write the partial differential equation, the boundary conditions and the initial conditions using only ξ. This transformation cannot be found by a single general procedure. In many cases, engineers developed sufficient understanding to propose the right choice. In other cases, one may have some idea of the form e.g. $\xi = f\,(r_D)\,/t_D$. In this case $\xi = r_D^2/4t_D$ appears to do the trick. Of course the factor 4 is not important; it is just convenient. We like to find the function $\hat{p}_D\,(\xi)$ instead of $p_D\,(r_D, t_D)$. Mathematicians like to stress that $p_D\,(r_D, t_D)$ and $\hat{p}_D\,(\xi)$ are different functions and should therefore not be indicated by the same symbol p_D. However, for convenience we omit the "$\hat{\ }$" on $\hat{p}_D\,(\xi)$ and do use the same symbol. We rewrite Eq. (3.32)

$$\frac{1}{r_D}\frac{d}{d\xi}\left(r_D\frac{dp_D}{d\xi}\frac{\partial\xi}{\partial r_D}\right)\frac{\partial\xi}{\partial r_D} = \frac{dp_D}{d\xi}\frac{\partial\xi}{\partial t_D}. \tag{3.44}$$

We use that $\frac{\partial\xi}{\partial t_D} = -r_D^2/4t_D^2$ and $\frac{\partial\xi}{\partial r_D} = r_D/2t_D$. Substitution into (3.44) leads to

$$\frac{1}{r_D}\frac{\partial}{\partial\xi}\left(r_D\frac{dp_D}{d\xi}\frac{r_D}{2t_D}\right)\frac{r_D}{2t_D} = -\frac{dp_D}{d\xi}\frac{r_D^2}{4t^2{}_D}, \tag{3.45}$$

$$\frac{\partial}{\partial\xi}\left(\xi\frac{dp_D}{d\xi}\right) = -\xi\frac{dp_D}{d\xi}. \tag{3.46}$$

To reduce the radial diffusion equation to an ordinary differential equation with $\xi = r_D^2/4t_D$ is called the Boltzmann transformation. Such a name (Boltzmann) is tied up with the transformation which shows once again that finding the appropriate combination of variables making up ξ is not easy. We use $y = \xi\frac{dp_D}{d\xi}$ and the equation can be written as $\frac{\partial y}{\partial\xi} = -y$ or $y = A\exp\,(-\xi)$.

   In addition, the boundary conditions and initial condition can be written in terms of ξ. Indeed, the initial condition for $t = 0$ reads $p_D\,(\xi \to \infty) = 0$ and also boundary condition $p_D\,(r_D \to \infty, t_D) = 0$ reads for the new function $p_D\,(\xi \to \infty) = 0$. This is one of the features of a similarity solution that one of the boundary conditions in the new independent variable ξ coincides with the initial condition. Finally, we can rewrite boundary condition (3.33)

$$\lim_{r_D \to 0}\left(r_D\frac{\partial p_D}{\partial r_D}\right) = 1,$$

$$\lim_{\xi \to 0}\left(2\xi\frac{\partial p_D}{\partial\xi}\right) = 1.$$

It follows that $A = -\frac{1}{2}$. Subsequently, we obtain $\frac{dp_D}{\partial\xi} = y/\xi$

$$\frac{dp_D}{\partial\xi} = \frac{1}{2}\frac{\exp\,(-\xi)}{\xi}.$$

The equation is integrated once, and we obtain

$$p_D = \frac{1}{2}\int_a^\xi\frac{\exp\,(-u)}{u}du.$$

Engineers sometimes write $\int_a^\xi\frac{\exp(-\xi)}{\xi}d\xi$, but mathematicians dislike this because the upper boundary ξ is not the same as the $\xi's$ in the integrand. The lower boundary a is an integration

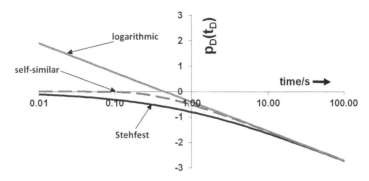

Figure 3.1 The dimensionless draw-down pressure as a function of the dimensionless time for the exact solution, the self similar solution and the logarithmic solution.

constant. The initial condition and one of the boundary conditions demands that $p_D(\xi \to \infty) = 0$. From this we find that $a \to \infty$. Consequently, the solution satisfying all boundary conditions and the initial condition reads

$$p_D(\xi) = -\frac{1}{2}\int_\xi^\infty \frac{\exp(-u)}{u}\,du =: -\frac{1}{2}E_1(\xi), \tag{3.47}$$

$$p_D(r_D, t_D) = -\frac{1}{2}E_1\left(\frac{r_D^2}{4t_D}\right) \approx \frac{1}{2}\left(\gamma + \ln\frac{r_D^2}{4t_D}\right), \tag{3.48}$$

where $E_1(\xi)$ is the exponential integral [4, p. 228]. We also find in Abramowitz [4, p. 229] that

$$E_1(z) = -\gamma - \ln(z) - \sum_{n=1}^\infty \frac{(-1)^n z^n}{nn!}.$$

Note that expression (3.48) agrees with expression (3.43) obtained above. A comparison of results is shown in Figure 3.1.

We note that it is not necessary for self-similar formulation to include the initial condition and the self-similar solution describes the long time behavior.

3.1.9   THE DIMENSIONAL DRAW-DOWN PRESSURE

We obtain the full dimensional draw-down pressure by using $p = p_R p_D + p_i$ with $p_R = \frac{qB_o\mu}{2\pi kH}$. Moreover, we use $t = t_R t_D$ with $t_R = \frac{\phi c_{eff}\mu r_w^2}{k}$. Subsequently, we obtain from $p_D(r_D = 1, t_D) = \frac{1}{2}\left(\gamma + \ln\frac{1}{4t_D}\right) = -\frac{1}{2}\ln\frac{4t_D}{\exp(\gamma)}$ for the well flow pressure $\tilde{p}_{wf}$ at the well

$$\tilde{p}_{wf} = p_i - \frac{qB_o\mu}{4\pi kH}\ln\frac{4t_D}{\exp(\gamma)} = p_i - \frac{qB_o\mu}{4\pi kH}\ln\frac{4kt}{\phi c_{eff}\mu r_w^2 \exp(\gamma)},$$

$$\tilde{p}_{wf} = p_i - \frac{qB_o\mu\ln(10)}{4\pi kH}\log\frac{4t_D}{\exp(\gamma)} := p_i - m_H\log\frac{4t_D}{\exp(\gamma)}.$$

In addition, all kinds of extra resistances occur near a well, e.g., because the drilling mud has been clogging the pores. Such a resistance is proportional to the flow rate q and leads to an additional pressure drop. This pressure drop is given as

$$\Delta p_{well} = \frac{qB_o\mu}{2\pi kH}S.$$

This expression defines the skin factor S. Adding this additional resistance leads to the real well flow pressure p_{wf} as opposed to the ideal well flow pressure $\widetilde{p}_{wf}$ used above.

$$p_{wf} = p_i - \frac{qB_o\mu}{4\pi kH}\left(\ln\frac{4t_D}{\exp(\gamma)} + \frac{1}{2}S\right) = p_i - \frac{qB_o\mu}{4\pi kH}\left(\ln\frac{4kt}{\phi c_{eff}\mu r_w^2 \exp(\gamma)} + \frac{1}{2}S\right), \quad (3.49)$$

$$p_{wf} = p_i - \frac{qB_o\mu\ln(10)}{4\pi kH}\left(\log\frac{4t_D}{\exp(\gamma)} + \frac{1}{2}\frac{S}{\ln 10}\right) := p_i - m_H\left(\log\frac{4t_D}{\exp(\gamma)} + \frac{1}{2}\frac{S}{\ln 10}\right).$$

3.2 PRESSURE BUILD UP

One of the methods [122] used for years to determine the permeability height product in a reservoir is to shut in the well and observe how quickly the pressure rises. The method combined with draw-down, which is described above, also determines the skin, i.e., the extra resistance factor near the well, the average reservoir pressure and gives an estimate of the volume of the reservoir from which oil is obtained. Therefore, you will spend the next year quite some time to interpret the build up tests. What we like to do today is give you the basic background for all of this. The most important part of this background is how to handle time-dependent boundary conditions. This can be done with superposition. In almost all engineering courses, there is a vague reference to the principle of superposition, something you ought to know. I found this always very confusing, when I was a student. For instance, you may know the solution of one well producing into an infinite field and add the solution of another well at some distance from this well to simulate a boundary condition of constant potential or a boundary condition of no flow. In other cases, we deal with time-dependent boundary conditions. All of this requires different ways of superposition. Hopefully, I am able to avoid the vagueness here by including an appendix, which I copied more or less from [137, section 10.5-4].

3.2.1 SUPERPOSITION

The mathematical background for superposition is given in appendix 3.D. The idea of superposition is that we can add the solutions of two separate differential equations to obtain the solution of a more complex problem. Or, more generally, to add the solutions of n separate differential equations to obtain the solution of a more complex problem. In our case, we have an interest in the effect of a time-dependent boundary condition. Indeed, if one has produced for a while and subsequently shut in the well at time t_p we are changing the boundary condition at the well at a certain time $t = t_p$. Duhamel's theorem [137] stated in appendix 3.D gives the result. I like to avoid to prove his theorem. From the engineering point of view, we like to state the superposition principle (what a word) as follows: the pressure at time $t_p + \Delta t$ that occurs in a well that has been shut in after it had been producing q during time t_p is equivalent to the sum of the pressures of two separate problems that do not involve time dependent boundary conditions. These problems are the problem of a well that has been producing q during time $t_p + \Delta t$ and the problem of a well that has been producing $-q$ between time t_p and time $t_p + \Delta t$. This is illustrated in Figure 3.2.

The dimensionless draw-down pressure at the well is given by $p_D(r_D = 1, t_D)$. We abbreviate this for convenience as $p_D(t_D) \approx -\frac{1}{2}\ln\frac{4t_D}{\exp(\gamma)}$, which is easily obtained from Eq. (3.48) for $r_D = 1$. Therefore, the dimensionless pressure p_D^s that occurs after production from a well with rate q during time t_p and subsequently the well is shut in during time Δt is given by (see also appendix 3.D).

$$p_D^s = p_D\left(t_{pD} + \Delta t_D\right) - p_D\left(\Delta t_D\right) = -\frac{1}{2}\ln\frac{t_{pD} + \Delta t_D}{\Delta t_D}.$$

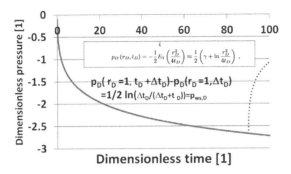

Figure 3.2 Owing to linearity of the transient pressure equation, a build up can be constructed from the difference between two shifted draw-downs.

We obtain the full dimensional build-up pressure by using $p = p_R p_D + p_i$ with $p_R = \frac{qB_o\mu}{2\pi kH}$. Subsequently we obtain

$$p^s = p_i - \frac{qB_o\mu}{4\pi kH}\ln\frac{t_p + \Delta t}{\Delta t},$$

$$p^s = p_i - \frac{qB_o\mu\ln(10)}{4\pi kH}\log\frac{t_p + \Delta t}{\Delta t},$$

where log means the logarithm with base 10. Note that $\frac{t_{pD}+\Delta t_D}{\Delta t_D} = \frac{t_p+\Delta t}{\Delta t}$. The factor $\frac{qB_o\mu\ln(10)}{4\pi kH}$ is usually abbreviated as m_H. The skin factor drops out of the well shut in pressure. Indeed, the skin acts both during the production of $+q$ and the production of $-q$ and appears in identical terms with different signs. These terms cancel. The only reason to show this conversion to the log base 10, is that you will find this procedure in many of the old files of the company. One of the prime objectives of pressure transient testing is to determine the reservoir deliverability, given by the product of permeability and net formation pay (kH). The parameter, $m_H = \frac{qB_o\mu\ln(10)}{4\pi kH}$ is inversely proportional to this quantity. One way of obtaining this parameter is by plotting the build-up response against $\frac{\Delta t}{t_p+\Delta t}$ on semi-logarithmic paper. The slope of the curve plotted this way will yield a value for and consequently gives an estimate for the formation deliverability. An equivalent technique was first introduced by Horner [121, 122], who suggested to use as the appropriate time axis $\ln((t + \Delta t)/\Delta t)$. The equivalent time thus introduced is often referred to as Horner time.

Alternatively, one would just like to plot the well shut pressure versus $\ln\frac{t_p+\Delta t}{\Delta t}$. The slope $\frac{qB_o\mu}{4\pi kH}$ contains the unknown parameter product kH. The viscosity μ and the formation volume factor B_o can be measured in the laboratory. One can use the draw-down equation (3.49) to estimate the skin. In general, you will have some idea about the parameters occurring in the argument of the logarithm of (3.49), i.e., $4kt/\left(\phi c_{eff}\mu r_w^2 \exp(\gamma)\right)$. If one has an idea about the layer height one can estimate k and the effective compressibility c_{eff} is about twice the fluid compressibility. Any error of a factor of 2 only contributes $\ln(2) = 0.693\,15$. A bad skin factor is 5. Therefore, once one has used the buildup to determine the permeability, one can use the draw-down to estimate the skin.

The pressure as a function of time, which occurs after producing at a constant rate q during time t_p and subsequent well shut in is shown in Figure 3.2

Most of the analysis techniques developed for pressure transient analysis are based on the draw-down response indicated in Figure 3.2). The draw-down response (under the assumptions stated above) is reproduced in the build-up curve as shown in Figure 3.2. The "extrapolated" draw-down response (i.e., the lower dashed line, see Figure 3.1) cannot be obtained by direct measurement from the build-up data. In our case, we used for the draw-down response the pressure behavior in an infinite reservoir. However, usually the pressure in a well "feels" the boundaries a long time before $t = t_p$. In such a more general case, the difference between the shut-in pressure and the build-up pressure (indicated in Figure 3.2) is analyzed. Initially, i.e., for short build-up times, this represents a good approximation. However, when the build-up time is of the order of the producing time, t_p, the deviation from the "drawdown response" becomes significant (i.e., the difference between the dashed curves). This is most easily demonstrated by studying the time derivative of the transient pressure response of a radial well in an infinitely acting reservoir.

3.2.2 TIME DERIVATIVES OF PRESSURE RESPONSE

The time derivative of the well shut in pressure $p^s = p_i - \frac{qB_o\mu}{4\pi kH} \ln \frac{t_p + \Delta t}{\Delta t} = p^s = p_i - \frac{qB_o\mu}{4\pi kH} \ln \frac{t}{t - t_p}$
is given by

$$\frac{\partial p_s}{\partial t} = -\frac{qB_o\mu}{4\pi kH}\left(\frac{1}{t} - \frac{1}{t - t_p}\right) = \frac{qB_o\mu}{4\pi kH}\frac{1}{t}\frac{t_p}{(t - t_p)}.$$

In the same way, we find for the draw-down pressure derivative

$$\frac{\partial p_{wf}}{\partial t} = -\frac{qB_o\mu}{4\pi kH}\frac{1}{t} \text{ or equivalently } \frac{\partial p_{wf}}{\partial \ln t} = -\frac{qB_o\mu}{4\pi kH}. \tag{3.50}$$

The derivatives become of comparable magnitude as $\Delta t = t - t_p$ becomes of the order of t_p. In such a case, our uncertainty in the draw-down curve disturbs our analysis.

3.2.3 PRACTICAL LIMITATIONS OF PRESSURE BUILD UP TESTING

For a pressure transient test producing time is often limited by flare restrictions (for gas wells) or production storage capacity. Moreover, pressure gauges are generally rented from service companies, again limiting the time allotted to the production period. In order to make a larger part of the build-up suitable for analysis, a correction procedure is introduced.

3.3 FORMULATION IN A BOUNDED RESERVOIR

Dietz proposed a self-similar solution for the situation that we are withdrawing at a constant rate q_i at a number of wells at $\mathbf{r}_i$ in a bounded reservoir. He proposed that the long time behavior of the solution could be characterized by $\frac{\partial p}{\partial t} = constant$. We consider a reservoir with the $z-$direction perpendicular to the layer direction. We like to write the equation for the long time behavior of

$$\varphi c_{eff}\frac{\partial P}{\partial t} = \frac{\partial}{\partial x}\left(\frac{k}{\mu}\frac{\partial \phi}{\partial x}\right) + \frac{\partial}{\partial y}\left(\frac{k}{\mu}\frac{\partial \phi}{\partial y}\right) + \frac{\partial}{\partial z}\left(\frac{k}{\mu}\frac{\partial \phi}{\partial z}\right) - \sum_{i=1}^{n} q_i B_{oi}\delta\left(\mathbf{r} - \mathbf{r}_i\right),$$

where we have added the term $+\sum_{i=1}^{n} q_i\delta\left(\mathbf{r} - \mathbf{r}_i\right)$ to represent n production or injection wells at rate q_i at position $\mathbf{r}_i$. The delta function $\delta\left(\mathbf{r} - \mathbf{r}_i\right)$ indicate that the wells act as point sources (sinks). Note that we can write $\delta\left(\mathbf{r} - \mathbf{r}_i\right) = \delta\left(x - x_i\right)\delta\left(y - y_i\right)\delta\left(z - z_i\right)$. We relate a positive sign of q to production and a negative sign to injection. The total rate $\sum_{i=1}^{n} q_i B_{oi}$ in the wells has to come from the oil expansion in the reservoir due to the pressure decrease. In other words, we have

$$\varphi c_{eff} V \frac{\partial P}{\partial t} = -\sum_{i=1}^{n} q_i B_{oi},$$

where V is the volume of the reservoir. Therefore, we like to write the equation as

$$-\frac{\sum_{i=1}^{n} q_i B_{oi}}{V} = \frac{\partial}{\partial x}\left(\frac{k}{\mu}\frac{\partial \phi}{\partial x}\right) + \frac{\partial}{\partial y}\left(\frac{k}{\mu}\frac{\partial \phi}{\partial y}\right) + \frac{\partial}{\partial z}\left(\frac{k}{\mu}\frac{\partial \phi}{\partial z}\right) - \sum_{i=1}^{n} q_i B_{oi} \delta(x-x_i)\,\delta(y-y_i)\,\delta(z-z_i).$$

After integration over the height and using the vertical equilibrium approximation, i.e., the assumption that the potential ϕ does not depend on the cross-dip direction coordinate z, we obtain an equation of the same form

$$-\frac{\sum_{i=1}^{n} \frac{q_i}{H} B_{oi}}{A} = \frac{\partial}{\partial x}\left(\frac{k}{\mu}\frac{\partial \phi}{\partial x}\right) + \frac{\partial}{\partial y}\left(\frac{k}{\mu}\frac{\partial \phi}{\partial y}\right) - \sum_{i=1}^{n} \frac{q_i}{H} B_{oi} \delta(x-x_i)\,\delta(y-y_i), \qquad (3.51)$$

but now $\frac{q_i}{H}$ has units $\mathrm{m^3/m/s}$, i.e., the production per unit layer height. From now on we use the notation $\delta(\mathbf{r} - \mathbf{r}_i) = \delta(x-x_i)\,\delta(y-y_i)$. A is the area of the reservoir. The position coordinate $\mathbf{r}$ has only components (x, y). With a horizontal reservoir in the radial setting we replace the potential ϕ by the pressure p and obtain

$$-\frac{\sum_{i=1}^{n} \frac{q_i}{H} B_{oi}}{A} = \frac{1}{r}\frac{\partial}{\partial r}\left(\frac{kr}{\mu}\frac{\partial p}{\partial r}\right) - \sum_{i=1}^{n} \frac{q_i}{H} B_{oi} \delta(\mathbf{r} - \mathbf{r}_i). \qquad (3.52)$$

For a single well, we like to solve the equation

$$-\frac{1}{r}\frac{\partial}{\partial r}\left(\frac{kr}{\mu}\frac{\partial p}{\partial r}\right) = \frac{qB_o}{AH}, \qquad (3.53)$$

with the boundary conditions that for $r = r_{well}$ Eq. (3.53) is satisfied and that at the boundary, i.e., at $r = r_e$ there is zero pressure gradient $\frac{\partial p}{\partial r} = 0$ corresponding to no flow, i.e.,

$$-\frac{1}{r_{well}}\frac{\partial}{\partial r}\left(\frac{kr}{\mu}\frac{\partial p}{\partial r}\right)_{r_{well}} = \frac{qB_o}{AH}, \qquad (3.54)$$

$$\left(\frac{\partial p}{\partial r}\right)_{r_e} = 0. \qquad (3.55)$$

Integration of Eq. (3.53) leads to

$$\left(\frac{k}{\mu}r\frac{\partial p}{\partial r}\right) = -\frac{qB_o}{2AH}\left(r^2 - r_e^2\right),$$

where we have used boundary condition (3.55). After one more integration step we obtain

$$p - p_{wf} = -\frac{q\mu B_o\left(r^2 - r_w^2\right)}{4kAH} + \frac{q\mu B_o r_e^2}{2kAH}\ln\frac{r}{r_w},$$

where we have introduced the well flow pressure p_{wf} at the well at $r = r_w$ for convenience. As the surface area $A = \pi \left(r_e^2 - r_w^2 \right)$, we can write the equation as

$$p - p_{wf} = \frac{q\mu B_o \left(r^2 - r_w^2 \right)}{4\pi k \left(r_e^2 - r_w^2 \right) H} - \frac{q\mu B_o r_e^2}{2\pi k \left(r_e^2 - r_w^2 \right) H} \ln \frac{r}{r_w},$$

The average pressure is obtained by integrating the above expression after multiplying with $2\pi r$ and dividing by $\pi \left(r_e^2 - r_w^2 \right)$. This is a straightforward but tedious operation. We obtain after using Maple

$$\frac{q\mu B_o r_e^2}{2\pi^2 k \left(r_e^2 - r_w^2 \right)^2 H} \int_{r_w}^{r_e} 2\pi r \ln \frac{r}{r_w} dr ,$$

and

$$\bar{p} - p_{wf} = -\frac{1}{8\pi} B_o \mu \frac{q}{Hk} - \frac{1}{4\pi} \frac{q\mu B_o r_e^2}{k \left(r_e^2 - r_w^2 \right) H} + \frac{1}{4\pi} q\mu B_o r_e^2 \frac{2 r_e^2 \ln \frac{r_e}{r_w}}{k \left(r_e^2 - r_w^2 \right)^2 H}.$$

We use $r_e \gg r_w$ and obtain

$$\bar{p} - p_{wf} = \frac{q\mu B_o}{2\pi k H} \left(\ln \frac{r_e}{r_w} - \frac{3}{4} \right). \tag{3.56}$$

In non-circular domains, the same type of equation holds except that we have to insert a geometric factor. Traditionally, this is done by rewriting (3.56) as follows:

$$\bar{p} - p_{wf} = \frac{q\mu B_o}{2\pi k H} \left(\frac{1}{2} \ln \frac{\pi r_e^2}{\pi r_w^2 \exp \left(\frac{3}{2} \right)} \right) = \frac{q\mu B_o}{2\pi k H} \left(\frac{1}{2} \ln \frac{A}{\pi r_w^2 \exp \left(\frac{3}{2} \right)} \right).$$

This again is rewritten and for historical reasons, we include the exponential of Euler's constant $\exp(\gamma) = \exp(0.5772)$, i.e.,

$$\bar{p} - p_{wf} = \frac{q\mu B_o}{2\pi k H} \left(\frac{1}{2} \ln \frac{4A}{\exp(\gamma) C_A r_w^2} + S \right), \tag{3.57}$$

where we have also included a skin factor S, which takes care of the additional resistance due to well damage, e.g., by the drilling mud or by sand redeposited during oil production. The factor C_A in the denominator is the so-called Dietz-shape-factor. It is $C_A = 31.6$ for a circular reservoir. They can be found in [64, p. 151]. To mention a few of the values, we have $C_A = 30.9$ for a square field with the well in the center, $C_A = 22.6$ for a rectangle with a ratio $2 : 1$ of the sides and the well in the middle.

3.4   NON-DARCY FLOW

We have seen that Forchheimer equation reads

$$\mathbf{grad}(p + \rho gz + \frac{1}{2}\rho v^2) = - \left(\frac{\mu \mathbf{u}}{k} + \beta \rho \mathbf{u} |\mathbf{u}| \right). \tag{3.58}$$

In the radial setting and ignoring gravity and the velocity contribution to the potential, we obtain noting that the velocity u toward the production well is negative

$$\rho \frac{\partial p}{\partial r} = - \left(\frac{\mu}{k} \rho u - \beta \left(\rho u \right)^2 \right) ,$$

$$\frac{\rho k}{\mu} \frac{\partial p}{\partial r} = \left(\frac{\rho q}{2\pi r H} + \frac{k\beta}{\mu} \left(\frac{\rho q}{2\pi r H} \right)^2 \right). \tag{3.59}$$

We note that $\rho u = -\rho q/(2\pi r H)$. In a quasi steady state approximation the mass flow $\rho q = \rho_{sc}q_{sc} = cons\tan t$. As most of the effect of non-Darcy flow occurs near the well we can as well use the quasi-steady state approximation to estimate the effect of Non-Darcy flow. We use again the Kirchoff integral transform

$$m(p) = \left(\frac{\mu}{\rho k}\right)_{ref} \int_{p_{ref}}^{p} \left(\frac{\rho k}{\mu}\right) dp',$$

and

$$\frac{\partial m}{\partial r} = \left(\frac{\mu}{\rho k}\right)_{ref} \left(\frac{\rho k}{\mu}\right) \frac{\partial p}{\partial r}.$$

We find from Eq. (3.59)

$$\left(\frac{\rho k}{\mu}\right)_{ref} \frac{\partial m}{\partial r} = \left(\frac{\rho_{sc}q_{sc}}{(2\pi r H)} + \frac{k\beta}{\mu}\left(\frac{\rho_{sc}q_{sc}}{(2\pi r H)}\right)^2\right),$$

$$\frac{\partial m}{\partial r} = \left(\left(\frac{\mu}{\rho k}\right)_{ref}\frac{\rho_{sc}q_{sc}}{(2\pi r H)} + \frac{k\beta}{\mu}\left(\frac{\mu}{\rho k}\right)_{ref}\left(\frac{\rho_{sc}q_{sc}}{(2\pi r H)}\right)^2\right).$$

We can as well assume that the permeability in the reference state is equal to the permeability at any other pressure. Next we make the approximation that the viscosity is only weakly dependent on the pressure and that we can choose an average value $\overline{\mu}$. We obtain after integration that

$$m(r) - m(r_{well}) = \left(\frac{\mu}{\rho k}\right)_{ref}\frac{\rho_{sc}q_{sc}}{(2\pi H)}\ln\left(\frac{r}{r_{well}}\right)$$
$$+ \frac{\beta}{\overline{\mu}}\left(\frac{\mu}{\rho}\right)_{ref}\left(\frac{\rho_{sc}q_{sc}}{(2\pi H)}\right)^2\left(\frac{1}{r_{well}} - \frac{1}{r}\right).$$

We like to define this in terms of $\Delta m = \left(\frac{\mu}{\rho k}\right)_{ref}\frac{\rho_{sc}q_{sc}}{(2\pi H)}\left(\ln\left(\frac{r}{r_{well}}\right) + S_{dyn}\right)$, where S_{dyn} is the dynamic skin factor. We obtain

$$m(r) - m(r_{well}) = \left(\frac{\mu}{\rho k}\right)_{ref}\frac{\rho_{sc}q_{sc}}{(2\pi H)}\left(\ln\left(\frac{r}{r_{well}}\right)\right.$$
$$+ \left.\frac{k_{ref}\beta}{\overline{\mu}}\left(\frac{\rho_{sc}q_{sc}}{(2\pi H)}\right)\left(\frac{1}{r_{well}} - \frac{1}{r}\right)\right).$$

Because usually $r \gg r_w$ we define the dynamic skin S_{dyn} as

$$S_{dyn} = \frac{k_{ref}\beta}{\overline{\mu}}\left(\frac{\rho_{sc}q_{sc}}{(2\pi r_{well}H)}\right) := Dq_{sc},$$

$$D = \frac{k_{ref}\beta}{\overline{\mu}}\left(\frac{\rho_{sc}}{(2\pi r_{well}H)}\right),$$

where D got the misnomer "turbulence factor". The dynamic skin has, however, nothing to do with turbulence and is the result of inertia forces, i.e., the acceleration and deceleration of the gas while it moves through the pore bodies and the pore necks. The total skin is equal to the Darcy skin (static skin) plus the dynamic skin, i.e., $S = S_{dyn} + S_{Darcy}$.

3.A   ABOUT BOUNDARY CONDITION AT $r = r_{eD}$

- For a reservoir bounded at $r_D = r_{eD}$, where it has a no flow condition we need to find implement the boundary condition into

$$p_D(r_D, s) = A(s) K_o(r_D \sqrt{s}) + B(s) I_o(r_D \sqrt{s}).$$

We use (see [4, page 376]) that $I_o'(z) = I_1(z)$ and $K_o'(z) = -K_1(z)$. Consequently we obtain that

$$A(s) K_1(r_{eD} \sqrt{s}) - B(s) I_1(r_{eD} \sqrt{s}) = 0.$$

The boundary condition at the well reads for $\left(r_D \frac{\partial \widetilde{p_D}}{\partial r_D} \right)_{r_D = 1} = \frac{1}{s}$ or for $\left(\frac{\partial \widetilde{p_D}}{\partial r_D} \right)_{r_D = 1} = \frac{1}{s}$

$$-A(s) K_1(\sqrt{s}) + B(s) I_1(\sqrt{s}) = \frac{1}{s^{\frac{3}{2}}}.$$

- and we obtain

$$A(s) = \frac{1}{s^{\frac{3}{2}} \left(-K_1(\sqrt{s}) + \frac{K_1(r_{eD}\sqrt{s})}{I_1(r_{eD}\sqrt{s})} I_1(\sqrt{s}) \right)}$$

$$B(s) = \frac{1}{s^{\frac{3}{2}} \left(-\frac{I_1(r_{eD}\sqrt{s})}{K_1(r_{eD}\sqrt{s})} K_1(\sqrt{s}) + I_1(\sqrt{s}) \right)}.$$

3.A.1   EXERCISE, STEHFEST ALGORITHM

About application of the Stehfest algorithm for obtaining a transient pressure profile

- Use $r_{eD} = 100$ and compute the dimensionless pressure difference near the well with the Stehfest algorithm. Compare to the solution in the infinite space, the self similar solution and the logarithmic solution in a single plot. Hint one needs to write a function to compute

$$p_D(r_D = 1, s) = \frac{K_o(\sqrt{s})}{s^{\frac{3}{2}} \left(-K_1(\sqrt{s}) + \frac{K_1(r_{eD}\sqrt{s})}{I_1(r_{eD}\sqrt{s})} I_1(\sqrt{s}) \right)}$$

$$+ \frac{I_o(\sqrt{s})}{s^{\frac{3}{2}} \left(-\frac{I_1(r_{eD}\sqrt{s})}{K_1(r_{eD}\sqrt{s})} K_1(\sqrt{s}) + I_1(\sqrt{s}) \right)}. \qquad (3.60)$$

Note that one can neglect $\frac{K_1(r_{eD}\sqrt{s})}{I_1(r_{eD}\sqrt{s})} I_1(\sqrt{s})$ with respect to $K_1(\sqrt{s})$ in the expression for $A(s)$ when the argument $r_{eD}\sqrt{s}$ becomes very large e.g. larger than fifty. Also when $r_{eD}\sqrt{s}$ becomes large we can disregard the term $B(s)$ altogether. In this case that $r_{eD}\sqrt{s}$ is large we get the same expression as in an infinite reservoir. All the Bessel function of integer order that are required are already programmed in Stehfest.xls. This function will be slightly more complicated but similar to the function Function lapl2(s).

- give also a plot of the fully dimensional well flow pressure $\widetilde{p}_{wf}$ versus $\ln(t)$, both for the infinite reservoir and the bounded reservoir. Use some reasonable numbers for the input data.

- compute the well shut in pressure after having produced for dimensionless time of $10,000$ both for the infinite reservoir and the bounded reservoir. Hint the dimensionless pressure is $p_D^s = p_D(t_{pD} + \Delta t_D) - p_D(\Delta t_D)$.

- can one derive an expression of the Laplace transform of the pressure build-up curve when one applies the boundary condition $\left(r_D \frac{\partial p_D}{\partial r_D}\right)_{r_D=1} = 1 - U\left(t_D - t_{pD}\right)$, where $U(x) = 0$ for $x < 0$ and $U(x) = 1$ for $x > 0$. Furthermore we use $U(x) = \frac{1}{2}$ for $x = 0$. The Laplace transform of $U(t-k) = \frac{1}{s}\exp(-ks)$. Note that this boundary condition corresponds to well-shut in. Use also that $L^{-1}\left[\exp(-bs)f(s)\right] = F(t-b)U(t-b)$
- make a choice of reasonable input data to compute the reference radius, time and pressure.
- explain (Figure 3.2) why the interpretation of the build-up pressure in a finite reservoir becomes difficult long after shut in. Hint plot the buildup in a finite reservoir and an infinite reservoir both as in Figure 3.2).
- show that one can obtain the permeability height product and the skin from a combination of a build-up plot and a draw-down plot. Hint this is explained in Dake "fundamentals of reservoir engineering".
- show analytically that for draw-down in a finite reservoir the pressure finally decreases proportionally to the time t. Hint in the limit that for small s, one finds the behavior for $t \to$ large. Does one also observe this in the with the Stehfest algorithm inverted Eq. (3.60)? Hint: taking the limit for small s is tedious. Perhaps MAPLE can help.
- try to find the volume of the reservoir from the numerical behavior of behavior of $\frac{\partial p_D}{\partial t_D}$. Hint use the reference pressure and reference time.

3.B ROCK COMPRESSIBILITY

This text was prepared in collaboration with K. Drabe (Geochem).

3.B.1 PHYSICAL MODEL

For two-phase flow problems concerning rock compressibility are usually ignored. We want to include rock compressibility in the model. Essentially, the problem has been solved by Verruijt. He uses, however, the assumption that the air and water flow are equal because his interest is not in airflow. Therefore, we give here the derivation for oil/ water flow. We use the following assumptions:

- we consider oil and water flow in a compressible medium,
- the overburden pressure is constant and equal to the sum of the fluid pressure and the grain pressure,
- the oil, water, solid and porous medium compressibility $(c_\alpha, \alpha = w, o, s, f)$ have been measured at constant mass.

$$c_\alpha = \frac{1}{\rho_\alpha}\left(\frac{\partial \rho_\alpha}{\partial P}\right)_T, \quad \alpha = o, w, \tag{3.61}$$

$$c_s = \frac{1}{\rho_s}\left(\frac{\partial \rho_s}{\partial P}\right)_T, \tag{3.62}$$

$$c_{f1} = \frac{1}{V_f}\left(\frac{\partial V_f}{\partial P}\right)_T \approx \frac{1}{\varphi V_b}\left(\frac{\partial V_b}{\partial P}\right)_T, \tag{3.63}$$

$$c_{f2} = \frac{1}{\varphi V_b}\left(\frac{\partial V_b}{\partial P}\right)_T, \tag{3.64}$$

where P is the fluid pressure. Here V_f and V_b are the pore volume and bulk volume, respectively. The approximation to relate to the bulk volume is convenient to obtain the pore compressibilities from measurements [64] but only correct in the limit of zero grain compressibility. We use two types of formation compressibilities, viz., a theoretical c_{f1} and a practical c_{f2} which lends itself to measurement in the laboratory. The minus sign for the formation compressibility is introduced as a change in fluid pressure is minus the change in grain pressure. This limits our model to the one-dimensional situation in the vertical direction only. It is straightforward to express the formation compressibility in terms of the porosity $\varphi = V_f/V_b$ (see [64] p. 26 and p. 98)

$$c_{f1} = \frac{1}{V_f}\left(\frac{\partial V_f}{\partial P}\right)_T = \frac{1}{V_f}\left(\frac{\partial V_b\varphi}{\partial P}\right)_T = \frac{V_b}{V_f}\left(\frac{\partial \varphi}{\partial P}\right)_T + \frac{\varphi}{V_f}\left(\frac{\partial (V_s+V_f)}{\partial P}\right)_T.$$

We denote the grain volume with V_s. This can be rearranged to

$$c_{f1}(1-\varphi) = \frac{1}{\varphi}\left(\frac{\partial \varphi}{\partial P}\right)_T + \frac{\varphi V_s}{V_f}\frac{V_b}{V_b}\frac{1}{V_s}\left(\frac{\partial V_s}{\partial P}\right)_T.$$

or rearranging once more we obtain:

$$c_{f1} = \frac{1}{\varphi(1-\varphi)}\left(\frac{\partial \varphi}{\partial P}\right)_T - \frac{1}{\rho_s}\left(\frac{\partial \rho_s}{\partial P}\right)_T = \frac{1}{\varphi(1-\varphi)}\left(\frac{\partial \varphi}{\partial P}\right)_T - c_s. \qquad (3.65)$$

It is useful to consider the definition

$$c_{f2} = \frac{1}{\varphi V_b}\left(\frac{\partial V_b}{\partial P}\right)_T = \frac{1}{\varphi V_b}\left(\frac{\partial (V_s+V_f)}{\partial P}\right)_T = c_{f1} - \frac{1-\varphi}{\varphi}c_s. \qquad (3.66)$$

- Compressibilities are small, i.e., $c_\alpha p_i << 1$ where p_i is the initial pressure,
- Darcy's law of two-phase flow is valid, where the interstitial velocity is with respect to the grain velocity,
- capillary forces are disregarded in the flow direction, viscous forces are disregarded in the X-dip direction,
- the medium is homogeneous with constant permeability, compressibility and initial porosity.

3.B.2   MASS BALANCE IN CONSTANT CONTROL VOLUME

The mass balance equations for the solid, water and oil for a constant volume read

$$\frac{\partial}{\partial t}\left((1-\varphi)\rho_s\right) = -\mathbf{div}\cdot\left((1-\varphi)\rho_s\mathbf{V_s}\right),$$

$$\frac{\partial}{\partial t}\left(\varphi S_w\rho_w\right) = -\mathbf{div}\cdot\left(\varphi\rho_w S_w\mathbf{V_w}\right),$$

$$\frac{\partial}{\partial t}\left(\varphi S_o\rho_o\right) = -\mathbf{div}\cdot\left(\varphi\rho_o S_o\mathbf{V_o}\right),$$

where V denotes the interstitial velocity. The interstitial velocity is with respect to a fixed frame of reference and not with respect to the moving grains [62] as we should use to be able to apply Darcy's law. We carry out the differentiation as a product of densities and other terms, i.e.,

$$\rho_s\frac{\partial}{\partial t}(1-\varphi) + (1-\varphi)\frac{\partial}{\partial t}\rho_s = -\rho_s\mathbf{div}\cdot\left((1-\varphi)\mathbf{V_s}\right) - (1-\varphi)\mathbf{V_s}\cdot\mathbf{grad}\left(\rho_s\right),$$

$$\rho_w\frac{\partial}{\partial t}(\varphi S_w) + \varphi S_w\frac{\partial}{\partial t}(\rho_w) = -\rho_w\mathbf{div}\cdot(\varphi S_w\mathbf{V_w}) - \varphi S_w\mathbf{V_w}\cdot\mathbf{grad}(\rho_w),$$

$$\rho_o\frac{\partial}{\partial t}(\varphi S_o) + \varphi S_o\frac{\partial}{\partial t}(\rho_o) = -\rho_o\mathbf{div}\cdot(\varphi S_o\mathbf{V_o}) - \varphi S_o\mathbf{V_o}\cdot\mathbf{grad}(\rho_o).$$

The gradient (density) terms are usually considered small and disregarded. Subsequently, we obtain, if we consider the densities to be function of the pressure under the condition that the temperature is constant,

$$\frac{\partial}{\partial t}(1-\varphi) + \frac{(1-\varphi)}{\rho_s}\left(\frac{d\rho_s}{dP}\right)_T \frac{\partial P}{\partial t} = -\mathbf{div}\cdot((1-\varphi)\mathbf{V}_s),$$

$$\frac{\partial}{\partial t}(\varphi S_w) + \frac{\varphi S_w}{\rho_w}\left(\frac{d\rho_w}{dP}\right)_T \frac{\partial P}{\partial t} = -\mathbf{div}\cdot(\varphi S_w \mathbf{V}_w),$$

$$\frac{\partial}{\partial t}(\varphi S_o) + \frac{\varphi S_o}{\rho_o}\left(\frac{d\rho_o}{dP}\right)_T \frac{\partial P}{\partial t} = -\mathbf{div}\cdot(\varphi S_o \mathbf{V}_o).$$

Therefore, by introducing the compressibilities, we obtain

$$\frac{\partial}{\partial t}(1-\varphi) + (1-\varphi)c_s \frac{\partial P}{\partial t} = -(1-\varphi)\mathbf{div}\cdot(\mathbf{V}_s) - \mathbf{V}_s\cdot\mathbf{grad}(1-\varphi), \qquad (3.67)$$

$$\frac{\partial}{\partial t}(\varphi S_w) + \varphi S_w c_w \frac{\partial P}{\partial t} = -\mathbf{div}\cdot(\varphi S_w(\mathbf{V}_w - \mathbf{V}_s)) - \mathbf{div}\cdot(\varphi S_w \mathbf{V}_s), \qquad (3.68)$$

$$\frac{\partial}{\partial t}(\varphi S_o) + \varphi S_o c_o \frac{\partial P}{\partial t} = -\mathbf{div}\cdot(\varphi S_o(\mathbf{V}_o - \mathbf{V}_s)) - \mathbf{div}\cdot(\varphi S_o \mathbf{V}_s), \qquad (3.69)$$

where we have also rewritten the balance equation for the solid and subtracted and added the solid velocity to the equation. We assert that $u_w = \varphi S_w(V_w - V_s)$ and $u_o = \varphi S_o(V_o - V_s)$ are the Darcy velocity of water and oil, respectively. Subsequently, we divide the equation for the solid by $1-\varphi$ and disregarding the grad$(1-\varphi)$ term we obtain the following equation

$$\frac{1}{1-\varphi}\left(\frac{d\varphi}{dP}\right)_T \frac{\partial P}{\partial t} - c_s \frac{\partial P}{\partial t} = \mathbf{div}\cdot(\mathbf{V}_s), \qquad (3.70)$$

which can be rearranged to

$$\varphi c_{f1}\frac{\partial P}{\partial t} - c_s(1-\varphi)\frac{\partial P}{\partial t} = \varphi c_{f2}\frac{\partial P}{\partial t} = \mathbf{div}\cdot(\mathbf{V}_s).$$

We add Eqs. (4.90) and (3.69) and define the liquid compressibility as $\varphi c_{liq} = \varphi(S_w c_w + S_o c_o)$ to obtain

$$\frac{\partial}{\partial t}(\varphi) + \varphi c_{liq}\frac{\partial P}{\partial t} = -\mathbf{div}\cdot(\mathbf{u}_w + \mathbf{u}_o) - \varphi\mathbf{div}\cdot(\mathbf{V}_s) + \mathbf{V}_s\mathbf{grad}\varphi,$$

or disregarding the grad (φ) term

$$\left(\frac{d\varphi}{dP}\right)_T \frac{\partial P}{\partial t} + \varphi c_{liq}\frac{\partial P}{\partial t} + \varphi\,\mathbf{div}(\mathbf{V}_s) = -\mathbf{div}\cdot(\mathbf{u}_w + \mathbf{u}_o). \qquad (3.71)$$

Substitution of Eq. (3.77) into Eq. (3.71) leads to

$$\varphi(1-\varphi)(c_{f1} + c_s)\frac{\partial P}{\partial t} + \varphi c_{liq}\frac{\partial P}{\partial t} + \varphi^2\left(c_{f1} - \frac{1-\varphi}{\varphi}c_s\right)\frac{\partial P}{\partial t} = -\mathbf{div}\cdot(\mathbf{u}_w + \mathbf{u}_o). \qquad (3.72)$$

Substitution of Eq. (3.63) into Eq. (3.72) leads to the final result

$$\left(\varphi c_{f1} + \varphi c_{liq}\right)\frac{\partial P}{\partial t} = -\mathbf{div}\cdot(\mathbf{u}_w + \mathbf{u}_o). \qquad (3.73)$$

3.C   EQUATIONS DISREGARDING THE GRAIN VELOCITY IN DARCY'S LAW

The mass balance Eqs. (3.77), (4.90), (3.69) for the solid, water and oil for a constant volume read

$$\frac{\partial}{\partial t}(1-\varphi) + (1-\varphi)c_s \frac{\partial P}{\partial t} = -(1-\varphi)\,\mathbf{div}\cdot(\mathbf{V}_s) - \mathbf{V}_s\cdot\mathbf{grad}\,(1-\varphi), \tag{3.74}$$

$$\frac{\partial}{\partial t}(\varphi S_w) + \varphi S_w\,c_w\,\frac{\partial P}{\partial t} = -\mathbf{div}\cdot(\varphi S_w \mathbf{V}_w) = -\mathbf{div}\cdot\mathbf{u}_w, \tag{3.75}$$

$$\frac{\partial}{\partial t}(\varphi S_o) + \varphi S_o\,c_o\,\frac{\partial P}{\partial t} = -\mathbf{div}\cdot(\varphi S_o \mathbf{V}_o) = -\mathbf{div}\cdot\mathbf{u}_o, \tag{3.76}$$

where we now assume that u_w, u_o can be calculated with Darcy's law without correcting for the grain velocity. Again we divide the equation for the solid by $1-\varphi$ and disregarding the grad(1-φ) term   we obtain the following equation

$$\frac{1}{1-\varphi}\left(\frac{d\varphi}{dP}\right)_T \frac{\partial P}{\partial t} - c_s \frac{\partial P}{\partial t} = \mathbf{div}\cdot(\mathbf{V}_s), \tag{3.77}$$

which can be rearranged as before to

$$\varphi c_{f1}\frac{\partial P}{\partial t} - c_s(1-\varphi)\frac{\partial P}{\partial t} = \varphi c_{f2}\frac{\partial P}{\partial t} = \mathbf{div}\cdot(\mathbf{V}_s).$$

We add Eqs. (4.90) and (3.69) and define the liquid compressibility as $\varphi c_{liq} = \varphi(S_w\,c_w + S_o\,c_o)$ to obtain

$$\frac{\partial}{\partial t}(\varphi) + \varphi c_{liq}\frac{\partial P}{\partial t} = -\mathbf{div}\cdot(\mathbf{u}_w + \mathbf{u}_o), \tag{3.78}$$

or

$$\left(\frac{d\varphi}{dP}\right)_T \frac{\partial P}{\partial t} + \varphi c_{liq}\frac{\partial P}{\partial t} = -\mathbf{div}\cdot(\mathbf{u}_w + \mathbf{u}_o). \tag{3.79}$$

Substitution of Eq. (3.65) into Eq. (3.71) leads to

$$\varphi(1-\varphi)(c_{f1}+c_s)\frac{\partial P}{\partial t} + \varphi c_{liq}\frac{\partial P}{\partial t} = -\mathbf{div}\cdot(\mathbf{u}_w + \mathbf{u}_o). \tag{3.80}$$

Disregarding the grain compressibility c_s, which is often considered negligible, we obtain

$$\big((1-\varphi)\varphi c_{f1} + \varphi c_{liq}\big)\frac{\partial P}{\partial t} = -\mathbf{div}\cdot(\mathbf{u}_w + \mathbf{u}_o). \tag{3.81}$$

3.D   SUPERPOSITION PRINCIPLE

Our interest is of course not restricted to the case of a constant production rate q at $r_D = 1$. We like to include the possibility that the production rate is a function of time $q(t)$. In order to avoid the cumbersome notation, we drop the subscript D in the equations.

The idea here is based on an application of the superposition principle, which is known as Duhamel's formula's [137, Section 10.5-4]. Let L be a homogeneous linear operator whose coefficients and derivatives do not involve the time variable t (e.g., $L = \frac{1}{r}\frac{\partial}{\partial r}\left(r\frac{\partial}{\partial r}\right)$). Let $p(r,t)$ be the solution of the initial value problem

$$Lp(r,t) + A_o(r)\frac{\partial^2 p}{\partial t^2} + A_1(r)\frac{\partial p}{\partial t} = 0 \quad \text{for} \quad 1 < r < r_e; t > 0,$$

$$p(r,t=0^+) = 0 \quad \left(\frac{\partial p}{\partial t}\right)_{t=0^+} = 0 \quad \text{for} \quad 1 < r < r_e,$$

$$\left(\alpha\frac{\partial}{\partial r} + \beta\right)p(r,t) = b(t) \quad \text{for} \quad r = 1, \tag{3.82}$$

with as many homogeneous linear boundary conditions as needed (3.33), (3.34). In our case, we have that $A_o(r) = 0$ and $A_1(r) = 1$. For the case with water influx (as opposed to pressure buildup) $\alpha = -(r)_{r=1}, \beta = 0$. Let $\tilde{p}(r,t)$ be the solution of the same problem with $b(t) \equiv 1$. Then

$$p(r,t) = \tilde{p}(r,t)b(0^+) + \int_0^t \tilde{p}(r,t-\ )b'(\ )d\ . \tag{3.83}$$

However, in our case, $b(0^+) = 1$. Note that we can interpret $b(t)$ as a measure of the "production rate". This measure assumes an initial value equal to one. As we shut in the wells, we change the value from one to zero. In other words, the sudden change in b at time   is equal to minus one or $b'(\ ) = -\delta(\ -\ _o)$, where $\delta(x)$ represents the Dirac delta function and $_o$ is the time of shut off. The delta function is the derivative of a unit step function. The Dirac delta function has the property that $\int_{-\infty}^{\infty} f(x)\delta(x-x_o)dx = f(x_o)$. Therefore, we obtain for water shut in

$$p(r,t) = \tilde{p}(r,t) - \tilde{p}(r,t-\ _o).$$

Engineers like to interpret this equation as the result of two actions. One is to produce for a time t and then start to inject at the same rate during production with the result of zero production and injection.

We numerically approximate the integral with the trapezium rule of integration and use a center difference scheme to approximate the derivative of $b(t)$.

3.E   LAPLACE INVERSION WITH THE STEHFEST ALGORITHM [222]

The Stehfest algorithm is an excellent method for diffusional problems, which in general lead to monotonous solutions. An overview of Laplace inversion methods can be found in [66]. The algorithm is described in [221]. I am not repeating the proofs here except for saying that it is not easy to understand. I usually dislike the use of methods, for which I do not know how they are derived. Nonetheless, I have to make an exception for this case. I only will state the method.

An approximation $\widetilde{F}(t)$ is found from the equation

$$\widetilde{F(T)} = \frac{\ln 2}{T}\sum_{i=1}^{N} V_i f\left(i\frac{\ln 2}{T}\right),$$

where $f(s)$ is the Laplace transformed function, N is even and

$$V_i = (-1)^{\frac{N}{2}+i}\sum_{k=\frac{i+1}{2}}^{\min(i,\frac{N}{2})}\frac{k^{\frac{N}{2}}(2k!)}{(\frac{N}{2}-k)!k!(k-1)!(i-k)!(2k-i)!}.$$

Note that $k = \frac{i+1}{2}$ is equal to 1 for $i = 2$ etc. The algorithm is implemented in the spreadsheet "stehfest.xls".

3.F   EXCEL NUMERICAL LAPLACE INVERSION PROGRAMME

It is useful to be able to use numerical Laplace inversion to obtain some solutions. This can be done using EXCEL. This section describes how to implement such a program in EXCEL using Eq. (3.41).

The value of $n_{row} = n_{column} = 4$, means that the time label indicating the top of the column where the time is displayed is written in cell D4; the top of column in which the result $p_D(t)$ is displayed is E4. The numerical program can give the Laplace inverted pressure $P_D(t)$ is NDERIV =0, but also its derivative if NDERIV =1, or any higher derivatives. The stehfest inversion algorithm uses a representation of a sum of base functions in Laplace space to be inverted that can consist of ns terms. In this case we chose $ns = 10$. If one uses too few terms, one does not get an accurate representation of the Laplace inverted function. However, the algorithm becomes unstable if one chooses ns too high. Optimization of ns makes Laplace inversion rather an art than a technique. It only works out fine for diffusion problems, not for convection-dominated problems. The other parameters are connected to the output. num is the number of time points for which the inverted Laplace transform function is calculated. The Laplace inverted function is given between the time values t_{min} and t_{max}. The choice is that the time points are separated by a constant amount (addmul = 1) or on a logarithmic scale Figure 3.3.

To insert the knobs (a) On the Developer tab, in the controls group, click "Insert", and then under form controls click button, (b) click the worksheet location where one wants the upper-left corner of the button to appear, (c) Assign a macro to the button and then click OK. The button clear D:E has as macro ""sub clear(); Columns ("D:E").Select; Selection. ClearContents; Range("D1").Select; End Sub "". The macro Stehfest is more complicated and is given in Figure 3.4

The Stehfest coefficients are calculated in Figure 3.5.

	A	B	C
1	You need to program F(s) in Visual Basic in lapl(s)		
2	or some other function that must be called by Stehfest		
3			
4	nrow	4	first output row
5	ncol	4	first output column
6	NDERIV	0	order of derivative required
7	ns	10	number of Stehfest coefficients
8			
9	num	100	number of output data
10	tmin	0.01	minimum time
11	tmax	100	maximum time
12	addmul	2	if =1 additive, otherwise multiplicative
13			i.e tnew=told + Δt or tnew = told*factor
14			
15		clear D:E	stehfest

Figure 3.3 The steering parameters of the EXCEL program for Laplace inversion. Please be sure that the variables, "nrow", "ncol", "NDERIV", "ns", "num", "tmin", "tmax","addmul" in column B are named)(Figure 2.3) in the sheet "data". The names of the cells in column B are indicated in column A.

```
Sub stehfest()
' note that the Stehfest Laplace inversion algorithm is particularly suitable
' for diffusion problems.
' In cases that it really matters you like to check limiting formula analytically
  Dim j As Integer ' running dummy counter
  Dim i As Integer ' running dummy counter
  Dim ns As Integer ' number of terms in Stehfest series e.g. 8
  Dim num As Integer 'number of times for which output is required
  Dim nrow As Integer ' first output row
  Dim ncol As Integer ' first output column
  Dim NDERIV As Integer ' NDERIV^th derivative of the function required
'                          zeroth derivative is just the function
  Dim addmul As Integer 'if =1 then time = time + dt , otherwise time = time * factor
  Dim t As Double ' time variable
  Dim arg As Double ' argument for F(s)
  Dim tmin As Double 'minimum time for which output is required
  Dim tmax As Double ' maximum time for which output is required
  Dim factor As Double ' factor between output times
  Dim hulp As Double ' any intermediate computation
  Sheets("data").Select
  With Sheets("data") ' this allows reading from the worksheet
'     with this formulation you can move the cells up and down and
'     add rows without affecting the reading
    nrow = .Range("nrow"): ncol = .Range("ncol")
    num = .Range("num"):   tmin = .Range("tmin")
    tmax = .Range("tmax"):   ns = .Range("ns")
    addmul = .Range("addmul"): NDERIV = .Range("NDERIV")
  End With
  hulp = (Log(tmax) - Log(tmin)) / num:   factor = Exp(hulp):   ' factor for multiplicative spacing
  Cells(nrow, ncol).Value = "time" ' put "time" in cell (nrow, ncol)
  Cells(nrow, ncol + 1).Value = "F(t)" ' put "time" in cell (nrow, ncol+1)
  Call STEHFCF(ns, VSTEH) ' calculate Stehfest coefficients
  t = tmin / factor
  For j = 0 To num
    If addmul = 1 Then 'additive
        t = tmin + (tmax - tmin) * j / num
    End If
    If addmul <> 1 Then
        t = t * factor 'multiplicative
    End If
    For i = 1 To ns
          arg = (Log(2) * i) / (t)
             FLP(i) = lapl2(arg) ' or any other function of (arg = s) to be defined below
    Next i
  Call LPLINV(FLP, ns, VSTEH, t, NDERIV, F) ' find Stehfest coefficients
  Cells(nrow + j + 1, ncol).Value = t ' put time in row nrow+j+1, and in column ncol
  Cells(nrow + j + 1, ncol + 1).Value = F ' put result F in column ncol +1
  Next j
End Sub
```

Figure 3.4 The subroutine Stehfest Core program to calculate via the module the Laplace inverted function -1 / (s * Sqr(s)) * ko / k1, where k0 and k1 are modified Bessel functions.

The subroutine calls a subroutines that calculate the Stehfest coefficients and the required modified Bessel functions, K0, K1, I0 and I1. These functions are also tabulated in [2] (Figures 3.6–3.8).

3.F.1   ALTERNATIVE INVERSION TECHNIQUES

The importance of Laplace transformation was already stressed in 1949 by Hurst and Van Everdingen [238]. If analytical inversion is possible through the Bromwich inversion integral [1] (which is complicated) or when the inverse Laplace transform is given in a table [3, 54, 102, 170] this is of course preferable over numerical inversion. Analytical Laplace inversion can also be obtained when using MAPLE. Hassanzadeh and Pooladi-Darvish [115] tested various methods and concluded that the Stehfest method leads to accurate results for many diffusion-dominated problems, but that the Fourier inversion technique was the most powerful but also the most computationally expensive.

```
        Sub LPLINV(FLP, ns, VSTEH, t, NDERIV, F)
Rem
Rem     this subroutine returns the nth derivative of f
Rem     using the Stehfest algorithm
Rem   input
Rem     FLP    : FLP(i ln2/t) (FLP (s) = L(f(t)))
Rem     NS     : number of coefficients in Stehfest algorithm
Rem     VSTEH  : coefficients in Stehfest algorithm
Rem     tijd   : t, time at which f is evaluated
Rem     NDERIV : n, order of derivative
Rem     schale is scale factor
Rem   output
Rem F: F (t)
        SCHALE = Log(2#) / t
        F = 0#
        For i = 1 To ns
          s = i * SCHALE
          F = F + FLP(i) * VSTEH(i) * (s ^ NDERIV)
        Next i
        F = F * SCHALE
        End Sub
Rem
```

```
        Sub STEHFCF(ns, VSTEH)
Rem         Dim VSTEH(1000) As Double
        Dim FF(0 To 50) As Double
        Dim k As Integer
        Dim j As Integer
        Dim i As Integer
        Dim M1 As Integer
        Dim M2 As Integer
        Dim jtok As Double
        Dim sum As Double
        k = ns / 2
        FF(0) = 1#
        For m = 1 To ns
          FF(m) = FF(m - 1) * m
        Next m
        For i = 1 To ns
          sum = 0#
          a = (-1) ^ (k + i) ' -1 for k+i is odd and +1 for k+i is even
          M1 = Int((i + 1) / 2) ' note that for basic integers (1+2)/2=2 and for FORTRAN (1+2)/2=1
          M2 = min(i, k) ' minimu of i and k
          For j = M1 To M2 'determine coefficients
            jtok = j ^ k
            sum = sum + jtok * FF(2 * j) / (FF(k - j) * FF(j) * FF(j - 1) _
                           * FF(i - j) * FF(2 * j - i))

          Next j
          VSTEH(i) = a * sum ' Stehfest coefficients
        Next i
        End Sub
```

Figure 3.5   The subroutine Stehfest Core program requires the computation of the Stehfest coefficients.

```
Function Kzero(x)
Dim a0 As Double: Dim a1 As Double: Dim a2 As Double
Dim a3 As Double: Dim a4 As Double: Dim a5 As Double
Dim a6 As Double: Dim eul As Double: Dim y As Double
Dim y2 As Double: Dim y3 As Double: Dim y4 As Double
Dim y5 As Double: Dim y6 As Double: Dim part1 As Double

If x <= 2 And x > 0 Then
    y = x * x / 4#: y2 = y * y: y3 = y * y2
    y4 = y * y3:  y5 = y * y4: y6 = y * y5
    eul = 0.57721566:  a0 = 0.4227842: a1 = 0.23069756
    a2 = 0.0348859: a3 = 0.00262698: a4 = 0.0001075
    a5 = 0.0000074
    part1 = a0 * y + a1 * y2 + a2 * y3 + a3 * y4 _
            + a4 * y5 + a5 * y6
    Kzero = -Log(x / 2#) * Izero(x) - eul + part1
End If
If x > 2 Then
    y = 2# / x: y2 = y * y: y3 = y * y2
    y4 = y * y3: y5 = y * y4: y6 = y * y5
    a0 = 1.25331414:    a1 = -0.07832358: a2 = 0.02189568
    a3 = -0.01062446: a4 = 0.00587872:   a5 = -0.0025154
    a6 = 0.00053208
    part1 = a0 + a1 * y + a2 * y2 + a3 * y3 _
            + a4 * y4 + a5 * y5 + a6 * y6
    Kzero = part1 * Exp(-x) / Sqr(x)
End If
End Function
```

Figure 3.6   An auxillary function is the modified Bessel function K0.

```
Function Kone(x)
Dim a0 As Double: Dim a1 As Double: Dim a2 As Double
Dim a3 As Double: Dim a4 As Double: Dim a5 As Double
Dim a6 As Double: Dim y As Double: Dim part1 As Double

If x <= 2 And x > 0 Then
    y = x * x / 4#
    y2 = y * y: y3 = y * y2:   y4 = y * y3
    y5 = y * y4: y6 = y * y5: eul = 0.57721566
    a0 = 0.15443144:  a1 = -0.67278579: a2 = -0.18156897
    a3 = -0.01919402: a4 = -0.00110404: a5 = -0.00004686
    part1 = a0 * y + a1 * y2 + a2 * y3 + a3 * y4 _
            + a4 * y5 + a5 * y6
    Kone = Log(x / 2#) * Ione(x) + (1# + part1) / x
End If
If x > 2 Then
    y = 2# / x: y2 = y * y: y3 = y * y2
    y4 = y * y3: y5 = y * y4:  y6 = y * y5
    a0 = 1.25331414:  a1 = 0.23498619:     a2 = -0.0365562
    a3 = 0.01504268:  a4 = -0.00780353:   a5 = 0.00325614
    a6 = -0.00068245
    part1 = a0 + a1 * y + a2 * y2 + a3 * y3 _
            + a4 * y4 + a5 * y5 + a6 * y6
    Kone = part1 * Exp(-x) / Sqr(x)
End If
End Function
Function lapl2(s)
Dim ko As Double
Dim k1 As Double
ko = Kzero(Sqr(s))
k1 = Kone(Sqr(s))
lapl2 = -1# / (s * Sqr(s)) * ko / k1
End Function
```

Figure 3.7   An auxillary function is the modified Bessel function K1.

```
Function Izero(x)
Dim a0 As Double: Dim a1 As Double: Dim a2 As Double
Dim a3 As Double: Dim a4 As Double: Dim a5 As Double
Dim a6 As Double: Dim a7 As Double: Dim a8 As Double
Dim hulp As Double: Dim t As Double: Dim y As Double
Dim y2 As Double: Dim y3 As Double: Dim y4 As Double
Dim y5 As Double: Dim y6 As Double: Dim y7 As Double
Dim y8 As Double
If x <= 3.75 Then
    t = x / 3.75: y = t * t: y2 = y * y
    y3 = y * y2: y4 = y * y3: y5 = y * y4
    y6 = y * y5: a0 = 3.5156229:    a1 = 3.0899424
    a2 = 1.2067492: a3 = 0.2659732: a4 = 0.0360768
    a5 = 0.0045813
    Izero = 1# + a0 * y + a1 * y2 + a2 * y3 + a3 * y4 _
            + a4 * y5 + a5 * y6
End If
If x > 3.75 Then
    y = 1# / x: y2 = y * y: y3 = y * y2
    y4 = y * y3: y5 = y * y4: y6 = y * y5
    y7 = y * y6: y8 = y * y7
    a0 = 0.39894228:   a1 = 0.01328592:   a2 = 0.00225319
    a3 = -0.00157565: a4 = 0.00916281:   a5 = -0.02057706
    a6 = 0.02635537: a7 = -0.01647633:    a8 = 0.00392377
       hulp = a0 + a1 * y + a2 * y2 + a3 * y3 _
              + a4 * y4 + a5 * y5 + a6 * y6 + a7 * y7 + a8 * y8
       Izero = hulp * Exp(x) / Sqr(x)
End If
End Function
Function Ione(x)
Dim a0 As Double: Dim a1 As Double: Dim a2 As Double
Dim a3 As Double: Dim a4 As Double: Dim a5 As Double
Dim t As Double: Dim y As Double: Dim y2 As Double
Dim y3 As Double: Dim y4 As Double: Dim y5 As Double
Dim y6 As Double: Dim y7 As Double: Dim y8 As Double
If x <= 3.75 Then
    t = x / 3.75:  y = t * t:  y2 = y * y
    y3 = y * y2: y4 = y * y3: y5 = y * y4
    y6 = y * y5: a0 = 0.87890594: a1 = 0.51498869
    a2 = 0.15084934:  a3 = 0.02658733:  a4 = 0.00301532
    a5 = 0.00032411: '
    Ione = (0.5 + a0 * y + a1 * y2 + a2 * y3 + a3 * y4 _
            + a4 * y5 + a5 * y6) * x
End If
If x > 3.75 Then
    y = 1# / x: y2 = y * y: y3 = y * y2
    y4 = y * y3: y5 = y * y4: y6 = y * y5
    y7 = y * y6: y8 = y * y7
    a0 = 0.39894228: a1 = -0.03988024: a2 = -0.00362018
    a3 = 0.00163801: a4 = -0.01031555: a5 = 0.02282967
    a6 = -0.02895312: a7 = 0.01787654: a8 = -0.00420059
       hulp = a0 + a1 * y + a2 * y2 + a3 * y3 _
              + a4 * y4 + a5 * y5 + a6 * y6 + a7 * y7 + a8 * y8
    Ione = Exp(x) * hulp / Sqr(x)
End If
End Function
```

Figure 3.8 The Bessel functions K need also the Bessel functions I0 and I1 for their numerical computation.

4 Two-Phase Flow

OBJECTIVE OF THIS CHAPTER

To enable the course participant, based on the treatment of simplified models of two-phase flow,

- To set up simple models of two-phase flow and find the solution of the ensuing model equations,
- to calculate cumulative oil production with the 1-D simplified models to show features of interest such as effect of wetting and viscosity,
- to use the models for partial validation of reservoir simulation results,
- recognition of salient features in reservoir simulation results,
- to use an Exergy Return on Exergy Invested (ERoEI) analysis to estimate the viability of an oil recovery process,
- to understand the effect of wetting on relative permeability behavior,
  - effect of wetting shown in Brooks-Corey relperms,
  - versatility of the Lomeland-Ebeltoft-Thomas (LET) approach,
- to understand the structure of water-oil displacement,
  - occurrence of rarefaction waves, constant state and shocks,
  - to understand the difference between dispersed models of two-phase flow and interface models of two-phase flow,
  - to understand the numerical procedures used in two-phase flow,
- to understand that for $M < G + 1$ a shock solution occurs where the interface has a constant angle with respect to layer direction, where M is the mobility ratio and G is the gravity number (gravity forces/viscous forces).

INTRODUCTION

Much of the oil is produced by water displacement [142, 205]. The water can originate from a connected water reservoir (natural water drive) or be injected. Another part of the conventional oil is produced by natural gas-drive (solution-gas or gas cap drive). In all these cases, at least two phases are flowing through the porous medium. Two-phase flow is, therefore, an important subject in reservoir engineering. We confine our examples to two-phase flow of oil and water. They also form the basis of many enhanced oil recovery methods [75, 76, 140]. Much of the basic principles are also applicable to oil-gas flow. Relevant notions are summarized in the glossary.

Models of flow in porous media relevant for the operations related to the recovery of hydrocarbons are Darcy's law [65, 124, 157] and the component conservation laws. The model equations are conventionally solved with the use of reservoir simulators. Most basic principles of reservoir engineering involving numerical computations can be illustrated with EXCEL. Our focus is, however, on a subset of these two-phase flow models, considering isothermal flow of a single component in each phase. These models form a framework from which many relevant aspects can be derived. All the same to keep the model equations tractable, we must make a number of simplifying assumptions (see the glossary for clarification), i.e.,

1. We consider flow in a layer of length L with a constant thickness $H << L$, bounded above and below respectively by the impermeable cap- and base rock (Figure 4.1),
2. the liquids (and the rock) are approximately incompressible,

DOI: 10.1201/9781003168386-4

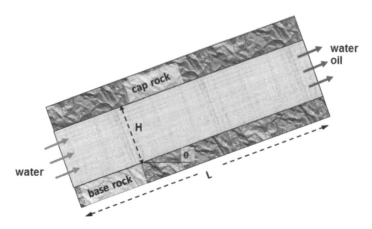

Figure 4.1 Vertical cross-section of oil bearing layer. Injected or edge water enters from the left. Oil and water are produced at the right.

3. the permeability and the porosity in the reservoir are constant,
4. there exists a water saturation below which water cannot flow easily in the porous medium, because fully filled water pores are disconnected. In fully water-wet media, still water flow occurs through water films bounding the porous medium. This saturation is called the connate water saturation S_{wc}. In practice we consider the initial water saturation in the oil reservoir is equal to the connate water saturation S_{wc}. There exists also an oil saturation called the residual oil saturation S_{or} below which oil cannot flow,
5. both the water and oil phase flow independently of each other and behave according to a generalized form of Darcy's law. The water flows due to a water pressure gradient **grad** P_w and gravity $\rho_w\mathbf{g}$; the oil flows due to an oil pressure gradient **grad** P_o and gravity $\rho_o\mathbf{g}$,
6. the capillary pressure $P_o - P_w = P_c$ between the phases is a function only of the saturation, i.e., the fraction of one of the fluids in the pores averaged over a representative elementary volume,
7. for each phase there is instantaneous equilibrium between gravity forces and capillary forces in the "cross-dip" direction. This means first that the viscous forces in the cross-dip direction can be neglected. Secondly, the saturation distribution has attained the equilibrium distribution subject to a given value of the capillary pressure at the boundary of the reservoir adjacent to the base rock. These assumptions are known as the Vertical Equilibrium (VE) assumption in the American Literature and as the Dupuit approximation in the European literature,
8. in the dip (layer) direction we disregard (as much as possible) the capillary forces and we take only the viscous forces into account.

Remark: a number of the assumptions, e.g., constant permeability, can be relaxed in more advanced models but this would, for a first acquaintance with the theory, be didactically confusing.

We will work out the assumptions above and give an illustration based on the Schoonebeek field. We will discuss the assumptions made above in the context of the conditions in the Schoonebeek field.

1. The Schoonebeek field [233] is an example of the many fields for which the layer thickness H is much smaller than the square root of its area (Figure 4.1). A completely

sealing fault segregates the reservoir in two parts. The South-West part where primary production is caused by solution gas drive and the rest of the field where an edge water drive is active. The largest part of the reservoir consists of unconsolidated (beach) sand. The net thickness varies between 9 and 30 m and the top is between 720 and 840 m below sea level. The permeability k varies from 10 Darcy at the top of the reservoir to one Darcy at the bottom, the porosity φ is about 30%. The reservoir contains paraffinic oil with an in-situ viscosity μ between 160 and 180 $[cP]$. The density of the oil is about 860 [kg/m^3]. The average dip of the reservoir is about 5.5° . The field is the largest on-shore oil field in Europe with a STOIIP of one billion (10^9) barrel.

2. the assumption that the reservoir liquids and the rock are incompressible (total compressibility $c_t \approx 0$) implies, according to the model assumption, that each pressure change has an immediate effect in the whole reservoir [27]. This assumption is well justified for the conditions considered by us. Indeed, incompressibility can only be justified if the time scale $\tau_1 \sim L/u$ at which the reservoir (in practice time of breakthrough is several months) is filled with water is much bigger than the time with which a pressure pulse "diffuses" in the reservoir $\tau_2 \sim \varphi \mu c_t L^2 /k$. For Schoonebeek conditions with well distances of a few hundred meters, this appears to be an acceptable approximation. In these equations, u denotes the Darcy velocity.

3. The permeability and porosity of the reservoir are not constant. The reservoir is built up out of layers with different permeability. Shale lenses, calcite layers, faults and areal heterogeneity will be the building blocks of a complex heterogeneous structure. The VE assumption [140] is often used to calculate effective or averaged properties comprising several layers [63]. Such average properties are called pseudo quantities, e.g., the pseudo relative permeability.

4. The concept of connate water is somewhat controversial. In water wet media, one imagines that water films remain continuous and that, given enough patience, it is possible to reach saturation values far below the water saturation encountered initially in reservoirs [178, 211]. This initial water saturation is formed by drainage though films during geologic times. Such a low connate water saturation is almost never achieved in the laboratory. In other words, the definition of the connate water saturation S_{wc} is rather vague. Residual oil saturation S_{or} is the oil that remains trapped at the pore scale level [241, 240]. In the presence of water, there are no oil films and the concept seems to be straightforward. In heterogeneous media, however, oil can be trapped in high permeable regions leading to much higher "remaining trapped" oil saturations than one would infer from small-scale core experiments. One must consider that cores are never homogeneous and large permeability variations are rather the rule than the exception. In other words, measured residual oil saturations are very core specific [211]. On average residual oil saturations appear to be less in intermediate-wet cores. In three-phase flow situations, i.e., in the presence of oil, water and gas the oil can also form films [45] and the same what has been said about the connate water saturation is now also true for the residual oil saturation. In spite of this, both the connate water saturation and residual oil saturation are parameters that are used to define relative permeability behavior and capillary pressure (see below).

5. Capillary pressures are discussed in Section 4.1. The capillary pressure is the difference between the non-wetting phase pressure and the wetting phase pressure. In our case, it is the difference between the oil pressure and the water pressure. In the absence of other data, we can express the capillary pressure by the following expression for the capillary

pressures [40, 149]

$$P_c = \sigma_{ow} \cos\theta\gamma\sqrt{\frac{\varphi}{k}}\left(\frac{\frac{1}{2}-S_{wc}}{1-S_{wc}}\right)^{\frac{1}{\lambda}}\left(\frac{S_w-S_{wc}+\varepsilon}{1-S_{wc}}\right)^{\frac{-1}{\lambda}},$$

where γ is a parameter that in many cases assumes values between 0.3 and 0.7. We suggest to use $\gamma = 0.5$ in all calculations where experimental data are lacking. Furthermore, the interfacial tension between oil and water is $\sigma_{ow} = 0.03$ [N/m] and as above for the sorting factor typical values are $0.2 < \lambda < 7$. Other symbols are explained in the glossary and list of symbols, with values given above. As the capillary pressure becomes infinity at $S_w = S_{wc}$, it is recommended to include a value $S_w = S_{wc} + \varepsilon(> 0)$ for computations (see, however, also Subsection 4.2).

6. in Darcy's law for two-phase flow we need to relate the Darcy velocity of a phase say the water phase u_w and the pressure gradient using the permeability of a phase [120, 124, 157]. The phase permeability depends on the distribution of fluids. The notion relative permeability is of help. We denote the phases with the symbol $\alpha = o, w$. The approximately valid (see, however, [183]) idea is that two immiscible fluids (far from the critical point) follow each their tortuous paths in the porous medium. For water wet media, the water will fill a network of smaller pores and the oil-phase a network of larger pores. In other words, for each water saturation, there is a set of channels in the porous medium filled with the water-phase and a complementary set of channels filled with the oil-phase. Momentum transfer between the oil- and the water-phase can be disregarded. This idea allows us to generalize the concept of permeability for one-phase to the concept of permeability for two phases. Oil flows through a porous medium that consists of rock and water with a permeability k_o, whereas water flows through a porous medium that consists of oil and rock with a permeability k_w. Within this concept, one expects that the permeability is only a function of the saturation (and not of the viscosity contrast between the fluids), i.e., $k_o = k_o(S_o)$ and $k_w = k_w(S_w)$. Therefore, we can write Darcy's law for the oil- and water-phase as follows: (z is opposite to the direction of the force of gravity)

$$\mathbf{u_o} = -\frac{k_o(S_w)}{\mu_o}(\mathbf{grad}P_o + \rho_o g\mathbf{e_z}),$$

$$\mathbf{u_w} = -\frac{k_w(S_w)}{\mu_w}(\mathbf{grad}P_w + \rho_w g\mathbf{e_z}), \tag{4.1}$$

where k_α a function only of the saturation. The pressure in the oil P_o is not equal to the pressure in the water P_w. We call the pressure difference between P_o and P_w the capillary pressure.

The permeabilities k_w and k_o must be measured as a function of the saturation. We can get some idea of this function. For a water saturation $S_w = 1$, the permeability of water will be (by definition) equal to the one-phase permeability k. For the same reason is $k_o(S_o = 1) = k$. Therefore it is convenient to introduce the relative permeabilities $k_{r\alpha}$ where $k_o = kk_{ro}$ and $k_w = kk_{rw}$. For the connate water saturation S_{wc} we must of course have $k_{rw}(S_{wc}) = 0$. In the same way, $k_{ro}(S_o = S_{or}) = 0$ at residual oil saturation S_{or}. As $S_w > S_{wc}$ the relative permeability k_{rw} increases monotonically with S_w. Also does the relative permeability k_{ro} increase monotonically with S_o.

There is, however, an asymmetry in the functional form of both functions that originates in the difference in wetting properties [58]. Small amounts of water have little

effect on the relative permeability of oil whereas small amounts of oil considerably reduce the relative permeability of water. An explanation for this effect can be that in a
water wet medium small amounts of water, corresponding to low water saturations have
the tendency to retreat in the small pores, which do not contribute much to the oil permeability. Small amounts of oil stay away from the small pores leading to a unfavorable
high surface energy. They occupy the large pores, which will now marginally contribute
to the water permeability.

An alternative but probably less realistic explanation is that oil forms bubbles that
get trapped in the center of the pores and thus form an obstruction for the water permeability. Water that retracts in the corners of the pore matrix has considerably less effect
on the relative oil permeability. Residual oil that forms globules in the middle of the
pores obstruct the flow of water and considerably decreases the relative permeability of
water.

Summarizing we expect that the relative permeabilities behave as shown in Figure 4.14.

7. the concept of vertical equilibrium [255] is explained in detail in Section 4.1.10. For
   relatively thin layers, it is possible to assume that viscous forces in the X-dip direction
   are negligible and therefore that we have equilibrium between gravity and capillary
   forces. It can be shown (see Section 4.1.10) that the transition zone h_{trans}, i.e., where
   the water saturation changes from values close to one minus the residual oil saturation
   to values close to the connate water saturation, will be practically given by

$$h_{trans} \approx \frac{1}{3}\sigma \cos\theta \sqrt{\frac{\varphi}{k}} / (\Delta\rho g \cos\vartheta),$$

where θ is the contact angle, ϑ is the dip angle, $\Delta\rho$ is the density difference between
the heaviest phase (e.g., water) and the lightest phase (e.g., oil). In a gas oil system the
heaviest phase would be oil and the lightest phase would be gas. Moreover, ϑ is the dip
angle and g is the acceleration due to gravity. Other quantities can be found here above
or in the glossary. The factor of $\frac{1}{3}$ arises as an empirical factor.

If the transition zone height h_{trans} is much larger than the layer height H, saturations
will be more or less constant and the Buckley–Leverett model applies (see Section 4.3).
If h_{trans} is much smaller than the layer height H, an interface model (see Section 4.7.1)
is more appropriate (see also Figure 4.11).

4.1   CAPILLARY PRESSURE FUNCTION

Capillary effects in porous media play an important role due to the large pore-fluid interface
[228]. It can thus be expected that interfacial effects have a great influence on the behavior of
fluids in porous media. For two-phase flow, this interface changes continuously. Consequently,
two-phase flow are for an important part determined by interfacial effects. Flow in porous media
is an important application area of surface chemistry [5].

Interfacial phenomena affect constitutive flow relations in porous media in a fundamentally
complex manner. To keep the situation tractable we need to use simplifications. Part of this
simplification is the generalization of Darcy's law leading to the concepts of relative permeability
discussed above. The second part of the simplification entails the introduction of the capillary
pressure between nonwetting and wetting phases $P_c = P_o - P_w$, which is only considered to be
a function of the saturation. The understanding of the behavior of capillary pressure is closely
related to the notions of interfacial tension and capillary rise.

4.1.1 INTERFACIAL TENSION AND CAPILLARY RISE

For better understanding, we shall go into the theoretical background that have led to formulate Eq.(4.8). We shall therefore first summarize the basic principles of surface chemistry [176]. The occurrence of interfacial tension is illustrated with a frame with a rod on it [29]. Between the frame and the rod there is a soap film. The soap is only introduced to have a stable film, and what is said could equally well be illustrated, without soap albeit that without soap the film snaps immediately, which is experimentally unwieldy. (Figure 4.3).

The rod will move such that the surface of the soap film is decreased. We need to exert a force on the rod, which is proportional to the length of the rod, to keep it in place. The force per unit length is the interfacial tension. The interfacial tension has thus units [N/m]. A typical value for the interfacial between oil and water is $\sigma_{ow} = 30 \times 10^{-3}$ [N/m = J/m^2].

Referring back to Figure 4.2, we can express the contact angle in terms of the interfacial tension between the rock and the water σ_{ws}, the interfacial tension between the oil and the rock σ_{os}, and the interfacial tension between the water and the oil σ_{ow}. For this, we make a force balance [29] and note the interfacial tension is the force per unit length along the contact line between the solid, water and oil, $\sigma_{os} = \sigma_{ws} + \sigma_{wo} \cos\theta$ from which we derive Young-Laplace equation (see Exercise 4.1.3 and Eq. (4.3)),

$$\cos\theta = \frac{\sigma_{os} - \sigma_{ws}}{\sigma_{wo}}.$$

As two immiscible fluids are in contact, then there exists an interface between the fluids. This is shown schematically in Figure 4.4 [125]. The fluid element between the two blocks from which

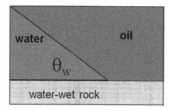
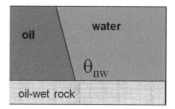

Figure 4.2 Contact angles for water wet and oil wet rocks.

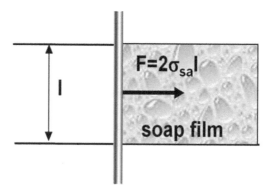

Figure 4.3 A soap film on a frame. On the freely moving rod a force is exerted, which is the product of the length of the rod wetted by the film multiplied by twice (the surface below and above) the interfacial tension.

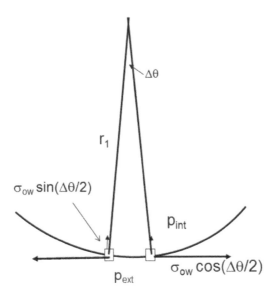

Figure 4.4 Force balance on element between the two squares due to interfacial tension.

the arrows originate is subjected to a horizontal and vertical force. The horizontal forces are balanced. The force in the vertical direction must be compensated by a higher pressure at the inside than at the outside. The force balance reads for small angles $\Delta\theta$

$$2\sigma_{ow}\sin(\Delta\theta/2) = (p_{int} - p_{ext})r_1\Delta\theta.$$

As the angle $\Delta\theta$ goes to zero, we can use $\sin(\Delta\theta/2) \approx \Delta\theta/2$. Therefore, we find that $(p_{int} - p_{ext}) = \sigma_{ow}/r_1$. The same argument is also true (if one wants to do this rigorously, it is quite some tedious algebra) perpendicular to Figure 4.4. Now the radius is r_2. Both contributions can be added, and therefore, when the interface is curved, the pressure on the concave (hollow) side is larger than on the convex side. The pressure difference between internal pressure and external pressure is known as the capillary pressure. It is given by Laplace's formula

$$P_c = \sigma\left(\frac{1}{r_1} + \frac{1}{r_2}\right), \tag{4.2}$$

where r_1, r_2 are two mutually perpendicular radii that can be drawn through the surface. Note, for a droplet of water in the oil, the water pressure is higher in the water than in the oil, whereas for a oil droplet in the water the oil pressure is higher in the oil.

4.1.2   EXERCISE, LAPLACE FORMULA

About the application of Laplace formula

- What pressure in atmospheres would be necessary to blow water at 25°C out of a sintered glass filter with a uniform pore diameter of 0.10 μm, [177]. Hint: the interfacial tension $\sigma_{wg} = 72\,\text{mN/m}$.
- two vertical cylindrical glass rods of radius 1 cm and with well-polished flat ends are separated by a 1 μm thick layer of water between the two ends. If the water meniscus is semicircular, all around the cylinder circumference, calculate the length the lower

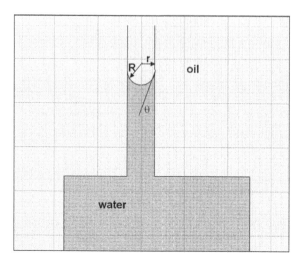

Figure 4.5 Capillary rise of water in a glass tube in vessel filled with oil. Capillary and gravity forces are in equilibrium (see text). The contact angle is θ.

cylinder may have before it is detached by gravity from the upper one. Assume the density of glass to be 2.5 g/cm^3 [177].

Consider a vertical tube in a liquid reservoir (Figure 4.5). There is equilibrium between capillary forces and gravity forces. The gravity force is $\Delta \rho g h$. There is capillary rise over a height h. In Figure 4.5 we have $R = r_1 = r_2$. From Eq. (4.2) it follows that $P_c = 2\frac{\sigma}{R}$. From r (the radius of the tube) $= R \cos \theta$ it follows

$$P_c = \frac{2\sigma \cos \theta}{r} = \Delta \rho g h.$$

4.1.3 EXERCISE, YOUNG'S LAW

About showing the difference between water-wet, oil-wet with finite contact angles and completely water-wet, considering Young's law

$$\cos \theta = \frac{\sigma_{os} - \sigma_{ws}}{\sigma_{ow}} \qquad (4.3)$$

and that the interfacial tension [N/m] can be interpreted as the surface energy [J/m^2]

- Considering that $|\cos \theta| \leq 1$, give the range of contact angles for $\sigma_{os} > \sigma_{ws}$ and the range of contact angles for which $\sigma_{os} < \sigma_{ws}$.
- what happens if the right side of Eq. (4.3) is bigger than one. What happens if the right side is smaller than minus one?
- if the rock is covered with a water layer, write Young's law (Eq. (4.3)) between the water/oil/gas phase.
- for which conditions do I observe an oil film on the water layer, and what is consequence for the relative oil-gas relative permeability?
- under what conditions is a rock completely water-wet. This happens for hydroxides; oxides, e.g., SiO$_2$ is completely hydrophobic, so why is it often stated that quartz is completely water-wet? Hint: consider the chemical reaction $SiO_2 + 2H_2O \rightarrow Si(OH)_4$.

- under what conditions does one expect in a completely water-wet medium that oil spreads on the grains in a porous medium, leading to very low oil saturations?
- given the values of the water-gas, water-oil and gas-oil interfacial tension, is it possible that spreading occurs on a water film covering the completely water-wet medium?
- what can be expected from water drive recovery from fractured reservoirs in an oil-wet matrix?
- why is in the definition of the USBM wetting index the capillary pressure for imbibition completely negative? Hint: after drainage, the grains are temporarily covered with the non-wetting phase.
- do you prefer a water-wet or intermediate-wet medium for low residual oil saturations? Hint: Inspect the end point relative permeabilities.

4.1.4 APPLICATION TO CONICAL TUBE; RELATION BETWEEN CAPILLARY PRESSURE AND SATURATION

We must be able to describe the capillary forces in a porous medium without considering the gravity forces, i.e., by considering a horizontal configuration. In a conical tube (Figure 4.6), the amount of non-wetting phase that can be injected in the tube depends on the applied pressure difference P_c.

Figure 4.6 consists of a horizontal tube in which a piston system has been mounted both from the left and from the right. The left piston displaces water through a semi-permeable filter, which can only transmit water (and thus no oil) and injects (withdraws) water in (from) the conically shaped tube. The right piston displaces oil through a filter that only transmits oil (and thus no water) and injects (withdraws) oil in (from) the conically shaped tube. Displacement of the right piston to the left increases the oil content of the tube and displacement of the left piston to the right increases the water content in the tube.

The work delivered for this is thus $(P_o - P_w)dV = P_c dV$. The work is used to increase the surface free-energy that is required to dewet the glass tube with oil. Therefore the capillary pressure is in the ideal reversible case a measure for the surface free energy. If we denote the interfacial area between fluid α and fluid β by $A_{\alpha\beta}$, then we obtain in the reversible case [168]

$$(P_o - P_w)dV = \sigma_{os}dA_{os} + \sigma_{ws}dA_{ws} + \sigma_{ow}dA_{ow}$$
$$= (\sigma_{os} - \sigma_{ws})dA_{os} + \sigma_{ow}dA_{ow}$$
$$= \sigma_{ow}\cos\theta dA_{os} + \sigma_{ow}dA_{ow},$$

where we used that $dA_{ws} = -dA_{os}$ and the Young Laplace equation.
It therefore it follows for the capillary pressure P_c that

$$P_c = \sigma_{ow}\cos\theta \frac{dA_{os}}{dV} + \sigma_{ow}\frac{dA_{ow}}{dV}. \tag{4.4}$$

We notice that the capilary pressure is not necessarily proportional to $\cos\theta$ or the cosine of the contact angle θ. We can make a plot of the capillary pressure versus the wetting phase saturation.

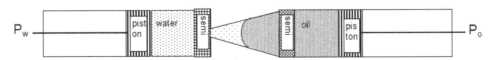

Figure 4.6 Piston setup to force oil into a tap-shaped tube. The work performed is equal to the increase in surface energy.

This plot is characteristic for the behavior of capillary pressure in porous media. This relationship is usually given by the capillary pressure versus the fraction of say the wetting fluid (saturation) in the porous medium. In practice, there will be always losses due to the irreversible nature of the drainage and imbibition processes in porous media. Part of the external work will be used to compensate for the heat losses. Consequently, the capillary pressure for drainage will be always higher and for imbibition will be always lower than the reversible values.

4.1.5   RELATION BETWEEN THE PORE RADIUS AND THE SQUARE ROOT OF THE PERMEABILITY DIVIDED BY THE POROSITY

Consider a porous medium that consists of bundle of straight capillary tubes. Such a model of the porous medium is called the capillary bundle model [77, 78]. It ignores tortuosity effects. We have shown a schematic picture of the capillary bundle model in Figure 4.7.

On the right, we show a hexagonal unit cell with area A. We only need to consider the unit cell. We assume that all cylinders have a radius r. Using Poiseuille flow we know that the total flow rate Q in a unit cell, i.e., through a single capillary tube is given by

$$Q = -\frac{\pi r^4}{8\mu} \frac{\Delta \phi}{\Delta L}.$$

In this equation, we use μ to denote the viscosity, ϕ to denote the potential $\phi = (P + \rho g z)$, L the length of the tubes. Furthermore, we use that the hexagonal unit cell has an area A. The cross-section of the tube is given by πr^2. Therefore, the porosity $\varphi = \frac{\pi r^2}{A}$. The Darcy velocity $u = \frac{Q}{A}$ reads

$$u = -\frac{\pi r^4}{8\mu A} \frac{\Delta \phi}{\Delta L} = -\frac{\varphi r^2}{8\mu} \frac{\Delta \phi}{\Delta L}.$$

Comparison to Darcy's law $u = -\frac{k}{\mu} \frac{\Delta \phi}{\Delta L}$ leads to the identification

$$k = \frac{\varphi r^2}{8}. \tag{4.5}$$

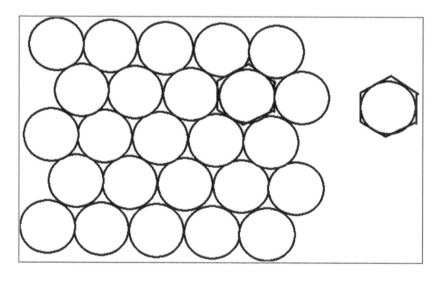

Figure 4.7   Cross-section through the capillary bundle model.

Note that this relation is based on the pore throat radius and not on the grain size radius (cf. Eq. (2.14)). Using the grain size radius overestimates the permeability by some factor of 5. Indeed, the pore throat diameter is much smaller than the pore diameter.

4.1.6   NON-DIMENSIONALIZING THE CAPILLARY PRESSURE

Apparently there exists some relation between the characteristic radius r of the pore size and $\sqrt{\frac{k}{\varphi}}$. This relation was used by Leverett to assume with Laplace's formula (see Eq. (4.2)) that $P_c \sim \frac{\sigma}{r}\cos\theta \sim \sigma\cos\theta\sqrt{\frac{\varphi}{k}}$, where σ is the interfacial tension and θ is the contact angle (Figure 4.3). The term $\sigma\cos\theta\sqrt{\frac{\varphi}{k}}$ has units $[Pa]$ and the remainder is a dimensionless function. It has been the merit of Leverett [149] to set up this theory, which is relatively simple and sufficiently accurate for engineering applications. Leverett was the first to define a dimensionless capillary pressure function, which is now known under the name Leverett J function and can be used to compare pressure data to one another.

$$P_c = \sigma\cos(\theta)\sqrt{\left(\frac{\varphi}{k}\right)}J(S_w). \tag{4.6}$$

For a given lithology, the Leverett J-function $J(S_w)$ appears to be independent of porosity and permeability or to formulate it from the pessimistic point of view: Eq. (4.6) is not of general validity but only for a specific lithology type within the same formation. From the fundamental point of view, the fact that for a given lithology the Leverett $J-$ function is independent of permeability implies that the pore size distribution is except for a scaling factor very similar. Furthermore, $J(S_w)$ is a non-increasing (usually monotonically decreasing) function of S_w. The success of the formulation indeed indicates that the pore-size distribution within a certain lithologic type is comparable except for a constant factor. The Corey [58] expression for capillary pressures is given by

$$P_c = P_{cb}\left(\frac{S_w - S_{wc}}{1 - S_{wc}}\right)^{\frac{-1}{\lambda}}, \tag{4.7}$$

where P_{cb} is the bubbling factor, and λ is the sorting factor, which are parameters. Combining with the Leverett J-function, we obtain the following expression for the capillary pressures

$$P_c = \sigma_{ow}\gamma\sqrt{\frac{\varphi}{k}}\left(\frac{\frac{1}{2} - S_{wc}}{1 - S_{wc}}\right)^{\frac{1}{\lambda}}\left(\frac{S_w - S_{wc}}{1 - S_{wc}}\right)^{\frac{-1}{\lambda}}, \tag{4.8}$$

where γ is a parameter that in many cases assumes values between 0.3 and 0.7. We suggest to use $\gamma = 0.5$ in all calculations where experimental data are lacking. As the water saturation decreases water retreats in the very small crevices. We also define the constant term in Eq. (4.8) as the bubbling pressure P_{cb},

$$P_{cb} = \sigma_{ow}\gamma\sqrt{\frac{\varphi}{k}}\left(\frac{\frac{1}{2} - S_{wc}}{1 - S_{wc}}\right)^{\frac{1}{\lambda}}. \tag{4.9}$$

The capillary pressure increases sharply as the water saturation decreases below the connate water saturation. However, it does not become infinity in water-wet media, as is suggested by Eq. (4.8). What happens is that there is no longer a connected system that involves pores, fully saturated with water, but the water retracts in the crevices on the surface of the pores. At higher capillary pressures, it will only penetrate in the deeper part of the crevices, thus decreasing the water saturation. It is possible to define a capillary pressure beyond the residual non-wetting

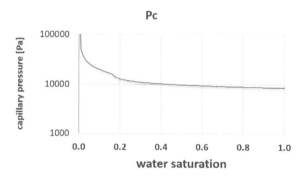

Figure 4.8 Capillary pressure at saturations below the connate water saturation. This is just a convenient way to represent the capillary pressure over the full saturation range including for $S_w < S_{wc}$, but more precise representations would lead to more complicated expressions.

saturations, but one needs to be aware of the fact that there is no longer a network of fully water-filled pores. However, the water forms still a continuous phase, which includes water at the pore surface (Figure 4.8).

It is convenient to define an auxiliary parameter S_{co} slightly above the connate water saturation. We define the capillary pressure [145]

$$P_c = \left[\begin{array}{ll} P_{cb}\, S_{we}^{(-1/\lambda)} & \text{for} \quad S_w > S_{co} \\ P_{cba}\, S_w^{(-1/fracdim)} & \text{for} \quad S_w \le S_{co}, \end{array} \right. \tag{4.10}$$

where $S_{we} = (S_w - S_{wc})/(1 - S_{wc} - S_{or})$ and

$$P_{cb} = \gamma \sigma_{ow} \sqrt{\frac{\varphi}{k}} \left(\frac{\frac{1}{2} - S_{wc} - S_{or}}{1 - S_{or} - S_{wc}} \right)^{\frac{1}{\lambda}}$$

$$P_{cba} = P_{cb} S_{co}^{\frac{1}{fracdim}} \left(\frac{S_{co} - S_{wc}}{1 - S_{wc} - S_{or}} \right)^{\frac{1}{\lambda}}. \tag{4.11}$$

The surface shows fractal pore surface behavior [145], and this is reflected in capillary pressure behavior below the connate water saturation. The connate water saturation is defined as the saturation below which we have no longer a network of fully water-filled pores, but that there are connecting bridges of surface water [204]. Therefore, water can still flow below the connate water saturation, but partly the water flows in crevices along the pore wall. Admittedly, our formulation above avoids infinite capillary pressures, which is useful, but is not an accurate representation of the capillary pressure below the connate water saturation. This aspect needs more fundamental research. In Eq. (4.10), we take account the residual oil saturation [145]. The situation for oil near the residual oil saturation is different as the oil can no longer flow below the residual oil saturation because the oil avoids the crevices on the pore surface.

4.1.7 EXERCISE, RATIO GRAIN DIAMETER/PORE THROAT DIAMETER

About the ratio of the grain diameter and the pore-throat diameter using $k = \varphi/8r^2$ and the Carman-Kozeny relation

- Take reasonable values for S_{wc}. Use $\gamma = 1/2$ and $k = 0.987 \times 10^{-12}$ [m²] and draw the capillary pressure curve (log-scale) versus the water saturation for five values of $0.1 < \lambda < 10$.
- consider a droplet trapped in front of pore throat. What pressure drop ΔP is required to move the drop into the pore, given that the pore body radius $r_b >> r_t$ is much larger than the throat radius r_t.
- what is the viscous pressure drop ΔP over the length of the drop Δl, in terms of the viscosity μ, the permeability k inferred from Darcy's law? Ignore gravity forces.
- for equilibrium the capillary pressure $P_c = \Delta P \sim (\sigma_{ow} \sqrt{\varphi/k})$. Use that the grain radius is roughly five times the throat radius, and the expression of the permeability $k = \varphi r_t^2/8$ to find the expression for the capillary number: N_{vc} = viscous/capillary; often N_{vc} is quoted to be $\mu u/(\varphi \sigma_{ow})$; which combination of constants is taken equal to one to find $N_{vc} = \mu u/(\varphi \sigma_{ow})$?
- use the Carmen-Kozeny relation $k \approx 4/150\varphi^3/(1-\varphi)^2 r_p^2 \sim \varphi r_t^2/8$ to relate the throat radius r_t to the grain radius r_p.

4.1.8   THREE-PHASE CAPILLARY PRESSURES

The capillary pressure between the phases is a function only of the saturations, i.e., the fraction of one of the fluids in the pores averaged over a representative elementary volume. In addition, it is assumed (see [16]) that the water-oil capillary pressure $p_o - p_w = P_c^{ow}(S_w)$ depends only on the water saturation and that the gas-oil capillary pressure $p_g - p_o = P_c^{go}(S_g)$ only depends on the gas saturation. Therefore, $p_g - p_w = P_c^{ow}(S_w) + P_c^{go}(S_g)$. In the absence of separate capillary pressure measurements, we use

$$P_c^{ow}(S_w) \quad = \sigma\gamma\sqrt{\frac{\varphi}{k}}\left(\frac{\frac{1}{2}-S_{wc}}{1-S_{wc}}\right)^{1/\lambda}\left(\frac{S_w-S_{wc}}{1-S_{wc}}\right)^{-1/\lambda} \tag{4.12}$$

$$P_c^{go}(S_g) \qquad = P_c^{ow}(S_o+S_w) \quad . \tag{4.13}$$

Here σ is the interfacial tension, φ is the porosity and k is the formation permeability. For the interfacial tension we take $\sigma \approx 0.03 N/m$. The adjustment parameter γ takes values $0.2 < \gamma < 0.7$ (see Bear [18], Figure 9.2.6 on p. 448) taken from reference [201]. We use $\gamma = 0.5$. In writing that the capillary pressure is given by Eq. (4.12), we assume zero contact angles, i.e., $\sigma \cos(\theta) = \sigma$. We assume that, as already mentioned, the sorting factor has typical values $0.2 < \lambda < 7$. The interfacial tensions between water and oil and between oil and gas are taken equal to avoid capillary pressure discontinuity effects, but it shows that the empirical approach adopted here is inconsistent Figure 4.9.

4.1.9   EXPERIMENTAL SET UP AND MEASUREMENTS OF
CAPILLARY PRESSURE

A setup for measuring the capillary pressure is given in Figure 4.9 [8, 60, 79, 197]. The results that can be obtained with such a setup are plotted in Figure 4.10. Leverett assumed that the petrophysical relation for the capillary pressure that holds under static conditions is also valid under dynamic conditions. In the right of Figure 4.10, we show the primary drainage curve, which is obtained if a water-filled sample is gradually filled with oil. Also shown in the imbibition curve, which is obtained if a core filled with oil at connate water saturation is gradually filled with water. The secondary drainage curve is obtained when the sample filled with water at residual oil saturation is gradually filled with oil. If we reverse from drainage to imbibition before residual

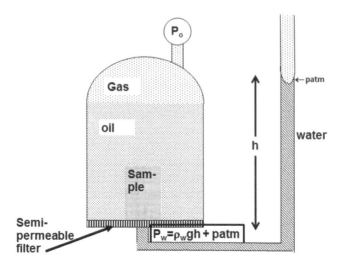

Figure 4.9 Schematic of setup to measure capillary pressure $P_c = P_o - P_w$.

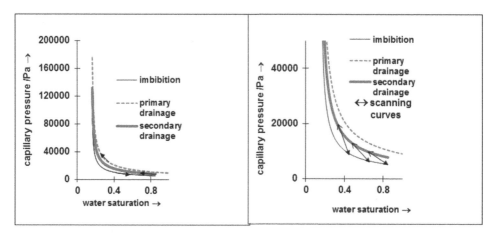

Figure 4.10 Primary drainage, imbibition and secondary drainage capillary pressure curves. The arrows in the left indicate the direction of saturation. In the right Figure the arrows indicate the imbibition scanning curves.

saturations are obtained, we obtain the imbibition scanning curves. Completely analogously, we obtain the drainage scanning curves if we reverse from imbibition to drainage. In summary, the capillary pressure is not only a function of the saturation.

4.1.10 CROSS-DIP CAPILLARY EQUILIBRIUM

All assumptions that have been treated up till now are also used in computer simulation programs, with the exception of the assumption that the liquids and rock are incompressible. Often the assertion of incompressibility is acceptable. In spite of this assumption, it appears that the model equations are too complex to be solved (semi-)analytically. By this term we mean without resorting to finite difference or finite element schemes; even for (semi)analytical models, the computer computations are almost always useful. The virtue of semi-analytical solutions is that they give us a better understanding of the physical model and the model consequences. In particular, the analytical solution does not suffer from the large-scale mixing in the large grid blocks. (Semi-) analytical solutions should always be used to validate simulation results and are indispensable for a balanced interpretation of the results.

The Dupuit approximation [67] allows us to find simple model equations that may lead to semi-analytical solutions [20, 255]. It consists of two basic assumptions. First, we introduce the approximation that the viscous forces in the cross-dip direction are negligible. Second, the changes in the reservoir are considered to occur sufficiently slowly such that the saturation has attained an equilibrium distribution in the X-dip direction. Such an equilibrium distribution requires a constraint, which is usually a given capillary pressure value at the boundary between the reservoir and the base rock.

We note that zero viscous forces, according to Darcy's law, would imply zero cross-dip velocity [255]. The velocity in the cross-dip direction is only one order of magnitude smaller than the velocity in the dip direction. It is outside the scope of this monograph to give a complete explanation.

The Dupuit approximation [73] disregards viscous forces in the X-dip direction such that there is in this direction equilibrium between capillary and gravity forces

$$P_c = \sigma \cos(\theta) \sqrt{(\frac{\varphi}{k})} J(S_w) = \Delta \rho g h(S_w) \cos \vartheta,$$

where h is the height from the phreatic level ($P_c = 0$ surface). The dip angle is ϑ. The Leverett J-function assumes values between the a nonzero threshold (when the medium is filled with water and residual oil) and increases with decreasing water saturation until the highest value at connate water saturation.

On average, the Leverett-J-function changes with one third between residual oil saturation and "near" connate water saturation. The word "near" is used because as explained above the definition of the connate water saturation is rather vague. We call the height over which the J-function changes between these values h_{trans}. Therefore

$$h_{trans} \approx \frac{1}{3} \sigma \cos(\theta) \sqrt{\frac{\varphi}{k}} / (\Delta \rho g \cos \vartheta), \tag{4.14}$$

where ϑ is the dip angle. This can be understood as follows: for gravity and capillary pressure equilibrium, we must have that both the oil and water pressure vary according to hydrostatics, i.e., for $\frac{dP_\alpha}{dz} = \rho_\alpha g \cos \vartheta$ with $\alpha =$ oil, water. Subtracting we find for the capillary pressure $P_c = P_o - P_w$ the differential equation $\frac{dP_c}{dz} = \Delta \rho g \cos \vartheta$ in obvious notation. The solution of this differential equation is $P_c = P_c^o + \Delta \rho g z \cos \vartheta$. Without loss of generality, we can define the height $z = 0$ as the point for which $P_c = 0$, i.e., the phreatic surface. Thus, we can write $P_c = \Delta \rho g z \cos \vartheta$ and this can be inverted to $z = h(S_w) = \frac{P_c}{\Delta \rho g \cos \vartheta}$. We can thus plot (from the $P_c - S_w$) the height $h(S_w)$ as a function of saturation (Figure 4.11). When we have capillary pressure/gravity equilibrium in the X-dip direction, then we can give the saturation as a function of the height. The distance h_1 of the bottom of the layer with respect to the level for which the capillary pressure is zero will be different on each x- location in the dip direction. With h_f we denote the height of the capillary fringe with respect to the level where the capillary pressure is zero. In Figure 4.11, we have shown such a saturation distribution on a location in the dip direction and for the case that the transition zone h_{trans} and the layer thickness H are of the same order of magnitude.

In this course, we largely confine ourselves to two extremes (Figure 4.11) namely one in which the capillary forces completely dominate, i.e., $H << \frac{1}{3} \sigma \cos(\theta) \sqrt{(\frac{\varphi}{k})} / (\Delta \rho g \cos \vartheta) = h_{trans}$ and one in which the gravity force completely dominates, i.e., $H >> \frac{1}{3} \sigma \cos(\theta) \sqrt{(\frac{\varphi}{k})} / (\Delta \rho g \cos \vartheta)$.

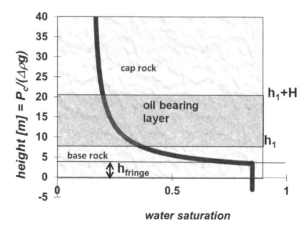

Figure 4.11 Height from phreatic level ($p_c = 0$) versus saturation assuming vertical equilibrium. This means that the saturation depends on the $p_c(S_w)/(\Delta\rho g \cos\vartheta)$. The height h_1 is the height of the bottom of the layer with respect to the phreatic level $p_c = 0$.

In the case that capillary forces dominate we expect no variation of the saturation in the "cross-dip" direction. We say that the water is completely dispersed in the X-dip direction. We can thus describe the water flow in the layer as a one-dimensional displacement process. The corresponding theory is called the theory of Buckley Leverett. [49]. In the case that gravity completely dominates, we expect that the length of the capillary transition zone h_{trans} between the water-phase at the bottom part of the reservoir and the oil-phase at the upper part of the reservoir is small with respect to the layer thickness H. We can thus assume a "sharp" interface between the water and the oil. We say that water is completely segregated from oil. This situation leads to the so-called interface models (e.g., the gravity tongue of Dietz [73]).

4.1.11   EXERCISE, CAPILLARY DESATURATION CURVE

About the capillary desaturation curve and required reduction in capillary pressure to significantly reduce the residual oil saturation considering Figure 4.12

- Calculate the capillary number of Forster, $N_{vc} = \mu u/(\varphi \sigma_{ow})$, for $\mu = 0.01$ (viscosity (Pa.s), $\varphi = 0.3$ (porosity, $u = 1$ (Darcy velocity) m/day and $\sigma_{ow} = 0.03$ N/m (interfacial tension) ,
- viewing the plot of the capillary desaturation curve (Figure 4.12), to which value should the interfacial tension drop to obtain a residual oil saturation of 15% (Figure 4.13).

4.2   RELATIVE PERMEABILITIES

There are a number of semi-empirical relations[1] [6] that one can use when (more the rule than the exception) experimental data of relative permeabilities are lacking. We disregard the viscosity dependence of the relative permeabilities [79]; however, the lubrication effect at high oil viscosities can become significant [120]. General observations are that in water-wet media, oil will occupy the larger pores and obstruct the flow of water, leading to low relative water permeabilities. A low

---

[1]Many aspects in this section are based on discussions with Hamidreza Salimi.

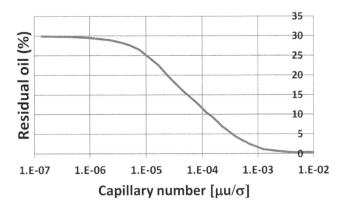

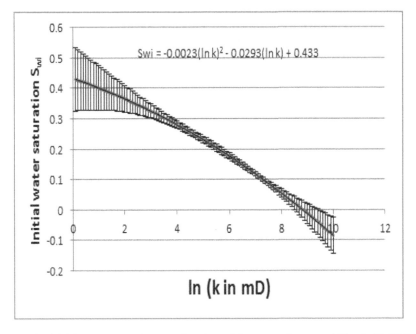

Figure 4.12 Typical capillary desaturation curve plots the residual oil saturation versus the capillary number $\mu u/(\varphi\sigma_{ow})$. The curve is remarkable as the residual oil saturation does not change much as the capillary number gets larger until the value is high enough and residual oil saturation starts to decrease. The capillary number can in practice not be sufficiently increased by increasing the viscosity, but it is possible to decrease the interfacial tension when a combination of the right surfactant, co-surfactant and salt concentration is used.

Figure 4.13 The initial water saturation as a function of the reservoir permeability (mD) [82]. The error bars are derived using the Bootstrap method on the initial data set recovered from [82].

relative water permeability leads to a favorable mobility ratio and more stable displacement experiment (see Section 4.7.2). In oil-wet media, oil will occupy the smaller pores and wet the pore walls of the larger pores, leading to a higher value of the relative water permeability [120]. Initial or connate water saturations in water-wet media are usually high, i.e., 20%–25%, whereas initial water saturations in oil-wet media are small 10%–15%. Consequently, residual oil saturations

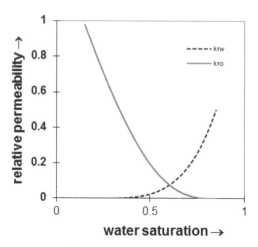

Figure 4.14 Relative permeability for water-wet systems. Note that the relative end point permeability for water is much lower than for oil.

in water-wet media are usually high, whereas they are low in oil-wet reservoirs. This leads to a lower ultimate recovery for water-wet media. The iso-permeability saturation point $S_{w,o=w}$ where the oil relative permeability is equal occurs at saturations below $S_{w,o=w} < 0.5$ for oil-wet media and at saturations $S_{w,o=w} > 0.5$ for water-wet media [118, 204] (Figure 4.14).

It is the purpose of this section to relate simple expressions of relative permeability to porous media properties, with the idea that they can be used to predict the consequence of these properties for 1-D recovery predictions by using the Buckley–Leverett theory described in Section 4.3.

For illustration of the viscosity effect, the oversimplified relations (so-called power law relative permeabilities) or Corey relative permeabilities can be used

$$k_{rw} = k'_{rw} S_{we}^{n_w}, \ k_{ro} = k'_{ro} S_{oe}^{n_o}, \tag{4.15}$$

where $S_{we} = (S_w - S_{wc})/(1 - S_{wc} - S_{or})$ and $S_{oe} = (S_o - S_{or})/(1 - S_{wc} - S_{or})$ and where k'_{rw} and k'_{ro} ($(k'_{rw} \neq k'_{ro})$) are the end-point permeabilities for water and oil, respectively. Moreover, the use of Eq. (4.15) requires an estimate of connate water saturation S_{wc} and of the residual oil saturation. We can find the connate water saturation from a plot of $\ln(k_{rw})$ versus the water saturation and find the water saturation at which the relative permeability dramatically decreases. Indeed, for water-wet media, the relative permeability is always above zero. The exponents n_w and n_o can assume values between 1 and 7. The oversimplified relative permeabilities in Eq. (4.15) do not show the wetting effect, which is present in Eq. (4.16)

$$k_{rw} = k'_{rw} S_{we}^{\frac{2+3\lambda}{\lambda}},$$

$$k_{ro} = k'_{ro}(1 - S_{we})^2 (1 - S_{we}^{\frac{2+\lambda}{\lambda}}),$$

$$S_{we} = \frac{S_w - S_{wc}}{1 - S_{wc} - S_{or}} \tag{4.16}$$

which are in fact a modified form of the Brooks-Corey relations [40]. In the original form, Brooks (personal communication) and Corey defined the effective saturation as $S_{we} = (S_w - S_{wc})/(1 - S_{wc})$ and used $k'_{rw} = k'_{ro} = 1$. Thus, the Brooks-Corey relative permeabilities in its original form describes primary drainage. In the equation, λ is the sorting factor. Typical values are $0.2 < \lambda < 7$. The parameter λ can be obtained experimentally by plotting the logarithm of log (S_{we}) versus logarithm of the capillary pressure log (P_c), where $P_c = P_{cb}S_e^{-1/\lambda}$. Such a plot would also be able to observe for which value of the connate water saturation S_{wc}, the log-log plot will be closest to a straight line. The determination of the sorting factor λ does not depend on whether the residual oil saturation is considered zero or has some finite value.

Procedure for Capillary Pressure Below the Connate Water Saturation

- Use parameters fracdim $= 2.5$, $S_{co} = S_{wc} + 0.01$
- $p_{cb} = \gamma\sigma_{ow}\sqrt{\varphi/k}((0.5 - S_{wc})/(1 - S_{wc}))^{(1/\lambda)}$
- $p_{cba} = p_{cb}s_{co}^{1/fracdim}((S_{co} - S_{wc})/(1 - S_{wc}))^{-1/\lambda}$
- $P_c = p_{cb}((S_w - S_{wc})/(1 - S_{wc}))^{-1/\lambda}U(S_w - S_{co}) + U(S_{co} - S_w)p_{cba}S_w^{fracdim}$, where $U(x) = 0$ for $x < 0, U(x) = 1$ for $x > 0$, and undetermined for $U(x = 0)$, but usually one takes $U(0) = 1/2$.

4.2.1   EXERCISE, BROOKS-COREY REL-PERMS

About showing that in the Brooks-Corey relative permeabilities $S_{or} = 0$ and $k'_{ro} = k'_{rw} = 1$, describes primary drainage and $S_{or} > 0$, e.g., $S_{or} = 0.3$ and $k'_{rw} < 1$, e.g. $k'_{rw} = 0.2$ describes secondary drainage. The exercise also illustrates the effect on the fractional flow functions.
Equation (4.16) can be used to describe primary and secondary drainage. For primary drainage we put $S_{or}=0$ and $k'_{rw}=k'_{ro} = 1$. For secondary drainage we may assign a value to S_{or} say $S_{or} = 0.25$. The end point permeability k'_{rw} is for example $k'_{rw} = 0.5$, whereas we keep the oil end point permeability equal to one, i.e., $k'_{ro} = 1.0$. A typical connate water saturation is $S_{wc} = 0.15$.

- Draw the Brooks-Corey relative permeabilities versus the water saturation for five values of $0.1 < \lambda < 10$.
- draw the fractional flow functions versus the water saturation for five values of $0.1 < \lambda < 10$, and a viscosity ratio M for oil/water of $M = 1$ and or $M = 10$.

Bootstrap Method

The Bootstrap method is apparently a method to statistically analyze small data sets. We consider [128, 194, 195]. The coefficient of variation $C_V = \sqrt{(var(k))/E(k)}$. The number of tests required for reliable statistics is $(10 C_v)^2$, see [128], p. 151, but here for $C_V << 0.1$ it does not apply. However, just to get the idea of how the method works: given 10 data points are normally distributed with average one and standard deviation of 0.01.

- Generate 3 (many are required, but just take three for an easy illustration) additional data sets with 10 data points, by randomly choosing 10 of the data points with replacement (putting them back).
- so there will be many duplicate data sets.
- determine the average (standard deviation) of each of the data sets to determine the standard deviation from the average (standard deviation) (Table 4.1).

4.2.2   LET RELATIVE PERMEABILITY MODEL

Spiteri et al. gave a new model of trapping and relative permeability hysteresis for all wetting conditions [218]. The parameters used here are borrowed from the so-called LET models of relative permeability named after its originators Lomeland, Ebeltoft, and Thomas (see, [82, 153,

Table 4.1

Data Set

1	0.990956
2	0.987208
3	1.016599
4	0.991087
5	0.997742
6	1.005716
7	0.99846
8	1.018087
9	1.005109
10	0.990013

155]). In this case, the connate water S_{wc} saturation is pragmatically similar to the initial water saturation that is established at high capillary pressure using a porous plate or centrifuge.

We will continue to discuss the modified form with nonzero residual oil saturation. The end point permeability k'_{rw} for water can be related to the initial water saturation S_{wi} or initial oil saturation S_{oi}. As an intermediate step, we relate the residual saturation to the initial water saturation by the empirical relation [155]

$$S_{or} = 2.0698S_{wi}^3 - 4.3857S_{wi}^2 + 2.1741S_{wi} + 0.1482. \tag{4.17}$$

The parameters are obtained by fitting Eq. (4.17) to data extracted from Figure 5 in [82]. We use Eq. (A9) from [155] for the end point relative water permeability

$$k'_{rw} = \frac{(1 - S_{or} - S_{wi})^{L_{wko}}}{(1 - S_{or} - S_{wi})^{L_{wko}} + E_{wko}S_{or}^{T_{wko}}}, \tag{4.18}$$

where $L_{wko} = 2.636$, $E_{wko} = 0.07667$, and $T_{wko} = 0.5413$. Note that $1 - S_{or} - S_{wi}$ is always positive for $0 < S_{wi} < 1$. The parameters are obtained by fitting Eq. (4.18) to data extracted from Figure 6 in [82]. The residual oil saturation S_{or} can also be expressed in terms of the USBM wettability index as shown in Figure 4.16.

The relative permeabilities using the LET correlation are

$$k_{rw} = k'_{rw} \frac{S_{we}^{L_w}}{S_{we}^{L_w} + E_w (1 - S_{we})^{T_w}} \tag{4.19}$$

$$k_{ro} = k'_{ro} \frac{(1 - S_{we})^{L_o}}{(1 - S_{we})^{L_o} + E_o S_{we}^{T_o}}, \tag{4.20}$$

where the effective water saturation S_{we} is given by

$$S_{we} = \frac{S_w - S_{wi}}{1 - S_{wi} - S_{or}}, \tag{4.21}$$

where $S_w > S_{wi}$. Only the residual oil saturation S_{or}, the initial water saturation S_{wi} and the end point permeabilities k'_{rw} (Eq. 4.18), $k'_{ro} \approx 1$ have a straightforward physical meaning. The other six parameters determine the shapes of the curves. Here L_w (L_o) determines the behavior at low effective water (oil, $S_{oe} = 1 - S_{we}$) saturations for k_{rw} (k_{ro}), whereas T_w (T_o) determines

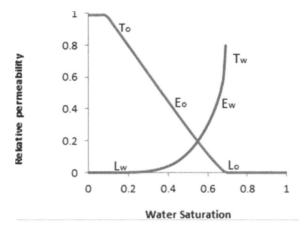

Figure 4.15 Relative oil and water relative permeability as a function of the parameters $L_w, L_o, E_w, E_o, T_w, T_o$ [82]. (See Eqs. (4.19, 4.20)). The figure has been obtained from the BSc (TU-Delft) thesis (2016) of Azzedine Toutou, "Enhanced Oil Recovery; The effect of wetting on the relative permeability behavior and oil recovery".

the behavior at high effective water (oil, $S_{oe} = 1 - S_{we}$) saturations. The parameters $E_w (E_o)$ determine the transition point between high- and low-effective saturations.

The end point wetting-phase permeability is derived from Figure 6 of [82] using the digitized plotter and using EUREQA^{TM} to obtain a regression expression. Unfortunately, this program EUREQA recently became part of a commercial program, which effectively precludes it from academic use.

4.2.3 ESTIMATE OF THE LET PARAMETERS

The LET relperms are given in Figure 4.15. The parameters can be obtained from data extracted from Figure 8 in [82].

4.2.4 EXERCISE, RESIDUAL OIL AND REL-PERM

About Residual oil saturation and relative permeability.

- make a plot of the residual oil saturation versus the initial water saturation.
- make a plot of the water end point permeability versus the residual oil saturation.

Ebeltoft et al. [82] plot in Figures 8, 11 and 12 the LET parameters as a function of the initial water saturations in this reference. Unfortunately, the authors have not indicated the scale in any of the y-axes. Figure 17 of [82] presents the calculated relative permeabilities. The SCAL parameters $L_w, L_o, E_w, E_o, T_w, T_o$ can, in principle, obtained from plots 15 and 16 of reference [82] (except for an absolute value on the y-axis). Assuming that the top of the plots in Figs. 15 and 16 of reference [82] is one (1) we try to match the relative permeability by assuming that the value of one must be replaced by a factor. These factors should preferably be simple numbers. We obtain a factor 8 for L_w, 2 for E_w, 8 for T_w, 6 for L_o, 4 for E_o and 8 for T_o. In this way, we derived the LET parameter values for initial water saturations equal to 0.1, 0.2, 0.3, 0.4 as shown in Table 4.2. These values can be used to study the relative permeabilities as a function of the wettability. The permeability can also be expressed as a function of the initial water saturation x.

$$\ln k(mD) = 1.00456717 + 0.5789093x^2 - 0.575279439x\tanh(12.4552153x^2) \qquad (4.22)$$

Table 4.2

LET Parameters for Various Initial Water Saturations
S_{wir}. **(See also [164].)**

Swir	0.1	0.2	0.3	0.4
L_w	3.52092	3.17219	2.66196	2.23261
E_w	2.39492	2.77307	3.01247	3.23192
T_w	0.71633	0.65202	0.63411	0.63933
L_o	3.22843	2.86294	2.45178	2.02538
E_o	2.39402	2.77307	3.01247	3.23192
T_o	1.09099	1.32197	1.47302	1.58395
S_{or}	0.13653	0.1845	0.19742	0.20295
k_{rw0}	0.93497	0.77807	0.56009	0.3331

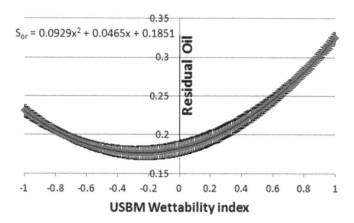

Figure 4.16 Residual Oil Saturation as a function of the USBM (United States Bureau of Mines) wetting index. [82].

4.3 THEORY OF BUCKLEY–LEVERETT

Buckley and Leverett [49] have derived a physically simple theory to describe the flow of water and oil. This theory makes it possible to predict the production behavior for the cases that the layer thickness is small with respect to the capillary transition zone length h_{trans}.

For incompressible fluids, the total production rate equals the injection rate in a 1D-setting. In other words, the theory allows us to estimate the fraction of oil and water in the fluids produced. The one extra assumption for the Buckley–Leverett theory with respect to what has been stated above is that conditions are such that capillary forces are the dominant force in the X-dip direction and consequently is the saturation approximately constant in the X-dip direction. These conditions may be reasonable in the laboratory but are almost never met in the field except for very low permeabilities and thin layer thicknesses. However, the Buckley–Leverett theory provides the best-case scenario (as to oil production) and can be of help in determining the maximum vertical grid size dimensions in numerical simulations.

1. We consider a layer of length L, constant thickness $H << L$ bounded from above and below respectively by impermeable cap- and base rock. The layer can be under a dip ϑ. Injection and production occur through a line drive configuration or through edge

water drive such that two-dimensional flow in a vertical cross-section occurs. The dip (layer) direction is indicated by "x" and the cross-dip (X-dip) direction is indicated by "y" (Figure 4.1),

2. the fluids and rock are considered incompressible,
3. the permeability and the porosity in the reservoir are considered constant. There exists a water saturation below which water cannot flow in the porous medium. This saturation is called the connate water saturation S_{wc}. The initial water saturation in the oil reservoir is equal to this connate water saturation. There exists also an oil saturation called the residual oil saturation S_{or} below which oil cannot flow,
4. both the water- and oil-phase flow independently of each other and behave according to a generalized form of Darcy's law,
5. in the dip direction, we disregard (as much as possible) the capillary forces and we take only the viscous and gravity forces into consideration.
6. the pressure gradient only depends only on the x-coordinate (in the dip direction) and is thus constant in the X-dip direction. Moreover, an equilibrium saturation has been attained balancing gravity and capillary forces in the X-dip direction subject to a given capillary pressure at the boundary between the reservoir and the base rock. Conditions are such that capillary forces dominate over gravity forces in the X-dip direction. Therefore, the saturation is approximately constant in the X-dip direction. This is also called completely dispersed flow. This reduces the physical problem to a problem in $1 - D$.

The physical simplicity of the model is unfortunately not matched with simple mathematics. There are a number of complex mathematical model implications. These difficulties largely arise due to disregarding the capillary pressure in the dip direction, which is "repaired" by the assumption of the occurrence of shocks. There are thus discontinuous saturation variations in the solution. We will therefore take a somewhat intermediate approach. We will specifically spell out where the problems are in Section 4.4.6. It is the combination of rarefaction, admissable shocks and constant states, which make seemingly simple and straightforward extensions of the theory [140] - such as for example for polymer drive [214, 215], surfactant flooding [51], foam displacement [141], nearly miscible displacement [219] and thermal recovery [50], - to a surmountable but time-consuming exercise. Fortunately, to solve the equations numerically is much easier and can be used as a starting point for obtaining the method of characteristics solution.

4.3.1   EXERCISE, VERTICAL UPSCALING RELATIVE PERMEABILITY

About constructing the fractional flow function from relative permeabilities.

Here we discuss constitutive relations. We consider a layer with a permeability k of 20 mDarcy and a height H of 5 meter. The distance between the row of injection and the row of production wells is 200 m. First we consider the situation where the reservoir is horizontal. The porosity φ is 0.3 [-]. The density difference $\Delta\rho$ between water and oil is [150kg/m^3], the acceleration due to gravity g is 9.81[m/s^2]. Make a choice of reasonable values for the viscosities of oil μ_o and water μ_w. The interfacial tension σ_{ow} between oil and water is $30 \times 10^{-3} \text{N/m}$. Include a value for the sorting factor λ, which assumes values between 0.2 (poor sorting) and very high values, e.g., 7 for well-sorted sands. The connate water saturation $S_{wc} = 0.15$ and the residual oil saturation S_{or} is also 0.15. The total Darcy velocity $u = 1 \ m/day$. Make a named list of variables in EXCEL by using the "Formulas-Create from Selection" command.[2] We use the so-called Corey relative permeabilities (see Eq. (4.16)) and the capillary pressures used in Eq. (4.8)

---

[2]It is expected that EXCEL commands will again slightly change in future versions of EXCEL, but the reader can in this case easily find the appropriate entry.

- Plot the Brooks-Corey relative permeability of oil and water versus saturation.
- plot the LET relative permeability of oil and water versus saturation for an initial water saturation of 0.4 and an initial water saturation of 0.1. Consider the original references on this subject [153, 154, 155].
- plot the quantity $\frac{P_c}{\Delta \rho g}$ versus saturation and vary the sorting factor. What can be concluded from these plots? Hint draw layers of various thicknesses with the bottom of the layer coinciding with the middle of the plot. Indicate the capillary transition zone.
- plot the fractional flow function f_w versus saturation, in the case that gravity and capillary forces can be disregarded.
- use insert-macro-module[3] to make the fractional flow function $f_w(S_w)$. Copy the text below and complete the missing statements at the dots.

Function $fw(sw)$
$swc = 0.15$
Rem connate water
$sor = 0.15$: Rem residual oil
$visrat = 0.5$
Rem $visrat = mu - w/mu - o$ = water viscosity / oil viscosity
$krw0 = 0.5$
Rem end point permeability for water
$kro0 = 1\#$
Rem end point permeability for oil
$se = ((sw - swc)/(1 - swc - sor))$
$kw = ...$
$ko = ...$
$fw = kw/(kw + ko * visrat)$
End Function

It is possible to use quantities defined in the worksheet in your basic program that defines the function([4]).

- Do you expect that the Buckley Leverett theory could be useful to describe the displacement process in the layer? In other words, is gravity segregation usually negligible for some practically relevant choices of the reservoir parameters? Hint: exceptionally in the Tambaredjo oil field [147] the oil density is very high (Shailesh Kisoensingh, private communication) and gravity segregation is very small.

4.4 MATERIAL BALANCE

The derivation of the material balance is only based on the assumption that the fluids and the rock are incompressible. We can therefore use the derivation for both the interface models as for the Buckley–Leverett Theory. Because we assume incompressible flow, we can instead of the mass balance also immediately derive the volume balance. This volume balance reads in words that

---

[3] or use any other software of your choice to program the fractional flow function.

[4] Sheets("data"). Select? With Sheets("data") 'this allows reading from the worksheet' swc = .Range("swc") End With, swc is a named variable in sheet in the sheet named "data". This allows to define data to be defined in your spreadsheet and use them in the module.

the sum of the amount that flows out of the volume element V and the amount that accumulates into the volume element must be zero. In symbols

$$\frac{d}{dt}\int_V \varphi S_\alpha dV = -\oint_S \mathbf{u}_\alpha.\mathbf{n}dS. \tag{4.23}$$

In this equation u_α is the "specific discharge", or the Darcy-velocity (volume flux) of phase α, i.e., oil and water. The vector $\mathbf{n}$ is perpendicular to the surface S that encloses the volume V. Furthermore, φ is the porosity and S_α is the saturation. To avoid the complex issue of averaging, we assume that the quantities in the integrand are already averaged over a representative elementary volume.

It is useful to remark that it is nowhere assumed that the dependent variables are continuous. If we rework this equation into its differential form, then the equation becomes of limited validity, i.e., it is only valid in the domains where the dependent variables (e.g., the saturation) are differentiable. We shall derive this differential equation for the one-dimensional case from the mass balance (4.23) stated above [148, 213, 249]. Subsequently, we will derive the general 3-D form using the divergence theorem (Integral theorem of Gauss). In one dimension, we can write Eq. (4.23) for a volume element $A\Delta x$ as follows (Figure 4.17)

$$\frac{d}{dt}\varphi S_\alpha A\Delta x = A(u_{\alpha,in} - u_{\alpha,out}). \tag{4.24}$$

where S_α denotes the average saturation in the block Δx. In the limit $\Delta x \to 0$ we can rewrite this equation as a partial differential equation, i.e.,

$$\varphi\frac{\partial S_\alpha}{\partial t} + \frac{\partial u_\alpha}{\partial x} = 0. \tag{4.25}$$

We can also write this equation in the general 3-D form. We use here the divergence theorem also called the Gauss integral theorem. This theorem states

$$\int_V \mathbf{div}\, \mathbf{F}dV = \oint_S \mathbf{F}.\mathbf{n}dS, \tag{4.26}$$

where $\mathbf{div}\, F = \frac{\partial F_x}{\partial x} + \frac{\partial F_y}{\partial y} + \frac{\partial F_z}{\partial z}$. We use Eq. (4.26) in Eq. (4.23)

$$\frac{d}{dt}\int_V \varphi S_\alpha dV = -\int_V \mathbf{div}\, \mathbf{u}_\alpha dV.$$

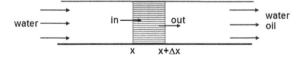

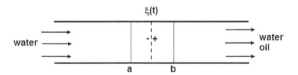

Figure 4.17 Control volumes for the derivation of the mass (volume) balance equations; top for the differential equation; bottom for the shock condition. The cross-sectional area is A (not shown).

The resulting equation is also valid for the dependent variables integrated over a representative elementary volume. Therefore, we can omit the integral sign and we find

$$\varphi \frac{\partial S_\alpha}{\partial t} + \mathbf{div}\, \mathbf{u}_\alpha = 0, \tag{4.27}$$

where S_α and $\mathbf{u}_\alpha$ are quantities averaged over the representative elementary volume. In the model of oil/water flow that we discuss here, the capillary forces will be omitted. Capillary forces act analogously to diffusion albeit with a saturation dependent diffusion constant. They make the solution smoother and get rid of discontinuities in the solution. When the capillary forces are omitted discontinuities can occur and Eqs. (4.25) and (4.27) are not valid at these discontinuities.

We shall now derive the volume balance at the discontinuities. The position where the discontinuity occurs (Figure 4.17) is denoted by $\xi(t)$ and find the equation for the velocity with which the discontinuity progresses [146]. The derivation is due to Lax. We write in one dimension the Eq. (4.24) over the interval (a, b), such that in this interval $a < \xi(t) < b$

$$\varphi \frac{d}{dt} \int_a^b S_\alpha dx = u_\alpha(a) - u_\alpha(b). \tag{4.28}$$

We split the integral on the left in a part in which we integrate from a to $\xi^-(t)$ and in a part in which we integrate from $\xi^+(t)$ to b to obtain

$$\varphi \frac{d}{dt} \left(\int_a^{\xi^-(t)} S_\alpha dx + \int_{\xi^+(t)}^b S_\alpha dx \right) = u_\alpha(a) - u_\alpha(b).$$

We subsequently apply Leibnitz's rule which states how to differentiate integrals, i.e.,

$$\frac{d}{d\lambda} \left(\int_{u(\lambda)}^{v(\lambda)} f(x, \lambda) dx \right) = \int_{u(\lambda)}^{v(\lambda)} \frac{\partial}{\partial \lambda} f(x, \lambda) dx$$
$$+ f(v, \lambda) \frac{\partial v}{\partial \lambda} - f(u, \lambda) \frac{\partial u}{\partial \lambda}. \tag{4.29}$$

We use the superscript $-$ to indicate a point that is just upstream of $\xi(t)$ and the superscript $+$ to indicate a point just downstream of $\xi(t)$. Note, a and b are constant in Eq. (4.28). We find with the introduced notation

$$\varphi(S_\alpha^- - S_\alpha^+) \frac{\partial \xi}{\partial t} + \int_a^{\xi^-(t)} \varphi \frac{\partial S_\alpha}{\partial t} dx + \int_{\xi^+(t)}^b \varphi \frac{\partial S_\alpha}{\partial t} dx$$
$$= u_\alpha(a) - u_\alpha(b). \tag{4.30}$$

In the domains $(a, \xi^-(t))$ and $(\xi^+(t), b)$ the saturation is differentiable (only not at $\xi(t)$) such that Eq. (4.24) holds. Therefore, we use Eq. (4.25) and substitute $-\partial u_\alpha/\partial x$ for $\varphi \partial S_\alpha/\partial t$ and we obtain

$$\varphi(S_\alpha^- - S_\alpha^+) \frac{\partial \xi}{\partial t} - \int_a^{\xi^-(t)} \frac{\partial u_\alpha}{\partial x} dx - \int_{\xi^+(t)}^b \frac{\partial u_\alpha}{\partial x} dx = u_\alpha(a) - u_\alpha(b).$$

The integrals are easy to evaluate and we find

$$\varphi(S_\alpha^- - S_\alpha^+) \frac{\partial \xi}{\partial t} - u_\alpha \big|_a^{\xi^-(t)} - u_\alpha \big|_{\xi^+(t)}^b = u_\alpha(a) - u_\alpha(b),$$

which can be rewritten as

$$\varphi(S_\alpha^- - S_\alpha^+)\frac{\partial \xi}{\partial t} + u_\alpha(a) - u_\alpha^- + u_\alpha^+ - u_\alpha(b) = u_\alpha(a) - u_\alpha(b),$$

and we obtain

$$\varphi(S_\alpha^- - S_\alpha^+)\frac{\partial \xi}{\partial t} = u_\alpha^- - u_\alpha^+. \tag{4.31}$$

Equations (4.24), (4.27) and (4.31) play an important part in the models to be discussed below.

4.4.1 SOLUTIONS OF THE THEORY OF BUCKLEY–LEVERETT

Summary of Material Balance Equations

Because we have assumed incompressibility and the saturation only varies in the dip direction, we use the 1-D material balance Eq. (4.24)

$$\varphi\frac{\partial S_\alpha}{\partial t} + \frac{\partial u_\alpha}{\partial x} = 0, \tag{4.32}$$

when the saturation and its derived quantities are differentiable. At discontinuities, we use

$$v_s = \frac{\partial \xi}{\partial t} = \frac{1}{\varphi}\frac{u_\alpha^- - u_\alpha^+}{S_\alpha^- - S_\alpha^+}. \tag{4.33}$$

To obtain Eq. (4.33), we have rearranged Eq. (4.31) to make it explicit in terms of the shock velocity.

Adding Eqs. (4.32) for the oil- and the water-phase leads to (using that $S_w + S_o = 1$ and thus that the derivative of one is zero) we obtain

$$\frac{\partial(u_o + u_w)}{\partial x} := \frac{\partial u}{\partial x} = 0, \tag{4.34}$$

where the total Darcy velocity $u = u_o + u_w$. The solution of this partial differential Eq. (4.34) is

$$u = u(t). \tag{4.35}$$

In the same way, we can also derive from Eq. (4.31) that the total Darcy velocity before and after the shock is the same. We add Eqs. (4.31) for $\alpha = o$ and for $\alpha = w$ and obtain

$$u^- = u_w^- + u_o^- = u_w^+ + u_o^+ = u^+.$$

We had also expected this because of our assumption of fluid incompressibility. Therefore, the total 1-D velocity u at each point is equal to the injection velocity $u_{inj}(t)$.

4.4.2 EQUATION OF MOTION (DARCY'S LAW) AND THE FRACTIONAL FLOW FUNCTION

It is convenient to write the Darcy velocity of one-phase (for instance, water) as a product of the total Darcy-velocity u and the fractional flow function $f_w = \frac{u_w}{u}$. We shall show that, in the absence of capillary forces, the fractional flow function only depends on the saturation. For the ease of notation, we write $\frac{k_\alpha(S_\alpha)}{\mu_\alpha} := \lambda_\alpha(S_\alpha)$. In the layer direction x we have

$$u_{ox} = -\lambda_o(\frac{\partial p}{\partial x} + \rho_o g \sin\vartheta), \ u_{wx} = -\lambda_w(\frac{\partial p}{\partial x} + \rho_w g \sin\vartheta). \tag{4.36}$$

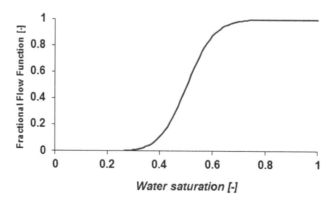

Figure 4.18   Fractional Flow Function versus Saturation in the absence of gravity.

We neglect the capillary forces and write that the water pressure equals the oil pressure, i.e., $P_w = P_o = p$. The fractional flow function is found by addition of the Eqs. (4.36) and elimination of the pressure gradient. We obtain for the total velocity

$$u = u_{wx} + u_{ox} = u_{wx} - \frac{\lambda_o}{\lambda_w} \lambda_w \left(\frac{\partial p}{\partial x} + \rho_w g \sin \vartheta \right) - \lambda_o (\rho_o - \rho_w) g \sin \vartheta. \qquad (4.37)$$

We resubstitute Darcy's law (4.1) for the water-phase and rearrange to obtain an expression for the fractional flow function $f_w = \frac{u_w}{u}$

$$f_w = \frac{\lambda_w}{\lambda_w + \lambda_o} - \frac{\lambda_w \lambda_o}{\lambda_w + \lambda_o} \left((\rho_w - \rho_o) g \sin \vartheta \right) \frac{1}{u}. \qquad (4.38)$$

With gravity terms included the fractional flow function versus saturation may lose its typical S-shaped form. The procedure to obtain a solution remains essentially the same, but it is sometimes helpful to obtain the numerical result before trying to write the analytical solution (see below). For horizontal reservoirs, the fractional flow function has an S-shape and is simply $f_w = \frac{\lambda_w}{\lambda_w + \lambda_o}$. For $S_w = S_{wc}$ we have $k_w = 0$ and therefore the fractional (water) flow function is also zero for $S_w = S_{wc}$. For $S_w = (1 - S_{or})$ we have that $k_o = 0$ and thus the fractional flow function is equal to one. For the relative permeabilities used we obtain an S-like curve with possibly more than one inflection point. If gravity is considered the fractional flow function can become larger than one and the shape becomes more complicated. The S-shaped like function without gravity term is shown in the left of Figure 4.18.

4.4.3   ANALYTICAL SOLUTION OF THE EQUATIONS

The solution of the model equations of Eqs. (4.25) and (4.33) consists of three parts, viz.,

1. The rarefaction wave,
2. constant state solution,
3. the shock wave.

The constant state part and the rarefaction part can be obtained from Eq. (4.25) by rewriting the equation in terms of an ordinary differential equation. We can achieve this by the use of a similarity transformation [235], in this case, a convenient combination of x and t, such that the equation and boundary condition can all be expressed in this new variable. Indeed we use $\eta = x/t$. We start with Eq. (4.25) for the aqueous-phase and substitute $u_w = u f_w$ to obtain

$$\frac{\partial S_w}{\partial t} + \frac{u}{\varphi} \frac{\partial f_w}{\partial x} = 0. \tag{4.39}$$

We use $\frac{\partial \eta}{\partial t} = -\frac{x}{t^2}$ and $\frac{\partial \eta}{\partial x} = \frac{1}{t}$. Substitution of $x/t = \eta$ in Eq. (4.39) leads to an ordinary differential equation. We use that $f_w = f_w(S_w)$, and thus $\frac{\partial f_w}{\partial \eta} = \frac{d f_w}{d S_w} \frac{\partial S_w}{\partial \eta}$ and obtain

$$-\eta \frac{d S_w}{d \eta} + \frac{u}{\varphi} \frac{d f_w}{d S_w} \frac{d S_w}{d \eta} = 0.$$

We assume for simplicity an initially uniform oil saturation and connate water saturation $S_w = S_{wc}$. We only inject water (no oil). The initial condition can be expressed in terms of η, i.e., $S_w(\eta \to \infty) = S_{wc}$. Water injection means that for $x = 0$ and thus for $\eta = 0$ we have the saturation $S_w = 1 - S_{or}$.

We observe that there are two solutions namely the constant state $\frac{d S_w}{d \eta} = 0$,

$$S_w = constant,$$

and the rarefaction wave

$$\eta = \frac{u}{\varphi} \frac{d f_w}{d S_w}. \tag{4.40}$$

Usually, a solution expresses the dependent variable as a function of the independent variable, i.e., $S_w = f(\eta)$ but the inversion of the solution where we express the independent variable as a function of the dependent variable (Eq. (4.40)) is not simple. Therefore, we prefer to give the solution where the independent variable is expressed as a function of the dependent variable. We take a value for the saturation and calculate the value of the right side of Eq. (4.40). Thus, we find $\eta = x/t$ where x denotes the position where we can find the given saturation at time t. One thus finds that a given saturation after two hours is twice as far from the injection point as after 1 hour (stretching principle).

Let us try to find qualitatively the behavior of the solution in the absence of gravity effects, i.e., a horizontal reservoir. The fractional flow function has an S-shaped form $f_w = \frac{\lambda_w}{\lambda_w + \lambda_o}$. If gravity would have been considered the fractional flow function can become larger than one and the shape becomes more complicated. In Figure 4.19 we sketch with the help of Eq. (4.40) the solution. We begin at $S_w = 1 - S_{or}$. From the plot we can determine the value of $\frac{d f_w}{d S_w}$ for this saturation. This value is zero. Therefore, also $\eta = x/t = 0$. This point is indicated as point A in Figure 4.19. When the water saturation becomes somewhat smaller than $(1 - S_{or})$ the value of $\frac{d f_w}{d S_w}$ becomes larger than zero and thus we find, for instance, point B in Figure 4.19 from $\frac{x}{t} = \frac{u}{\varphi} \frac{d f_w}{d S_w}$.

Furthermore $\frac{d f_w}{d S_w}$ versus S_w looks like a clock function, i.e., for the values $S_w = S_{wc}$ and for $S_w = (1 - S_{or})$ one finds the values zero whereas for a value in between it will be everywhere positive and one will find the maximum value of $\frac{d f_w}{d S_w}$.

This means that in Figure 4.19 one finds a maximum value for η at point C and subsequently η decreases for smaller values of S_w. One thus finds two values of S_w for every value of $\eta = x/t$. Apparently, something went wrong when we only consider Eq. (4.25) and disregard the occurrence of shocks. At point C, the derivative of the saturation versus η becomes infinite. A

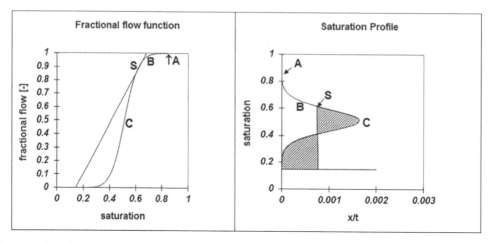

Figure 4.19 Construction of saturation profile from fractional flow function. The part from A to S is found from $x/t = u/\varphi \frac{df_w}{dS_w}$ and is called rarefaction part. The shock saturation is found from tangent line to the fractional flow function starting at $(S_{wc}, 0)$. Alternatively the (hatched) areas right and left (below) the shock line should be equal, to satisfy the mass balance. As pointed out below, this does not define the only possible solution.

high value of $dS_w/d\eta$ implies that it is no longer justified to disregard capillary forces. Stated more precisely, analogously to Fick's Law $(J = -D\frac{dc}{dx})$ as the capillary diffusion "constant" (D) goes to zero the saturation gradient $(\frac{dS}{dx})$ goes to infinity. Therefore, the capillary forces at the shock are not negligible ($0 \times \infty$ can be finite, see Figure 4.19).

Capillary forces are required to avoid a multivalued solution (physical nonsense) and thus to obtain shock-like behavior instead. It is a good example to show that mathematics and physical models have to be considered together. The problem with the solution shows that something is wrong with the physical model, i.e., it is not allowed to disregard capillary forces. We had, indeed, anticipated that a shock-like part needs to be included in the solution. Therefore, we gave a derivation for the velocity of the shock

$$v_s = \frac{\partial \xi}{\partial t} = \frac{u}{\varphi} \frac{f_\alpha^- - f_\alpha^+}{S_\alpha^- - S_\alpha^+}, \tag{4.41}$$

where we have substituted the definition of the fractional flow function. We use the notation v_s to denote the velocity of the shock. Summarizing we can say that the solution can be constructed from three solution parts, i.e., the constant state solution $S_w =$ constant, the rarefaction wave solution (4.40) and the shock solution (4.41).

4.4.4 CONSTRUCTION OF THE ANALYTICAL SOLUTION; REQUIREMENT OF THE ENTROPY CONDITION

Can the three solution parts build a unique single valued solution (for every x/t there is a single saturation value) that satisfies the initial condition $S_w(\eta \to \infty) = S_{wc}$ and the boundary condition $S_w(\eta = 0) = 1 - S_{or}$? The answer is "single valued" yes, "unique" no [234]. In Figure 4.20, we have drawn two single valued solutions. We can simply ascertain that these solutions can be constructed from our solution parts. In Figure 4.20, we have combined the $f_w - S_w$ plot and the $S_w - \frac{x}{t}$ plot. The common axis is the S_w axis. We consider the plots shown at the left. Above the

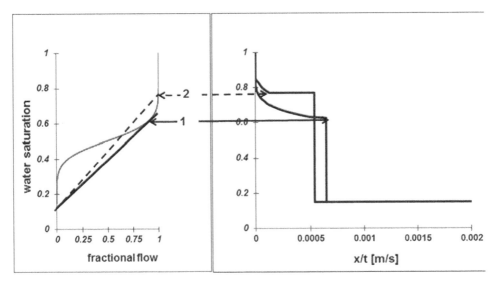

Figure 4.20 Tangent construction (1) according to Welge giving the right solution. The intersection construction (2) satisfies all requirements except for the entropy condition. Such a solution does not follow as a limit when the capillary forces tend to zero.

tangent line $(S_{wc}, 0) \rightarrow (S_w, f_w)$ we use the rarefaction wave solution $\frac{x}{t} = \frac{u}{\varphi} \frac{d f_w}{d S_w}$. With this part we construct the up-stream part of the shock. The shock saturation is found by the tangent line in the left part of Figure 4.19. We can find the position of the shock. To this end, we look at Figures 4.19 and 4.20. Point S in Figure 4.19 belongs both to the shock part and the rarefaction part. This means that the velocity of the shock $\frac{\partial \xi}{\partial t} \propto \frac{f_w^-}{S_w - S_{wc}}$ must be equal to the velocity $\frac{x}{t} = \eta \propto \frac{d f_w}{d S_w}$ with which the saturation point S moves. Thus, we equate η in Eq. (4.40) to v_s in Eq. (4.41). We divide on both sides by $\frac{u}{\varphi}$ and we obtain (Figure 4.19)

$$\frac{f_\alpha^- - f_\alpha^+}{S_\alpha^- - S_\alpha^+} = \left(\frac{d f_w}{d S_w} \right)_{shock}. \tag{4.42}$$

This equation is shown graphically in Figure 4.20 and indicated by connecting line (1). This is, however, not the only solution that can be constructed. Consider the solutions connected by line (2) in Figure 4.20. Here we have drawn the intersection line from point $(S_{wc}, 0)$ with the $f_w - S_w$ curve. If we consider only the upper intersection point, then we find the following solution. Above the highest intersection point of the line from $(S_{wc}, 0))$ with the fractional flow curve we use the rarefaction wave solution $\frac{x}{t} = \frac{u}{\varphi} \frac{d f_w}{d S_w}$. Thus, we construct the solution upstream of the shock. The intersection dotted line represents the shock in the same way as the tangent drawn line in the left part of Figure 4.20. In the right part of Figure 4.20, we observe that the velocity of the shock is not equal to the velocity of the rarefaction wave. Between the shock and the rarefaction wave, we have a constant state solution. The solution drawn in the right part of Figure 4.20 satisfies all the conditions that we require from the solution. It turns out that in the equation

$$\varphi \frac{\partial S_\alpha}{\partial t} + \frac{\partial u_\alpha}{\partial x} = \lim_{D \to 0} D \frac{\partial^2 S_\alpha}{\partial x^2}, \tag{4.43}$$

i.e., as the diffusion constant tends to zero this does not imply that we can disregard the right term altogether. The capillary diffusion term can be small but is positive. The diffusion coefficient

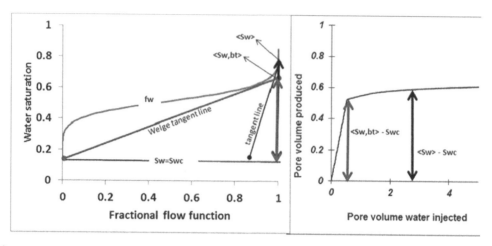

Figure 4.21 Extrapolation of line tangent to fractional flow curve at the saturation at the production point $S_w = S_{w,end}$ toward the line $f_w = 1$ gives the average saturation $< S_w >$ in the reservoir. The tangent line originating from the connate water saturation gives the average saturation at breakthrough $< S_{w,bt} >$. The pore volume of oil produced $N_{pd} = < S_w > -S_{wc}$. The pore volume of oil produced at breakthrough is $N_{pd,bt} = < S_{w,bt} > -S_{wc}$. The pore volume of water injected is $W_{id} = dS_w/df_w$.

must be positive as the local entropy production term $D(\partial S_\alpha/\partial x)^2$ is always positive. We can construct a traveling wave solution of Eq. (4.43) using the coordinate transformation $\xi = x - vt$ and obtain

$$\frac{\partial (u_\alpha - v\varphi S_\alpha)}{\partial \xi} = \lim_{D \to 0} D\frac{\partial^2 S_\alpha}{\partial \xi^2}.$$

It can be shown that this equation only has a solution if the boundary condition at $\xi \to -\infty$ is the saturation picked out by the tangent shock line. It is, however, outside the scope of this monograph.

The name entropy condition can be given the physical meaning by observing that according to (nonequilibrium) thermodynamics a positive entropy production term implies a positive diffusion constant. The entropy condition is known in Petroleum Engineering under the name "construction of Welge". Welge (a genius) intuitively picked out the right solution without fully understanding the problem. Therefore the left plot of Figure 4.20 is right and the right plot of Figure 4.20 is incorrect. Whenever you are dealing with a non-standard case, you have to understand the entropy condition, but this is outside the scope of this monograph. We observe that the tangent line from point $S_w = S_{wc}$, $f_w = 0$ to the $f_w - S_w$ curve satisfies precisely Eq. (4.42). The S_w value of the tangent point is the saturation value where the shock starts (Figure 4.21). Again the construction that uses the entropy condition is called the Welge construction. The reason that we emphasize this point is that a numerical scheme may or may not pick the right solution [16] (see page 152 of this reference). In the numerical schemes, which we discuss below. it turns out that we obtain the right solution.

4.4.5 EXERCISE, BUCKLEY LEVERETT PROFILE WITH EXCEL

Construction of Buckley Leverett Saturation profile

Use the assumptions as in exercise (4.3.1) Use also the same EXCEL sheet that you made in exercise (4.3.1) In particular, you need the macromodule for calculating the fractional flow function.

- Find the shock saturation. Take four cells next to each other. Put an assumed value say $S_w=0.7$ in the first cell. In the second cell, calculate $\frac{f_w}{S_w-S_{wc}}$. In the third cell, calculate $\frac{df_w}{dS_w}$. In the fourth cell, take the difference between the second and the third cell. Try to solve for the shock saturation by using Tools-Solver or Goal seek. Sketch graphically what you have been doing in the fractional flow function versus saturation plot. Use view-toolbars-drawing to get the menu that allows to draw lines in the fractional flow function. Click on the fractional flow plot. Click right on the fractional flow versus saturation plot and insert grid lines. Note that one has to reapply the solver every time that a change in one of the constants is made.
- plot the saturation profile. Make a column of saturation values going down from $(1-S_{or})$ to the shock saturation found above. Left of this column determine the $\frac{x}{t} = \frac{u}{\varphi}\frac{df_w}{dS_w}$ positions/time. Assume a value for t. Copy the last value (of the shock position) to one cell below and put S_{wc} right of it. Copy twice the value of the shock position and put again S_{wc} right of it. The two columns can be used to plot the saturation profile versus $\frac{x}{t}$.

4.4.6   DERIVATION OF THE SHOCK CONDITION

It is possible to prove that the Welge tangent [248] condition needs to be satisfied by considering that the shock must be represented by a traveling wave solution [146, 138].

Therefore, consider Eq. (4.43) for the water-phase, i.e.,

$$\varphi\frac{\partial S_w}{\partial t} + \frac{\partial u_w}{\partial x} = \lim_{\varepsilon\to0}\varepsilon\frac{\partial^2 S_w}{\partial x^2}. \tag{4.44}$$

The problem can be solved using the method of matched asymptotic expansions [236], Chapter 5. Disregarding the right side in Eq. (4.44), i.e., the diffusion term $\varepsilon\frac{\partial^2 S_w}{\partial x^2}$ defines the outer solution. Using for the inner part, the transformation $\xi = (x-vt)/\varepsilon$ to obtain

$$\varphi\frac{\partial S_w}{\partial t} - \frac{\varphi v}{\varepsilon}\frac{\partial S_w}{\partial \xi} + \frac{u}{\varepsilon}\frac{\partial f_w}{\partial \xi} = \frac{1}{\varepsilon}\frac{\partial^2 S_w}{\partial \xi^2}. \tag{4.45}$$

Collecting the largest terms in Eq. (4.46) we obtain

$$u\frac{\partial f_w}{\partial \xi} - v\varphi\frac{\partial S_w}{\partial \xi} = \frac{\partial^2 S_w}{\partial \xi^2}. \tag{4.46}$$

The solution of this equation can be obtained by integrating once, i.e.,

$$\frac{dS_w}{d\xi} = uf_w - \varphi S_w - (uf_w - \varphi S_w)_{\xi\to-\infty}, \tag{4.47}$$

where we use that due to the scaling with $\frac{1}{\varepsilon}$ the derivative far upstream, i.e., at $\xi \to -\infty$ is zero.

The saturation derivative $\frac{dS_w}{d\xi} < 0$ because if $\frac{dS_w}{d\xi} > 0$ we can never reach a lower saturation value than the upstream saturation. Using the Taylor expansion we obtain

$$\frac{dS_w}{d\xi} = \left(u\frac{df_w}{dS_w} - \varphi v\right)_{\xi\to-\infty} \left(S_w - S_w^{-\infty}\right) + h.o.t. \le 0, \tag{4.48}$$

where $S_w < S_w^{-\infty}$. In other words, $v \leqq \frac{u}{\varphi} \frac{df_w}{dS_w}$. However, if $v < \frac{u}{\varphi} \frac{df_w}{dS_w}$, we obtain a double-valued solution. Therefore, we must have

$$v = \frac{u}{\varphi} \frac{df_w}{dS_w}. \tag{4.49}$$

Implicit in the derivation we used that $\varepsilon > 0$ or the diffusion coefficient is positive. The diffusion coefficient must be positive, otherwise the entropy might decrease. Therefore, it is justified to call Eq. (4.49) an entropy condition, and it is not a consequence of mass conservation [64].

4.4.7 ANALYTICAL CALCULATION OF THE PRODUCTION BEHAVIOR [64]

A practical engineer is, apart from understanding of the physics in the theories he uses, primarily interested in the production and pressure behavior of the reservoir under consideration [63]. The saturation profile downstream of the shock front is characterized by a constant state. Before the breakthrough of the shock, we expect water-free oil production. For a layer with length L, we expect water breakthrough after $t_{bt} = L/v_s$. From Eqs. (4.40) and (4.41), we can derive that $v_s = \frac{u}{\varphi}(\frac{df_w}{dS_w})_{shock}$. We define the dimensionless cumulative water injection (cumulative injection at water breakthrough divided by the pore volume $W_{id,bt} = Aut_{bt}/(\varphi AL)$. We have put the pore surface A in the numerator and the denominator for reasons of easy understanding. If it is clear, erase the pore surface A both from the numerator and the denominator. The symbol W_{id} is used for the dimensionless cumulative water injection related to the pore volume. The symbol W_{iD} is reserved for the dimensionless cumulative water injection related to the movable porevolume. We obtain with the help of Eqs. (4.41) and (4.42)

$$W_{d,bt} = \frac{Aut_{bt}}{\varphi AL} = \frac{u\frac{L}{v_s}}{\varphi L} =$$

$$= \frac{1}{\frac{df_w}{dS_w}_{shock}} = \frac{S_{w,shock} - S_{wc}}{f_{w,shock} - f_w(S_{wc})(=0)}. \tag{4.50}$$

Upstream of the shock front Eq. (4.40) holds and can thus be used for production calculations. With $\eta = x/t$ we rewrite Eq.(4.40) as follows

$$\frac{x}{L} = \frac{uAt}{\varphi AL} \frac{df_w}{dS_w} := W_{id} \frac{df_w}{dS_w}. \tag{4.51}$$

In particular for $x = L$

$$W_{id} = \frac{1}{(\frac{df_w}{dS_w})_e}, \tag{4.52}$$

where the subscript e indicates the end point $x = L$ of the layer. The dimensionless oil production N_{pd} is equal to the product of the fractional flow function of oil $f_o = 1 - f_w$ and the dimensionless injection (=total fluid) production rate W_{id}. From this we can calculate the cumulative oil production

$$N_{pd} = \int_0^{W_{id}} (1 - f_w)dW_{id}. \tag{4.53}$$

For evaluation of the integral (4.53), we use integration by parts and Eq. (4.52) where we indicate with the subindex e the end of the layer (the production point)

$$N_{pd} = (1 - f_w)_e W_{id} + \int_{S_{wc}}^{S_{we}} dS_w = (1 - f_w)_e W_{id} + S_{we} - S_{wc}, \tag{4.54}$$

where we have used $W_{id}df_w = \frac{dS_w}{df_w}df_w = dS_w$. Also Eq. (4.54) lends itself for graphical interpretation in a $f_w - S_w$ diagram (Figure 4.21).

The intersection point of the tangent line to the fractional flow curve with the top boundary line of the plot $f_w = 1$ occurs at a saturation that we shall denote with $<S_w>$. This notation is used in the Petroleum Engineering literature because we want to convey that it is the average water saturation in the reservoir. Many of us use the notation $<X>$ also for ensemble average and the notation may be confusing. How can we ascertain that $<S_w>$ is indeed the average saturation? For this purpose, we consider the term $(1 - f_{we})W_{id}$. According to Eq. (4.52), we rewrite this term as $(1 - f_{we})W_{id} = (1 - f_{we})(dS_w/df_w)_e$. From the plot we read that this is equal to $(1 - f_{we})(<S_w> - S_{we})/(1 - f_{we}) = <S_w> - S_{we}$. We can thus write the dimensionless oil production simply as

$$N_{pd} = <S_w> - S_{wc}. \tag{4.55}$$

As the produced oil in the reservoir is replaced by water, it is straightforward to interpret $<S_w>$ indeed as the average water saturation. The procedure is really simple and leads to a graphical "laundry prescription" like put the middle turning knob on $40°C$ and press start.

4.4.8 EXERCISE, BUCKLEY LEVERETT PRODUCTION FILE

Graphical construction of the Buckley–Leverett production curve.

- Draw the fractional flow curve f_w as a function of saturation S_w (left part of Figure 4.19).
- draw the tangent line to this curve from the point $(S_w = S_{wc}, f_w = 0)$. Draw the tangent line all the way until it intersects with the line $f_w = 1$. In this way one finds the point $(< S_{w,bt} >, 1)$ (Figure 4.21).
- use Eq. (4.53) and find the dimensionless water production until breakthrough $W_{d,bt}$ with Eq. (4.52). Before this, i.e., for $W_d < W_{d,bt}$ we have that the dimensionless oil production $N_{pd} = W_{id}$, the dimensionless water injection.
- draw subsequently a number of tangent lines above tangent point. Determine $W_{id} = dS_w/df_w$. Use Eq. (4.54) for the determination of N_{pd} (Figure 4.21).

4.4.9 EXERCISE, ANALYTICAL BUCKLEY LEVERETT PRODUCTION CURVE

About the computation of the Buckley–Leverett production profiles with EXCEL

Use the assumptions as in exercise 4.3.1. Use also the same EXCEL sheet that you made in exercise 4.3.1. In particular, you need the macromodule (function) for calculating the fractional flow function.

- Make a new fractional flow function versus saturation plot with a saturation spacing $\Delta S_w \approx 0.01$. Use again the insert-shapes to get the menu that allows to draw lines in the fractional flow function,
- make a column of 200 saturation values from $S_w = 1 - S_{or}$ to the shock saturation. Calculate the corresponding fractional flow function in the column right to it. Calculate the dimensionless cumulative water injection $W_{id} = 1/\frac{df_w}{dS_w}$ in the column to the right by using numerical differentiation. In other words. $W_{id}(i) \approx (S_w(i+1) - S_w(i-1))/(f_w(i+1) - f_w(i-1))$,
- check graphically in the f_w versus S_w curve that the cumulative water injection calculation is about right for a few points. Click right (if you not already did this) on the fractional flow versus saturation plot and insert grid lines. Perform the numerical differentiation for one saturation value above the shock saturation in the plot to obtain

$W_{id} = 1/\frac{df_w}{dS_w}$. Draw a tangent line in the fractional flow function for this chosen saturation. Read the dimensionless production from the plot,

- calculate the dimensionless oil production $N_{pd} = (1 - f_w)W_{id} + S_{we} - S_{wc}$ in the column right to the column where you calculated W_{id}.

4.4.10 DETERMINATION OF RELATIVE PERMEABILITIES FROM PRODUCTION DATA AND PRESSURE MEASUREMENTS

For laboratory conditions, for which capillary forces in the flow direction can be disregarded, we can use a core experiment to determine relative permeabilities [224, 225]. The disregarding of capillary forces is almost always a problem as the capillary end effect (water does not want to leave the water wet core and piles up at the end of the core) can never be completely ignored. With today's computational power, the validity of the assumptions can be easily checked. Under favorable conditions, we can measure the water fraction in the produced fluids and the pressure drop along the core to estimate the relative permeabilities outside the shock region. The water fraction of produced fluids can be immediately used to construct the fractional flow function-saturation curve above the shock saturation. To extract the relative permeabilities, we need extra information that is found from the pressure drop in the core.

Fractional Flow Function from the Water Fraction in the Produced Fluids

The cumulative production of water $Q_w = Au_wt$ allows the calculation of the average water saturation in the core. Material balance considerations can be interpreted as that the cumulative water injected Aut minus the cumulative volume of water produced $Q_w(t)$ divided by the pore volume $AL\varphi$ equals the increase of the average water saturation $< S_w > -S_{wc}$, i.e.,

$$< S_w >= S_{wc} + \frac{Aut - Q_w(t)}{AL\varphi}, \tag{4.56}$$

where A is the cross-section of the core and L its length. To express the average saturation $< S_w(t) >$ to the saturation at the end point $S_w(L,t)$, we write

$$< S_w >= \frac{1}{L}\int_0^L S_w(x,t)dx.$$

Integrating by parts leads to

$$< S_w > = \frac{1}{L}[xS_w(x,t)]_0^L - \frac{1}{L}\int_{S_w(0,t)}^{S_w(L,t)} xdS_w$$

$$= S_w(L,t) - \frac{ut}{\varphi L}\int_{S_w(0,t)}^{S_w(L,t)} \frac{df_w}{dS_w}dS_w = S_w(L,t) - \frac{ut}{\varphi L}(f_w(L,t) - f_w(0,t)), \tag{4.57}$$

where we have used that $x = \frac{u}{\varphi} \frac{df_w}{dS_w} t$. Therefore, we obtain with $f_w(0,t)=1$ and Eq.(4.56)

$$S_w(L,t) = <S_w> + \frac{ut}{\varphi L}(f_w(L,t)-1)$$

$$= S_{wc} + \frac{Aut - Q_w(t)}{AL\varphi} + \frac{Aut}{AL\varphi}(f_w(L,t)-1) \quad \text{or}$$

$$S_w(L,t) = S_{wc} + \frac{1}{AL\varphi}\left(t\frac{dQ_w}{dt} - Q_w\right),$$

where we have used that

$$f_w(L,t) = \frac{u_w(L,t)}{u} = \frac{1}{Au}\frac{dQ_w}{dt}.$$

Therefore, the production data make it possible for us to determine the fractional flow function versus saturation.

4.4.11   DETERMINATION OF THE RELATIVE PERMEABILITIES BY ADDITIONAL MEASUREMENT OF THE PRESSURE DROP

This method is called the JBN[5] method for determination of the relative permeabilities; the JBN method neglects capillary forces and is thus only applicable for the exceptional cases that this is allowed.

For the relative permeabilities, we need some extra experimental information. This is provided by the pressure difference across the core as a function of time. We assume a dip angle ϑ (a zero dip angle corresponds to horizontal flow and a dip angle of 90 degrees corresponds to vertical flow from the top to the bottom; different from the dip angle in the layer used above where we have flow from the bottom to the top) . First, we note that the expression for the fractional flow function (disregarding capillary forces) in terms of the mobilities $\lambda_\alpha := k\frac{k_{r\alpha}}{\mu_\alpha}$ is given by

$$f_w = \frac{\lambda_w}{\lambda_w + \lambda_o} + \frac{\lambda_w \lambda_o}{\lambda_w + \lambda_o}\frac{(\rho_w - \rho_o)g\sin\vartheta}{u}, \tag{4.58}$$

where we have now a plus sign before the gravity dependent term as we are flowing form the top to the bottom. Dividing nominator and denominator by the mobility product $\lambda_w \lambda_o$, we obtain

$$f_w = \frac{\frac{1}{\lambda_o} + \frac{(\rho_w - \rho_o)g\sin\vartheta}{u}}{\frac{1}{\lambda_w} + \frac{1}{\lambda_o}}. \tag{4.59}$$

Darcy's law can be written as (again changing the sign before the gravity term)

$$u = u_w + u_o = -(\lambda_w + \lambda_o)\frac{\partial p}{\partial x} + (\lambda_w\rho_w + \lambda_o\rho_o)g\sin\vartheta.$$

Therefore, the pressure gradient is given by the expression

$$\frac{\partial p}{\partial x} = -\frac{u - g\sin\vartheta\,(\rho_w\lambda_w + \rho_o\lambda_o)}{\lambda_w + \lambda_o}. \tag{4.60}$$

---

[5] Johnson, Bossler, and Naumann (JBN).

and we obtain for the pressure $P(t)$ across the core

$$P(t) = -\int_0^L \frac{\partial p}{\partial x} dx = \int_0^{\varphi L/(ut)} \frac{u - g\sin\vartheta\,(\rho_w\lambda_w + \rho_o\lambda_o)}{(\lambda_w + \lambda_o)} \frac{ut}{\varphi} d\frac{df_w}{dS_w},$$

where we have used that $x = \frac{u}{\varphi}\frac{df_w}{dS_w}t$. Differentiation of the integrand leads to the term $P(t)/t$. Application of Leibnitz's rule also comprises differentiation of the boundaries), and we obtain the following expression for the time derivative of the pressure

$$\frac{dP}{dt} = \frac{P(t)}{t} + \frac{u - g\sin\vartheta\,(\rho_w\lambda_w + \rho_o\lambda_o)}{\lambda_w + \lambda_o}\frac{ut}{\varphi}\left(-\frac{\varphi L}{ut^2}\right)$$

$$= \frac{P(t)}{t} - \frac{u - g\sin\vartheta\,(\rho_w\lambda_w + \rho_o\lambda_o)}{\lambda_w + \lambda_o}\frac{L}{t}. \tag{4.61}$$

After rearrangement, we obtain

$$P(t) - t\frac{dP}{dt} = \frac{u - g\sin\vartheta\,(\rho_w\lambda_w + \rho_o\lambda_o)}{(\lambda_w + \lambda_o)}L.$$

From this we calculate

$$P(t) + \rho_w g\sin\vartheta L - t\frac{dP}{dt} = \frac{u - g\sin\vartheta\,(\rho_w\lambda_w + \rho_o\lambda_o) + g\sin\vartheta\rho_w\,(\lambda_w + \lambda_o)}{(\lambda_o + \lambda_o)}L$$

$$= \frac{u + g\sin\vartheta\,(\rho_w\lambda_o - \rho_o\lambda_o)}{(\lambda_w + \lambda_o)}L. \tag{4.62}$$

Substitution of Eq. (4.59) and rearranging, i.e.,

$$uf_w = \frac{\frac{u}{\lambda_o} + (\rho_w - \rho_o)g\sin\vartheta}{\frac{1}{\lambda_w} + \frac{1}{\lambda_o}} = \lambda_w\frac{u + \lambda_o(\rho_w - \rho_o)g\sin\vartheta}{\lambda_w + \lambda_o}$$

leads to

$$P(t) + \rho_w g\sin\vartheta L - t\frac{dP}{dt} = \frac{u + g\sin\vartheta\,(\rho_w\lambda_o - \rho_o\lambda_o)}{(\lambda_w + \lambda_o)}L = \frac{uf_w L}{\lambda_w}.$$

Using the definition for the mobility λ_w we obtain

$$\frac{1}{k_{rw}} = \frac{k}{\mu_w L u f_w(L,t)}\left(P(t) + \rho_w g\sin\vartheta L - t\frac{dP}{dt}\right) \tag{4.63}$$

$$\frac{1}{k_{ro}} = \frac{k}{\mu_o L u (1 - f_w(L,t))}\left(P(t) + \rho_o g\sin\vartheta L - t\frac{dP}{dt}\right), \tag{4.64}$$

where we have added the analogous expression for the oil relative permeability k_{ro}.

4.5 FINITE VOLUME APPROACH TO OBTAIN THE FINITE DIFFERENCE EQUATIONS FOR THE BUCKLEY LEVERETT PROBLEM

To derive the numerical schemes, we use here the finite volume approach [181]. We refer the reader interested in the finite element approach to [55]. The finite volume approach follows straightforwardly from the integrated mass balance equation. In many cases, however, the model equations are presented in finite difference form (obtained by converting the integrated mass balance to a differential equation). Here we describe the way "back", i.e., how to use the differential equations to obtain the numerical scheme. The approach is usually successful unless you rearranged the equation out of its conservational form. By conservational form, we mean that the structure of the equation is that the time derivative of something + the divergence of some other quantities equal a source term. In our case, the source term is zero. The equation (Eq. (4.39), repeated here as Eq. (4.65) for convenience, we need to solve reads

$$\varphi \frac{\partial S_w}{\partial t} + \frac{\partial u_w}{\partial x} = 0, \tag{4.65}$$

where we need to substitute Darcy's law for the Darcy velocity u_w. This substitution should be postponed as long as possible. We integrate over one grid block, which can be visualized as a cell of your spreadsheet. Therefore, we obtain

$$\varphi \frac{d}{dt} \int_{in}^{out} S_w dx + \int_{in}^{out} \frac{\partial u_w}{\partial x} dx = 0. \tag{4.66}$$

We define the average saturation in cell i as $S_{w,i} = \frac{1}{\Delta x} \int_{in}^{out} S_w dx$, where Δx is the length of the cell. To first-order accuracy, we obtain

$$S_{w,i}(t + \Delta t) = S_{w,i}(t) + \frac{1}{\varphi} \frac{\Delta t}{\Delta x}(u_{w,in} - u_{w,out}) \tag{4.67}$$

for the grid cell labeled i (Figure 4.22).

It is convenient in $1 - D$ problems of incompressible flow to express the Darcy water velocity u_w in terms of the total velocity u, which is irrespective of the permeability field $u = u_{inj}(t)$ independent of position and the fractional flow function. Similarly, in a radial geometry $2\pi rhu(r,t) = 2\pi r_w h u_{inj}(t) := q_{inj}(t)$, where r is the radius, r_w is the well radius, h is the layer height and $u_{inj}(t)$ and $q_{inj}(t)$ are the injection velocity and injection rate respectively. In linear geometry, we have according to Darcy's law

$$u_w = -\lambda_w (\frac{\partial P_w}{dx} + \rho_w g \sin \vartheta), \tag{4.68}$$

$$u_o = -\lambda_o (\frac{\partial P_o}{dx} + \rho_o g \sin \vartheta), \tag{4.69}$$

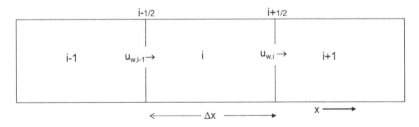

Figure 4.22 Labelling convention of flow into cell and out of cell. The properties u_w^{i-1} of and u_w^i are determined by the center of the cell from which the flow is coming. Patankar [181] indicates cell i by P, cell $(i-1)$ by W, cell $(i+1)$ by E, boundary $(i-\frac{1}{2})$ by wb and boundary $(i+\frac{1}{2})$ by eb.

where λ_α, P_α and ρ_α are the mobility, the pressure and the density of phase α respectively, ϑ is the dip angle, and g is the acceleration due to gravity. Therefore, we obtain

$$u = u_w + u_o = u_w - \lambda_o \left(\frac{\partial P_o}{\partial x} + \rho_o g \sin \vartheta \right)$$

$$= u_w - \frac{\lambda_o}{\lambda_w} \lambda_w \left(\frac{\partial P_w}{\partial x} + \rho_w g \sin \vartheta \right) - \lambda_o \left(\frac{\partial P_c}{\partial x} + (\rho_o - \rho_w) g \sin \vartheta \right). \qquad (4.70)$$

We can now resubstitute Darcy's equation to get rid of the pressure term. Rearranging the terms we can write explicitly for the Darcy's velocity of water

$$u_w = \frac{\lambda_w}{\lambda_w + \lambda_o} u - \frac{\lambda_w \lambda_o}{\lambda_w + \lambda_o} (\rho_w - \rho_o) g \sin \vartheta + \frac{\lambda_w \lambda_o}{\lambda_w + \lambda_o} \frac{\partial P_c}{\partial x} \qquad (4.71)$$

$$= u f_w(S_w) + \frac{\lambda_w \lambda_o}{\lambda_w + \lambda_o} \frac{\partial P_c}{\partial x}. \qquad (4.72)$$

We now use the so-called upstream weighting approximation (usw), which implies that $u_{w,in} = u f_{w,in}$ flows with the saturation value upstream, i.e., of cell $(i-1)$ and that $u_{w,out} = u f_{w,out}$ flows with the saturation properties of the cell considered, i.e., of cell i. For the situation considered here, it is obvious where the flow is coming from but for the $2-D$ case this may have to be determined in advance.

Remark: considering the westboundary (wb) of cell $[i = P]$, the flow $u_w = \frac{1}{2}(u_{w,W} + abs(u_{w,W}) + \frac{1}{2}(u_{w,P} - abs(u_{w,P})$, where W denotes cell $(i-1)$, automatically implements upstream weighting. In this nomenclature these cell $(i+1)$ is usually indicated as E.

A system that uses flows with the properties of the cell where it is flowing to become very unstable (you may want to try it). For the capillary pressure term, which behaves as a nonlinear diffusion term, we may or not may use upstream weighting term $\left(\frac{\lambda_w \lambda_o}{\lambda_w + \lambda_o} \right)_{eb}$. Therefore, for the flow out of the cell labeled i (through the east boundary) we obtain the numerical approximation-where eb denotes the mobility values are to be evaluated at the east boundary of cell $i = P$.

$$u_{w,i} = u f_w(S_{w,i}) + \left(\frac{\lambda_w \lambda_o}{\lambda_w + \lambda_o} \right)_{eb} \frac{P_c^{i+1} - P_c^i}{\Delta x}.$$

Including the capillary pressure term leads to an extra boundary condition so that the initial and boundary conditions are now

$$u_{w,i=0} = u_{w,\text{inj}}, \qquad (4.73)$$

$$P_c = P_{cb}, \qquad (4.74)$$

$$S_{w,i}(t = 0) = S_{wc}. \qquad (4.75)$$

The boundary condition $P_c = 0$ outside the last grid cell $i = N$, takes care of the capillary end effect, but introduces a discontinuity. This choice is physically appealing because outside the porous medium there is no capillary pressure. From the numerical point of view, a choice of $P_c = Pcb$ (see Eq. (4.9)) avoids a discontinuous capillary pressure. A full understanding of the behavior in the transition zone from inside the porous medium to outside the porous medium is beyond the scope of this book.

4.5.1   EXERCISE, NUMERICAL SOLUTION OF BUCKLEY LEVERETT PROBLEM

About implementation of the numerical solution of Buckley–Leverett problem in EXCEL.

- Use the EXCEL worksheet "Buckley" (see appendix B) to perform the numerical calculation. The description of the theory and EXCEL sheet are given in the appendix. Also understand the meaning of the input parameters in the data sheet,
- compare the analytical and numerical solution in a plot where we give the saturation versus x/t,
- compare the analytical and numerical solution in a plot of the dimensionless oil production versus the dimensionless water volume injected,
- compare the shock saturation obtained from the numerical calculations to the analytical shock saturation with the Welge tangent construction,
- what is preferable, also considering the production behavior: a water-wet medium with high residual oil saturations and low end point water permeability or an intermediate wet medium with low residual oil saturation and higher end point permeability?

4.6   VERTICAL EQUILIBRIUM AS A BASIS FOR UPSCALING OF RELATIVE PERMEABILITIES AND FRACTIONAL FLOW FUNCTIONS

The transition zone can be determined if one would be allowed to assume that the saturation distribution in the X-dip direction has come to equilibrium [63]. Equilibrium means equilibrium between gravity and capillary forces. However, gravity capillary equilibrium is, in practice, not established instantaneously, and the proposed upscaling method is theoretically incorrect. All the same a derivation of the relative permeability functions under the assumption of instantaneous vertical equilibrium is insightful and shows that upscaled oil (water) relative permeabilities tend to a more linear relationship with respect to the water saturation and the oil saturation. For convenience, we repeat the model assumptions for this case.

1. We consider a layer of length L, constant thickness $H << L$ bounded from above and below respectively by impermeable cap- and base rock. The layer can be under a dip ϑ. Injection and production occurs through a line drive configuration or through edge water drive such that the variations in the direction perpendicular to the paper can be disregarded. We say that two-dimensional flow in a vertical cross-section occurs. The dip (layer) direction is indicated by "x" and the cross-dip direction is indicated by "y" (Figure 4.1),
2. the fluids and rock are considered incompressible,
3. the permeability k and the porosity φ in the reservoir are considered constant. There exists a water saturation below which water cannot flow in the porous medium. This saturation is called the connate water saturation S_{wc}. The initial water saturation in the oil reservoir is equal to this connate water saturation. In a completely water-wet medium, water continues to flow in the water film covering the grains (see Exercise 4.1.3). The flow of water in the water film becomes so slow that it is no longer noticeable in the time frame of interest. The saturation at which this occurs can be conveniently called the connate water saturation. There exists also an oil saturation called the residual oil saturation S_{or} below which oil cannot flow, but oil does not flow in oil films and the oil is truly trapped at low saturations,
4. the water occupies the lower part of the layer and is separated by an interface of the oil that occupies the upper part of the reservoir. In the part where water flows oil is present

at residual oil saturation. In the part in which oil flows water is present at connate water saturation. The model remains two-dimensional,

5. the capillary pressure between the phases can be considered,
6. both the water- and oil-phase flow independently of each other and behave according to a generalized form of Darcy's law,
7. the pressure gradients in the water and the oil zone in the dip direction only depend on the coordinate in the dip direction. The capillary pressure $P_c(S_w) = \Delta \rho g(h(S_w) \cos \vartheta$ is in equilibrium with the gravity force, i.e., the equilibrium saturation distribution in the X-dip direction will be attained in a time that is short with respect to the rate of change of the saturation profile in the dip direction. Here the height $h(S_w)$ is with respect to the level where the capillary pressure is zero also called the phreatic level. The bottom of the reservoir adjacent to the base rock is at a height h_1 with respect to the phreatic level. This level serves as a boundary condition The VE approximation reduces the mathematical problem to a $1-D$ equation.

4.6.1 DAKE'S UPSCALING PROCEDURE FOR RELATIVE PERMEABILITIES

Dake [64], in his book proposed an upscaling procedure for relative permeabilities, which even includes a procedure for layered reservoirs. Here we describe the procedure for averaging in a single layer. As opposed to Dake, we give analytical expressions for Brooks-Corey permeabilities. We use capillary gravity equilibrium, which means that $P_c(S_w) = \Delta \rho g(h(S_w) \cos \vartheta$. If we measure the height $h(S_w)$ with respect to the level where the capillary pressure is zero and substitute Leverett's capillary pressure equation, i.e., Eq. (4.6) we obtain

$$h(S_w) = \frac{\gamma \sigma_{ow}}{\Delta \rho_{ow} g \cos \theta} \sqrt{\frac{\varphi}{k}} \left(\frac{\frac{1}{2} - S_{wc}}{1 - S_{wc} - S_{or}} \right)^{\frac{1}{\lambda}} \left(\frac{S_w - S_{wc}}{1 - S_{wc} - S_{or}} \right)^{\frac{-1}{\lambda}} := \frac{\varsigma^{\frac{1}{\lambda}}}{(S_w - S_{wc})^{\frac{1}{\lambda}}},$$

$$S_w = S_{wc} + \frac{\varsigma}{(h(S_w))^{\lambda}} , S_{we} = \frac{S_w - S_{wc}}{1 - S_{wc} - S_{or}} = \frac{1}{1 - S_{wc} - S_{or}} \frac{\varsigma}{(h(S_w))^{\lambda}}, \qquad (4.76)$$

where we usually take $\gamma = 0.5$ and ς is defined by the equation. This equation is valid for $S_{wc} \leqq S_w \leqq 1 - S_{or}$ and thus above the capillary fringe, i.e., for

$$h(S_w) \quad \geqq \frac{\gamma \sigma_{ow}}{\Delta \rho_{ow} g \cos \theta} \sqrt{\frac{\varphi}{k}} \left(\frac{\frac{1}{2} - S_{wc}}{1 - S_{wc} - S_{or}} \right)^{\frac{1}{\lambda}}. \qquad (4.77)$$

Below the capillary fringe, we will assume that $S_w = 1 - S_{or}$.

Turning to the conservation law in $2-D$, it can be written as

$$\varphi \frac{\partial S_w}{\partial t} + \frac{\partial u_{wx}}{\partial x} + \frac{\partial u_{wy}}{\partial y} = 0, \qquad (4.78)$$

which can be integrated over the height as

$$\frac{d}{dt} \int_{h_1}^{h_1+H} \varphi S_w dy + \frac{d}{dx} \int_{h_1}^{h_1+H} u_{wx} dy = 0, \qquad (4.79)$$

where h_1 denotes the height of the bottom of the layer with respect to the datum level where the capillary pressure is zero. It is assumed that the porosity is constant. First, the accumulation term is evaluated by substitution of Eq. (4.76). Eq. (4.76) is only valid above the capillary fringe, i.e., where the water saturation becomes less than $1 - S_{or}$. If we consider a layer that is entirely above the capillary fringe ($y > h_f$), we can write the average saturation as

$$\bar{S}_w = \frac{1}{H} \int_{h_1}^{h_1+H} S_w dy = S_{wc} + \frac{1}{H} \int_{h_1}^{h_1+H} \frac{\varsigma}{y^\lambda} dy$$

$$= S_{wc} - \frac{\varsigma}{H} \frac{(h_1+H)^{1-\lambda} - h_1^{1-\lambda}}{-1+\lambda}, \quad h_1 > h_f, \tag{4.80}$$

where y is the height with respect to the phreatic $P_c = 0$ level. Below $y = h_f$ the water saturation $S_w = 1 - S_{or}$, where h_f, the height of the capillary fringe, with $S_w = 1 - S_{or}$ (see Eq. (4.76)) is given by

$$h_f = \frac{\gamma \sigma_{ow}}{\Delta \rho_{ow} g \cos \theta} \sqrt{\frac{\varphi}{k}} \left(\frac{\frac{1}{2} - S_{wc}}{1 - S_{or} - S_{wc}} \right)^{\frac{1}{\lambda}}. \tag{4.81}$$

For large water saturations, we must split between a term below the capillary fringe h_f and a term above the capillary fringe, i.e.,

$$\bar{S}_w = \frac{1}{H} \int_{h_1}^{h_1+H} S_w dy = \frac{(h_f - h_1)}{H}(1 - S_{or}) + \frac{H + h_1 - h_f}{H} S_{wc} + \frac{1}{H} \int_{h_f}^{h_1+H} \frac{\varsigma}{y^\lambda} dy$$

$$= \frac{h_f - h_1}{H}(1 - S_{or}) + \frac{H + h_1 - h_f}{H} S_{wc} + \frac{\varsigma}{H} \frac{(h_1+H)^{1-\lambda} - h_f^{1-\lambda}}{1 - \lambda}, \quad h_1 < h_f. \tag{4.82}$$

Now we proceed to calculate the flow term in Eq. (4.79). First, for the VE approximation, which disregards viscous forces in the X-dip direction the pressure gradient in the flow direction is independent of the position in the cross-dip direction because

$$\frac{\partial^2 P_\alpha}{\partial y \partial x} = \frac{\partial^2 P_\alpha}{\partial x \partial y} = \frac{\partial}{\partial x}(-\rho_\alpha g \cos \vartheta) = 0.$$

In other words, one obtains

$$q_{wx} := \int_{h_1}^{h_1+H} u_{wx} dy = - \int_{h_1}^{h_1+H} \frac{kk_{rw}}{\mu_w} dy \left(\frac{\partial P_w}{\partial x} - \rho_w g \sin \vartheta \right) := -\Lambda_w \left(\frac{\partial P_w}{\partial x} - \rho_w g \sin \vartheta \right),$$

$$q_{ox} := \int_{h_1}^{h_1+H} u_{ox} dy = - \int_{h_1}^{h_1+H} \frac{kk_{ro}}{\mu_o} dy \left(\frac{\partial P_o}{\partial x} - \rho_o g \sin \vartheta \right) := -\Lambda_o \left(\frac{\partial P_o}{\partial x} - \rho_o g \sin \vartheta \right). \tag{4.83}$$

It is our purpose to express the integrated mobilities Λ_w and Λ_o in terms of h_1. Therefore one obtains, if the whole layer is above the capillary fringe (see Eqs. (4.76) and (4.16))

$$\Lambda_w = \int_{h_1}^{h_1+H} \frac{kk_{rw}}{\mu_w} dy = \frac{kk'_{rw}}{\mu_w} \int_{h_1}^{h_1+H} S_{we}^{\frac{2+3\lambda}{\lambda}} dy = \frac{kk'_{rw}}{\mu_w} \left(\frac{\varsigma}{1 - S_{wc} - S_{or}} \right)^{\frac{2+3\lambda}{\lambda}} \int_{h_1}^{h_1+H} \left(\frac{1}{y^\lambda} \right)^{\frac{2+3\lambda}{\lambda}} dy$$

$$= \frac{kk'_{rw}}{\mu_w} \left(\frac{\varsigma}{1 - S_{wc} - S_{or}} \right)^{\frac{2+3\lambda}{\lambda}} \frac{\left(\frac{1}{h_1} \right)^{3\lambda+1} - \left(\frac{1}{h_1+H} \right)^{3\lambda+1}}{(3\lambda + 1)}, \tag{4.84}$$

where we use that

$$\int_{h_1}^{h_1+H} \frac{1}{y^{2+3\lambda}} dy = \frac{\left(\frac{1}{h_1}\right)^{3\lambda+1} - \left(\frac{1}{h_1+H}\right)^{3\lambda+1}}{(3\lambda+1)}. \tag{4.85}$$

One obtains for $h_1 < h_f$, and $h_1 + H < h_f$.

$$\int_{h_1}^{h_1+H} \frac{kk_{rw}}{\mu_w} dy = (h_f - h_1)\frac{k'_{rw}}{\mu_w} + \frac{kk'_{rw}}{\mu_w}\left(\frac{\varsigma}{1-S_{wc}-S_{or}}\right)^{\frac{2+3\lambda}{\lambda}}$$

$$\times \frac{\left(\frac{1}{h_f}\right)^{3\lambda+1} - \left(\frac{1}{h_1+H}\right)^{3\lambda+1}}{(3\lambda+1)}. \tag{4.86}$$

For $h_1 + H \geqq h_f$ we obtain

$$\Lambda_w = \int_{h_1}^{h_1+H} \frac{kk_{rw}}{\mu_w} dy = H\frac{kk'_{rw}}{\mu_w}. \tag{4.87}$$

In the same way, one obtains for the oil mobility Λ_o and for $h_1 > h_f$ (see Eqs. (4.16) and Eq. (4.76))

$$\Lambda_o = \int_{h_1}^{h_1+H} \frac{kk_{ro}}{\mu_o} dy = \frac{kk'_{ro}}{\mu_o} \int_{h_1}^{h_1+H} (1-S_{we})^2 (1-S_{we}^{\frac{2+\lambda}{\lambda}}) dy$$

$$= \frac{kk'_{ro}}{\mu_o} \int_{h_1}^{h_1+H} \left(1 - \frac{\varsigma}{y^\lambda(1-S_{wc}-S_{or})}\right)^2 \left(1 - \left(\frac{\varsigma}{y^\lambda(1-S_{wc}-S_{or})}\right)^{\frac{2+\lambda}{\lambda}}\right) dy. \tag{4.88}$$

The integrand can be factorized, i.e.,

$$\Lambda_o = \frac{kk'_{ro}}{\mu_o} \int_{h_1}^{h_1+H} (1 - \frac{2\varsigma}{y^\lambda(1-S_{wc}-S_{or})} + \frac{\varsigma^2}{y^{2\lambda}(1-S_{wc}-S_{or})^2})$$

$$\times \left(1 - \left(\frac{\varsigma}{1-S_{wc}-S_{or}}\right)^{\frac{2+\lambda}{\lambda}} \frac{1}{y^{2+\lambda}}\right) dy.$$

This can be evaluated as

$$\Lambda_o = \frac{kk'_{ro}}{\mu_o} \int_{h_1}^{h_1+H} (1 - \frac{2\varsigma}{y^\lambda(1-S_{wc}-S_{or})} + \frac{\varsigma^2}{y^{2\lambda}(1-S_{wc}-S_{or})^2}) dy$$

$$- \frac{kk'_{ro}}{\mu_o} \int_{h_1}^{h_1+H} (1 - \frac{2\varsigma}{y^\lambda(1-S_{wc}-S_{or})} + \frac{\varsigma^2}{y^{2\lambda}(1-S_{wc}-S_{or})^2})$$

$$\times \left(\left(\frac{\varsigma}{1-S_{wc}-S_{or}}\right)^{\frac{2+\lambda}{\lambda}} \frac{1}{y^{2+\lambda}}\right) dy. \tag{4.89}$$

This equation can be expanded as

$$\Lambda_o = \frac{kk'_{ro}}{\mu_o} \int_{h_1}^{h_1+H} \left(1 - \frac{2\varsigma}{y^\lambda (1 - S_{wc} - S_{or})} + \frac{\varsigma^2}{y^{2\lambda} (1 - S_{wc} - S_{or})^2}\right) dy - \frac{kk'_{ro}}{\mu_o}$$

$$\times \int_{h_1}^{h_1+H} \left(\left(\frac{\varsigma}{1 - S_{wc} - S_{or}}\right)^{\frac{2+\lambda}{\lambda}} \frac{1}{y^{2+\lambda}} - \frac{2}{y^{2+2\lambda}} \left(\frac{\varsigma}{1 - S_{wc} - S_{or}}\right)^{\frac{2+2\lambda}{\lambda}}\right.$$

$$\left. + \left(\frac{\varsigma}{1 - S_{wc} - S_{or}}\right)^{\frac{2+3\lambda}{\lambda}} \frac{1}{y^{2+3\lambda}}\right) dy. \tag{4.90}$$

Upon integration one finds

$$\Lambda_o = \frac{kk'_{ro}}{\mu_o} \left[\left(y - \frac{2\varsigma}{(1-\lambda)y^{\lambda-1}(1 - S_{wc} - S_{or})} + \frac{\varsigma^2}{(1-2\lambda)y^{(2\lambda-1)}(1 - S_{wc} - S_{or})^2}\right)\right]_{y=h_1}^{y=h_1+H}$$

$$+ \frac{kk'_{ro}}{\mu_o} \left[-\left(\frac{\varsigma}{1 - S_{wc} - S_{or}}\right)^{\frac{2+\lambda}{\lambda}} \frac{1}{(\lambda+1)y^{1+\lambda}} + \frac{2}{(2\lambda+1)y^{1+2\lambda}} \left(\frac{\varsigma}{1 - S_{wc} - S_{or}}\right)^{\frac{2+2\lambda}{\lambda}}\right]_{y=h_1}^{y=h_1+H}$$

$$- \left[\left(\frac{\varsigma}{1 - S_{wc} - S_{or}}\right)^{\frac{2+3\lambda}{\lambda}} \frac{1}{(3\lambda+1)y^{1+3\lambda}}\right]_{y=h_1}^{y=h_1+H}. \tag{4.91}$$

For $h_1 < h_f$ one obtains

$$\Lambda_o = \frac{kk'_{ro}}{\mu_o} \left[\left(y + \frac{2\varsigma}{(1-\lambda)y^{\lambda-1}(1 - S_{wc} - S_{or})} + \frac{\varsigma^2}{(1-2\lambda)y^{(2\lambda-1)}(1 - S_{wc} - S_{or})^2}\right)\right]_{y=hf}^{y=h_1+H}$$

$$+ \frac{kk'_{ro}}{\mu_o} \left[-\left(\frac{\varsigma}{1 - S_{wc} - S_{or}}\right)^{\frac{2+\lambda}{\lambda}} \frac{1}{(\lambda+1)y^{1+\lambda}} + \frac{2}{(2\lambda+1)y^{1+2\lambda}} \left(\frac{\varsigma}{1 - S_{wc} - S_{or}}\right)^{\frac{2+2\lambda}{\lambda}}\right]_{y=hf}^{y=h_1+H}$$

$$- \left[\left(\frac{\varsigma}{1 - S_{wc} - S_{or}}\right)^{\frac{2+3\lambda}{\lambda}} \frac{1}{(3\lambda+1)y^{1+3\lambda}}\right]_{y=hf}^{y=h_1+H}. \tag{4.92}$$

The resulting layer averaged (upscaled) permeability is shown in Figure 4.23.
It is noted that the upscaled relative permeabilities tend more to straight lines. The definition of a fractional flow function $\mathcal{F}_w$ is useful, i.e.,

$$Q = q_{wx} + q_{ox} = q_{wx} - \frac{\Lambda_o}{\Lambda_w}\Lambda_w\left(\frac{\partial P_w}{\partial x} - \rho_w g \sin\vartheta\right) - \Lambda_o\left(\frac{\partial P_c}{\partial x} + (\rho_w - \rho_o)g\sin\vartheta\right) \tag{4.93}$$

$$\mathcal{F}_w = \frac{q_{wx}}{Q} = \frac{\Lambda_w}{\Lambda_w + \Lambda_o} + \frac{\Lambda_w\Lambda_o}{\Lambda_w + \Lambda_o}\frac{1}{Q}\left(\frac{\partial P_c}{\partial x} + (\rho_w - \rho_o)g\sin\vartheta\right). \tag{4.94}$$

4.6.2 EXERCISE, SORTING FACTOR DEPENDENCE

About sorting factor dependence of vertically averaged relative permeability; upscaling of Constitutive Relations using Vertical Equilibrium

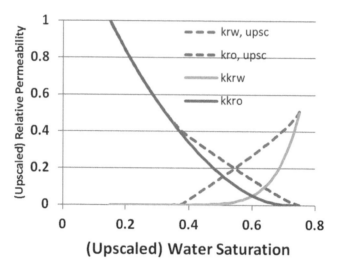

(Upscaled) Water Saturation

Figure 4.23 Upscaled (layer averaged) relative permeabilities compared to Brooks-Corey relative permeabilities ($\mu_o = \mu_w = 10^{-3}$[Pa.s], $k'_{rw} = 0.5, k'_{ro} = 1, S_{wc} = 0.15, S_{or} = 0.25, \lambda = 0.4, \gamma = 0.5, \sigma_{ow} = 0.03$[N/m$, \Delta\rho_{ow} = 200$[kg/m^3], $g\cos\vartheta = 9.81$m/s^2).

We consider a layer with a permeability k of 20 mDarcy and a height H of 5 m. The distance between the row of injection and the row of production wells is 200 m. First, we consider the situation where the reservoir is horizontal. The porosity φ is 0.3 [-]. The density difference $\Delta\rho$ between water and oil is [150kg/m^3], the acceleration due to gravity g is 9.81[m/s^2]. Make a choice of reasonable values for the viscosities of oil μ_o and water μ_w. The interfacial tension σ_{ow} between oil and water is 30×10^{-3}N/m. Include a value for the sorting factor λ, which assumes values between 0.2 (poor sorting) and very high values, e.g., 7 for well sorted sands. The connate water saturation $S_{wc} = 0.15$ and the residual oil saturation S_{or} is also 0.15. The total Darcy velocity $u = 1$ m/d.

Make a named list of variables in EXCEL by using the "Formulas-Create from Selection" command. We use the so-called Corey relative permeabilities (see Eq. (4.16) and the capillary pressures used in Eq. (4.8).

- Plot the over the height averaged relative permeability of oil and water versus the over the height averaged saturation,
- plot the averaged fractional flow function f_w versus the averaged saturation.

4.6.3   HOPMANS'S FORMULATION

We can also formulate the two-phase flow equations in terms of the wetting phase pressure and the non-wetting phase pressure [152]. By way of example, we consider the case that oil and water are injected from above at their respective pressures $P_{w,t}$ and $P_{o,t}$ and that oil and water are produced below at the respective pressures at the bottom $P_{w,b}$ and $P_{o,b}$. In other words, we consider vertical downward flow, where z increases in the downward direction. We assume incompressible flow. The capillary pressure $P_o - P_w = P_c$ satisfies the Brooks-Corey relation, i.e., Eq. (4.8), which can be conveniently simplified to

$$P_c = \sigma_{ow}\gamma\sqrt{\frac{\varphi}{k}}\left(\frac{\frac{1}{2} - S_{wc}}{1 - S_{wc}}\right)^{\frac{1}{\lambda}}\left(\frac{S_w - S_{wc}}{1 - S_{wc}}\right)^{\frac{-1}{\lambda}} := P_{cb}\left(\frac{S_w - S_{wc}}{1 - S_{wc}}\right)^{\frac{-1}{\lambda}},$$

which makes it easy to invert, i.e.,

$$S_w = S_{wc} + (1 - S_{wc}) \left(\frac{P_{cb}}{P_c} \right)^\lambda$$

$$\frac{dS_w}{dP_c} = -\lambda (1 - S_{wc}) \lambda \left(\frac{P_{cb}}{P_c} \right)^\lambda \frac{1}{P_c}. \tag{4.95}$$

The relative permeabilities can be straightforwardly expressed in terms of the capillary pressure instead of the saturation, i.e., $k_{rw}(P_c)$, $k_{ro}(P_c)$ as a function of the capillary pressure is given. Therefore, we can rewrite equations (4.25) for $\alpha = w, o$ as

$$\varphi \frac{dS_w}{dP_c} \frac{\partial P_c}{\partial t} = -\frac{\partial u_w}{\partial z}, \quad \varphi \frac{dS_o}{dP_c} \frac{\partial P_c}{\partial t} = -\frac{\partial u_o}{\partial z}$$

We use that $P_c = P_o - P_w$ and obtain the water equation and the oil equation

$$\varphi \frac{dS_w}{dP_c} \frac{\partial (P_o - P_w)}{\partial t} = -\frac{\partial u_w}{\partial z},$$

$$-\varphi \frac{dS_w}{dP_c} \frac{\partial (P_o - P_w)}{\partial t} = -\frac{\partial u_o}{\partial z},$$

where we use that $dS_o/dP_c = -dS_w/dP_c$.

Application of Darcy's law Eq. (4.1) leads to

$$\varphi \frac{dS_w}{dP_c} \frac{\partial (P_o - P_w)}{\partial t} = \frac{\partial}{\partial z} \left(\frac{kk_{rw}}{\mu_w} \left(\frac{\partial P_w}{\partial z} - \rho_w g \right) \right),$$

$$-\varphi \frac{dS_w}{dP_c} \frac{\partial (P_o - P_w)}{\partial t} = \frac{\partial}{\partial z} \left(\frac{kk_{ro}}{\mu_o} \left(\frac{\partial P_o}{\partial z} - \rho_o g \right) \right),$$

where we changed the plus sign in a minus sign to convert from upward flow to downward flow.

As initial conditions we can assume capillary gravity equilibrium

$$P_o = P_o^0 - \rho_o g (z - z_0)$$

$$P_w = P_w^0 - \rho_w g (z - z_0).$$

At $t = 0$, we change the pressure at the top and at the bottom to allow for a sudden change of saturation at the top. A transient capillary pressure profile develops, from which the saturation can be calculated. This formulation is easily implemented in COMSOL. The advantage is that it both incorporates capillary pressure and gravity in two-phase flow calculations in porous media.

4.7 PHYSICAL THEORY OF INTERFACE MODELS

4.7.1 DERIVATION OF INTERFACE EQUATION OF MOTION AND PRODUCTIONS FOR SEGREGATED FLOW

In many cases of practical interest is the transition zone length of two immiscible fluids (or miscible fluids) short with respect to the height of the reservoir. Under these conditions, we can assume that the flow of each fluid is confined to a well-defined part of the reservoir with an "abrupt" interface that separates both fluids from each other [61, 63, 91, 116]. It can also be used for stratified reservoirs [126]. This approximation is not only used for oil recovery by water but also for a description of the salt-freshwater interface near coastal area's. The transition zone for oil water flow is small for large gravity and small capillary forces. We repeat our assumptions stated above to make this description self-consistent. The one extra assumption for the "Interface Model" with respect to the "Buckley–Leverett model" is that conditions are such that gravity forces are dominant with respect to capillary forces. Consequently, the oil has been segregated into a zone with connate water saturation from a zone with water at residual oil saturation below. These conditions often occur in practice. Segregation is more complete in reservoirs with high permeabilities and large layer thicknesses.

Practical implications are, however, that the interface model provides a worst (as to recovery) case scenario. Numerical dispersion in most reservoir simulations often presents an over-optimistic picture of non-segregated (dispersed) oil and water when core-derived relative permeabilities are used. Setting the vertical grid size dimensions in numerical simulations such that the short transition zone is modeled adequately will be often computationally expensive. In these cases, it is suggested to use a linear relationship between relative permeability and saturation, i.e., use exponent $n_w = n_o = 1$ in the relative permeability relations (4.15), irrespective of measurements on cores. We summarize again the model assumptions,

1. We consider a layer of length L, constant thickness $H << L$ bounded from above and below respectively by impermeable cap- and base rock. The layer can be under a dip ϑ. Injection and production occurs through a line drive configuration or through edge water drive such that two-dimensional flow in a vertical cross-section occurs. The dip (layer) direction is indicated by "x" and the cross-dip direction is indicated by "y" (Figure 4.1),
2. the fluids and rock are considered incompressible,
3. the permeability and the porosity in the reservoir are considered constant. There exists a water saturation below which water cannot flow in the porous medium. This saturation is called the connate water saturation S_{wc}. The initial water saturation in the oil reservoir is equal to this connate water saturation. There exists also an oil saturation called the residual oil saturation S_{or} below which oil cannot flow,
4. the water occupies the lower part of the layer and is separated by an interface from the oil that occupies the upper part of the reservoir. In the part where water flows oil is present at residual oil saturation. In the part in which oil flows water is present at connate water saturation. The model remains two-dimensional,
5. the capillary pressure between the phases in particular at the interface are disregarded,
6. the oil and water phases behave both according to the (one-phase) Darcy's law, albeit with end point permeabilities, i.e., respectively at connate water saturation and residual oil saturation,
7. the pressure gradients in the water and the oil zone in the dip direction only depend on the coordinate in the dip direction and is thus constant in the cross-dip direction. This is the Dupuit approximation or Vertical Equilibrium approximation. Pressure continuity at the interface and the condition of hydrostatic equilibrium (disregarding viscous forces) in the X-dip direction relates the water and oil pressure gradients. It reduces the mathematical problem to a $1 - D$ equation.

Remark: this latter assumption is not an essential feature of an interface model. It helps in reducing to a $1 - D$ problem, which is convenient for computational purposes.

4.7.2   STATIONARY INTERFACE (MOBILITY NUMBER < GRAVITY NUMBER +1)

It turns out that a semi-stationary solution can only be found if the mobility ratio M is smaller than the gravity number G plus one, i.e., $M < G + 1$. The mobility ratio M, describes the ratio between the viscous forces in the oil flowing part of the reservoir divided by the viscous forces in

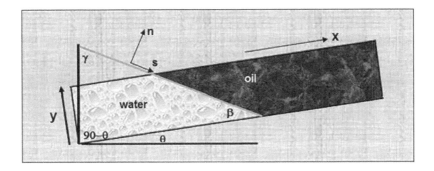

Figure 4.24 Definition of angles for the calculation of the slope of a stationary interface (M<G+1).

the water flowing part. The gravity number G is the ratio between gravity forces and the viscous forces in the flowing water-filled part of the reservoir.

We consider in Figure 4.24 a dipping layer, which makes an angle ϑ with the horizontal. The layer has a constant thickness H and a length L. The reservoir is originally filled with oil and water at connate water saturation. At time $t = 0$ water is injected downstream. The dimensions of the reservoir are such that the height of the capillary transition zone can be disregarded. We thus consider segregated flow. We shall show that for stable displacement the angle of the interface is constant, i.e., β =constant.

The derivation of the inclination of the interface is based upon:

- The equilibrium condition, i.e., the pressure is continuous on the interface. The fact that at every point of the interface the pressure in the oil equals the pressure in the water implies that the pressure gradient along the interface (s-direction) in both fluids is equal. This equilibrium condition was for the first time proposed by Edelman [84], i.e.,

$$\left(\frac{\partial P_o}{\partial s}\right) = \left(\frac{\partial P_w}{\partial s}\right), \tag{4.96}$$

- Darcy's law for single-phase flow in the domain of flowing water and in the domain of flowing oil,
- the existence of the quasi-steady state. The flow must be thus in the dip direction and it must be uniform as otherwise the interface is not quasi-stationary and it changes its shape.

Before we substitute Darcy's law, we must first express the angle γ in terms of ϑ and β. From Figure 4.24 follows that

$$\gamma + \beta + 90° - \vartheta = 180° \rightarrow \gamma = 90° - \beta + \vartheta.$$

We also observe that Δs has a component in the z-direction, which we denote by Δz

$$\Delta z = -\Delta s \cos \gamma = -\Delta s \sin(\beta - \vartheta).$$

Therefore, we can express Darcy's law with the potential $(\phi = p + \rho g z)$

$$u_{s\alpha} = -\frac{k_\alpha}{\mu_\alpha}\left(\frac{\partial p}{\partial s} - \rho_\alpha g \sin(\beta - \vartheta)\right),$$

where α denotes the phases (α=oil,water). This can be rewritten as

$$\frac{\partial p}{\partial s} = -\frac{\mu_\alpha}{k_\alpha} u_{s\alpha} + \rho_\alpha g \sin(\beta - \vartheta).$$

Substituted into the equilibrium condition (4.96) leads to

$$-\frac{\mu_w}{k_w} u_{sw} + \rho_w g \sin(\beta - \vartheta) = -\frac{\mu_o}{k_o} u_{so} + \rho_o g \sin(\beta - \vartheta).$$

We consider uniform flow in the x-direction and thus

$$u_s = u_x \cos(\beta).$$

Moreover we use that: $\sin(\beta - \vartheta) = \sin\beta \cos\vartheta - \cos\beta \sin\vartheta$, and thus

$$-\frac{\mu_w}{k_w} u_{xw} + \frac{\mu_o}{k_o} u_{xo} = -g\Delta\rho_{wo} \sin\vartheta \left(\frac{\tan\beta}{\tan\vartheta} - 1\right).$$

For a stationary interface, we must have that $u_{xw} = u_{xo} = u$ and we can write the equation with $dy/dx = -\tan\beta$ as

$$\left(\frac{\mu_o}{k_o} - \frac{\mu_w}{k_w}\right)u = g\Delta\rho_{wo} \sin\vartheta \left(\frac{dy/dx}{\tan\vartheta} + 1\right).$$

Furthermore, we define the dimensionless numbers:

- $M=$ mobility of displacing fluid (water) /mobility of displace fluid (oil) $= \frac{k_w/\mu_w}{k_o/\mu_o}$ (Mobility ratio),
- $G=$ gravity/viscous force$= \frac{g\Delta\rho_{wo}\sin\vartheta}{u\mu_w/k_w}$ (Gravity number).

When calculating the gravity number, one must note that k_w is the permeability of water and thus $k_w = kk'_{rw}$ where k'_{rw} is the relative end point permeability.

Substitution of the dimensionless numbers lead to an expression of the slope of the interface

$$\frac{dy}{dx} = \frac{M - 1 - G}{G} \tan(\vartheta). \tag{4.97}$$

It is obvious that for a stable interface the condition $dy/dx \leq 0$ must hold; otherwise the water-phase is above the oil-phase. In other words, $M \leq 1 + G$. This means that we can achieve a stable displacement with a small value of M. High viscosity of the displacing fluid is thus useful for efficient oil recovery. This is thus also the reason that one adds viscosifying agents (polymers) to the injection water: it lowers the mobility ratio. It is, therefore, practical to first calculate the mobility ratio to assess whether the displacement process is stable. $M < 1$ is a favorable mobility ratio, $M > 1$ is an unfavorable mobility ratio. From Eq. (4.97) one can quantify the condition for which a stationary interface can in theory be achieved, i.e., $dy/dx = 0 \rightarrow (M-1) = G$. We like to emphasize that in this case the stationary interface will only be formed after an infinitely long time. For $M > G+1$, a stationary interface cannot occur. The highest velocity at which the interface can become stationary (even if this is after an infinitely long time) is called the critical velocity u_{cr}. From the theory of viscous fingering (outside the scope of this monograph), it appears that the condition ($M \leq G+1$) can be derived as a condition of stability. Therefore the critical velocity can be interpreted as the velocity at which the interface is just stable. Therefore, one uses both the terms stability and stationarity of the interface. It is a matter of taste whether one wishes to use the term stability in this case. Instability means sensitive to small perturbations, i.e., viscous fingering, which only occurs in the absence of gravity.

For the critical velocity, we have

$$u_{cr} = \Delta\rho_{wo}g\sin\vartheta \Big/ \left(\frac{\mu_o}{k_o} - \frac{\mu_w}{k_w}\right).$$

When u_{cr} would be negative ($M < 1$) then this means (for our case $\rho_w > \rho_o$)) unconditional stability. In this case, indeed both the gravity forces and the viscous forces act to increase stability. If we would reverse the flow direction (negative velocity), then we would have destabilizing viscous forces leading again to a critical velocity. For unfavorable situations ($M > 1$) $\mu_o \gg \mu_w$ there will always be a positive value for u_{cr}. The critical velocity leads to an elucidating expression of the gravity number

$$G = \frac{u_{cr}(M-1)}{u}.$$

4.7.3 EXERCISE, INTERFACE ANGLE CALCULATIONS

About interface angle calculations for $M < G+1$.

We consider a layer with a distance between the row of injection and the row of production wells of 400 m. The reservoir dip is $5°$. The permeability k of the reservoir is 5 Darcy and the height H is 20 m. The porosity φ is 0.3 [-]. The density difference $\Delta\rho$ between water and oil is [200kg/m^3], the acceleration due to gravity g is 9.81[m/s^2]. We use for the viscosity of oil $\mu_o = 0.001$[Pa.s] and for the water viscosity $\mu_w = 0.001$[Pa.s] . The interfacial tension σ_{ow} between oil and water is 30×10^{-3} [N/m].

The connate water saturation $S_{wc} = 0.15$ and the residual oil saturation S_{or} is also 0.15. The total Darcy velocity $u = 1$ m/d. The end point relative permeabilities are 0.5 and 0.95 for water and oil respectively. Make a named list of variables in EXCEL by using the "Formulas-Create from Selection" command.

- Calculate the mobility ratio M and the gravity number G, i.e.,

$$M = \frac{kk'_{rw}}{\mu_w}\frac{\mu_o}{kk'_{ro}} \qquad G = \frac{kk'_{rw}g\Delta\rho\sin\vartheta}{u\mu_w}. \tag{4.98}$$

- calculate the angle that the water-oil interface will eventually make with respect to the layer direction,
- calculate the angle that the water-oil interface would eventually make if the layer were horizontal ($\vartheta = 0$).
  Interpret the formula

$$dy/dx = (M - 1 - G)/G, \tag{4.99}$$

where $M = (k'_{rw}/\mu_w)/(k'_{ro}/\mu_o)$ and $G = kk'_{rw}\Delta\rho/(u\mu_w)$,
- show that viscous fingering can occur in a horizontal reservoir for $M > 1$ and for $M > G+1$ in a tilted reservoir.
- for $M > G+1$ we have water above oil; $\rightarrow$ unstable,
- $\Delta S\partial_t h = \partial_x Q_w = Q\partial_x(\Lambda_w/(\Lambda_w + \Lambda_o))$, where Λ_w, Λ_o are the mobilities integrated over the height; The numerical solution with EXCEL is completely analogous to the solution for the Buckley–Leverett problem.

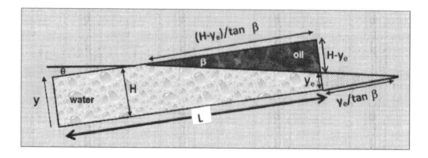

Figure 4.25 Geometric considerations for the recovery calculations of a stationary interface.

4.7.4 PRODUCTION BEHAVIOR FOR STATIONARY SOLUTION, I.E., $M < G + 1$

The production behavior in the case that $M < G + 1$ can be found form simple geometric considerations. We denote the width, length and height of the reservoir with w, L, en H. This means that the total movable oil volume is equal to $wHL\varphi(1 - S_{or} - S_{wc})$. For the situation shown in Figure 4.25, the volume swept by water can be expressed by the dimensionless production related to the total flowing pore volume in the reservoir. It is indeed equal to the total initial movable oil volume minus the movable pore volume of oil still present in the upper rectangular triangle in the reservoir

$$N_{pD} = 1 - \frac{(H - y_e)^2}{2HL\tan\beta}.$$

The amount of water that was injected up till now can be determined by ignoring the presence of the production well. The volume of water that flows beyond the well is given by $\frac{y_e}{2}\frac{y_e}{\tan\beta}w\varphi(1 - S_{wc} - S_{or})$. The cumulative water injection is therefore

$$W_{iD} = N_{pD} + \frac{y_e^2}{2HL\tan\beta}.$$

At breakthrough $(y_e = 0)$ we have

$$N_{pD,bt} = 1 - \frac{H}{2L\tan\beta}.$$

In this derivation, we have assumed that $\tan\beta > H/L$.

4.8 NON-STATIONARY INTERFACE

We want to describe the interface behavior without invoking an assumption of stationarity. This description is therefore valid for both conditions $M < G + 1$ and $M > G + 1$. For $M < G + 1$ we find a shock-like behavior, somewhat spread out by diffusion. This solution tends to the stationary solution found above. For $M > G + 1$, we find rarefaction behavior, i.e., the interface spreads out, and no stationary solution is attained.

We consider in Figure 4.26 a dipping layer, which makes an angle ϑ with the horizontal and has a height H and a length L. The reservoir is originally filled with oil and water at the connate water saturation. At time $t = 0$ water is injected upstream. The dimensions of the reservoir are such that the height of the capillary transition zone is negligible We thus consider segregated flow. Also we consider the non-stationary case for which $M > G + 1$ i.e., the slope of the interface is no longer constant.

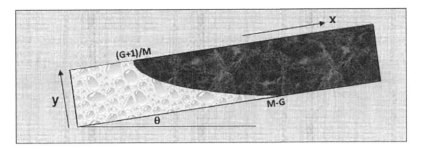

Figure 4.26 Oil displacement by water according to interface models with dominating viscous forces and unfavorable mobility ratios. The top of the interface moves at a speed proportional to $(G+1)/M$, whereas the bottom of the interface moves at a speed proportional to $M-G$, where M and G are the mobility ratio and the gravity number respectively.

The derivation of the shape of the interface depends on

- The incompressibility of the fluids, i.e., the volume balance Eq. (4.27),
- neglecting capillary pressure between the phases,
- the Dupuit-approximation, i.e., we disregard the viscous forces in the X-dip direction. Because we also disregard the capillary forces this means hydrostatic equilibrium in the X-dip direction,
- the approximate equilibrium condition, i.e., from the exact equilibrium condition that the pressure derivative along the interface on both sides of the interface is equal leads with the help of the Dietz-Dupuit approximation to an approximate equilibrium condition for the pressure derivatives in the flow direction,
- Darcy's Law.

4.8.1   THE VOLUME BALANCE IN THE FORM OF AN INTERFACE EQUATION

We can derive an interface equation of motion for incompressible flows. The procedure is to integrate Eq. (4.27) over the height to obtain

$$\varphi \int_0^H \frac{\partial S_\alpha}{\partial t} dy + \int_0^H \mathbf{div}\, u_\alpha dy = 0. \tag{4.100}$$

We evaluate first the accumulation term. Because the integration boundaries are constant (see also Eq. (4.29)), integration and differentiation can be swapped. Subsequently, we write the integral as a sum of an integral from the base rock to the interface and an integral from the interface to the caprock,

$$\varphi \int_0^H \frac{\partial S_\alpha}{\partial t} dy = \varphi \frac{d}{dt} \int_0^H S_\alpha dy$$
$$= \varphi \frac{d}{dt} \int_0^{h(x,t)} S_\alpha dy + \varphi \frac{d}{dt} \int_{h(x,t)}^H S_\alpha dy. \tag{4.101}$$

We apply Leibnitz's rule (4.29) and note that for the water-phase we have the following saturation values: for $0 \le x < h(x,t)$ we have that $S_w = 1 - S_{or}$ and for $h(x,t) < x \le H$ the water saturation is equal to the connate water saturation, i.e., $S_w = S_{wc}$. Therefore, one finds

$$\varphi \frac{d}{dt} \int_0^{h(x,t)} S_w dy + \varphi \frac{d}{dt} \int_{h(x,t)}^{H} S_w dy$$

$$= \varphi \frac{\partial h(x,t)}{\partial t}(1 - S_{or}) - \varphi \frac{\partial h(x,t)}{\partial t} S_{wc}$$

$$= \varphi(1 - S_{wc} - S_{or}) \frac{\partial h(x,t)}{\partial t}. \qquad (4.102)$$

For the convective part of Eq. (4.100) one obtains

$$\int_0^H \mathbf{div}\, \mathbf{u_w} dy = \int_0^H (\frac{\partial u_{wx}}{\partial x} + \frac{\partial u_{wy}}{\partial y}) dy. \qquad (4.103)$$

The first integral on the right side can be written as (exchange differentiation and integration) the partial derivative of the total discharge $[m^2/s]$ $Q_{wx} := \int_0^H u_{wx} dy$ to x. The second integral of the right side leads to $u_{wy}|_0^H = 0$. Substitution of the results of Eqs. (4.102) and (4.103) into Eq. (4.100) leads to the following equation of motion of the interface, i.e.,

$$\varphi(1 - S_{wc} - S_{or}) \frac{\partial h(x,t)}{\partial t} + \frac{\partial Q_{wx}}{\partial x} = 0. \qquad (4.104)$$

The same equation can be more easily found based on physical intuition by considering the mass (volume) balance over a control volume. (Figure 4.27). The hatched control volume below receives $Q_w(x)$. It releases $(Q_w(x + \Delta x))$. The difference between these flows is used to displace the interface, i.e., contributes the additional amount of water in the volume, indicated in Figure 4.27 by $\Delta x(h(t + \Delta t) - h(t))$. This (2D) volume can be approximated by a parallelogram. The surface area of the parallelogram is equal to $\Delta x(h(t + \Delta t) - h(t))$. The porosity is φ and the saturation in the parallelogram becomes one minus the residual oil saturation $1 - S_{or}$ instead of the original connate water saturation S_{wc}. Therefore the amount of water displaced per unit time is equal to $\varphi(1 - S_{wc} - S_{or})\varphi \frac{\partial h}{\partial t}\Delta x$. This leads then again to Eq. (4.104). Intuitive derivations that involve moving interfaces are error-prone. A mistake is easily made. It is preferred to validate an intuitive derivation with a rigorous mathematical derivation.

It is emphasized that Eq. (4.104) is found by the assumption that an interface exists and that the fluids are incompressible. It is valid independent of other assumptions, e.g., Darcy's law or the Dupuit approximation.

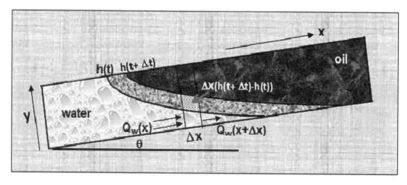

Figure 4.27 Control volume for the derivation of the mass balance equation for the non-stationary interface.

4.8.2 DIETZ-DUPUIT-APPROXIMATION

The Dupuit-approximation (vertical equilibrium) presupposes that viscous forces in the X-dip direction can be neglected. As capillary forces and gravity forces come to equilibrium, we obtain Eq. (4.14). When we additionally disregard capillary forces as we do for interface models, we have hydrostatic equilibrium ($p + \rho g y \cos(\vartheta)$ = constant) in the X-dip direction. A corollary of this is that the pressure gradient in the layer direction $\frac{\partial p}{\partial x}$ does not vary in the cross-dip direction, i.e., $\frac{\partial^2 p}{\partial x \partial y} = 0$. If one likes to read a more sophisticated approach of vertical equilibrium, we like to refer the reader to reference [255]. A seemingly straightforward corollary of hydrostatic equilibrium in the X-dip direction is that also the Darcy velocity in the X-dip direction would be zero. Such a corollary would be completely incorrect. Following reference [255], we can understand the reason for this. It appears that in the dimensionless equation the effective X-dip permeability is an order (L/H) larger than the "effective" permeability in the layer direction. The reason for being able to neglect the "effective permeability" in the X-dip direction is that the viscous force $\mu u/k_{eff}$ in the X-dip direction is small because of the high value of k_{eff} and thus does not imply that the Darcy-velocity u in the X-dip direction is zero.

   The Dupuit approximation leads to a substantial simplification for the calculation of the discharge Q_{wx}. Indeed, the computation of

$$Q_{wx} = \int_0^H u_{wx} dy = \int_0^{h(x,t)} u_{wx} dy$$
$$= - \int_0^{h(x,t)} \frac{k_w}{\mu_w} \left(\frac{\partial p}{\partial x} + \rho_w g \sin(\vartheta) \right) dy \qquad (4.105)$$

when $\frac{\partial p}{\partial x}$ depends on y is not easily possible while the corollary of the Dupuit approximation that $(\frac{\partial p}{\partial x} + \rho_w g \sin(\vartheta))$ constant is in de "cross-dip" direction "y" leads to

$$Q_{wx} = - \int_0^{h(x,t)} \frac{k_w}{\mu_w} dy \left(\frac{\partial p}{\partial x} + \rho_w g \sin(\vartheta) \right),$$

which as will be shown below lends itself to analytical solutions.

   The one-time application of disregarding the viscous forces in the "cross-dip" direction is also known as the Dupuit-approximation. We need to derive also a relation between the potential gradient in the dip direction $(\frac{\partial p}{\partial x} + \rho_w g \sin(\vartheta))$ below the interface and the gradient $(\frac{\partial p}{\partial x} + \rho_o g \sin(\vartheta))$ above the interface. De Josselin de Jong [68] derives the condition from the pressure equilibrium relation at the interface (4.96) and continuity of the Darcy velocities perpendicular to the interface. This leads to a rather complicated model equation. Here we derive this equation from the equilibrium Eq. (4.96) on the interface and subsequently we disregard once again the viscous forces in the cross-dip direction. This double application of disregarding the viscous forces in the X-dip direction is known as the Dietz-Dupuit-approximation. It leads to a relatively simple model equation.

4.8.3 APPROXIMATE EQUILIBRIUM EQUATION

In the equilibrium Eq. (4.96), we can write the coordinate along the interface s in terms of the coordinates x and y:

$$\frac{\partial P_o}{\partial x} \frac{\partial x}{\partial s} + \frac{\partial P_o}{\partial y} \frac{\partial y}{\partial s} = \frac{\partial P_w}{\partial x} \frac{\partial x}{\partial s} + \frac{\partial P_w}{\partial y} \frac{\partial y}{\partial s}.$$

We divide both the left and right side by $\frac{\partial x}{\partial s}$ and observe that $\frac{\partial y}{\partial x} = \frac{\partial h(x,t)}{\partial x}$. Introducing the approximation that there is hydrostatic equilibrium in the X-dip direction, i.e., $\frac{\partial P_\alpha}{\partial y} = -\rho_\alpha g \cos \vartheta$ leads to the following approximate equilibrium equation,

$$\frac{\partial P_w}{\partial x} - \frac{\partial P_o}{\partial x} = \Delta \rho g \cos \vartheta \frac{\partial h(x,t)}{\partial x}. \tag{4.106}$$

This equation was first derived by Dietz (1953).

4.8.4   DERIVATION OF FLOW RATE Q_{WX} FROM DARCY'S LAW

The approximate equilibrium Eq. (4.106) allows the elimination of the pressure gradient from expression (4.105) for the discharge Q_{wx}. This is completely analogous to the derivation of the fractional flow functions in Eqs. (4.38).

$$Q_{ox} = -\int_{h(x,t)}^{H} \lambda_o dy \left(\frac{\partial P_o}{\partial x} + \rho_o g \sin \vartheta \right) := -\Lambda_o \left(\frac{\partial P_o}{\partial x} + \rho_o g \sin \vartheta \right),$$

$$Q_{wx} = -\int_{0}^{h(x,t)} \lambda_w dy \left(\frac{\partial P_w}{\partial x} + \rho_w g \sin \vartheta \right) := -\Lambda_w \left(\frac{\partial P_w}{\partial x} + \rho_w g \sin \vartheta \right), \tag{4.107}$$

where we have introduced $\Lambda_w(h(x,t)) = \int_0^{h(x,t)} \lambda_w dy$ en $\Lambda_o(h(x,t)) = \int_{h(x,t)}^{H} \lambda_o dy$ for the ease of notation. For a homogeneous reservoir this can be simplified to $\Lambda_w = h(x,t)\lambda_w$ and $\Lambda_o = (H - h(x,t))\lambda_o$. In all cases, we can write $\Lambda_\alpha = \Lambda_\alpha(h)$. We have disregarded capillary forces, and we cannot use $P_w = P_o = p$ because the water and the oil pressure refer to different domains, i.e., P_w in the domain $0 < y < h(x,t)$ and P_o in the domain $h(x,t) < y < H$. We add Eqs. (4.107) and observe that incompressibility (completely analogous to the Eqs. (4.34) leads to $Q_{wx} + Q_{wo} = Q(t)$, i.e., the total discharge of oil and water in the flow direction depends only on the time (injection velocity) and not of the position. The following elimination procedure is the same as used in the Buckley–Leverett Theory. We use Eq. (4.107)

$$Q = Q_{wx} + Q_{ox} = Q_{wx} - \frac{\Lambda_o}{\Lambda_w} \Lambda_w \left(\frac{\partial P_w}{\partial x} + \rho_w g \sin \vartheta \right)$$

$$+ \Lambda_o(\rho_w - \rho_o)g \sin \vartheta - \Lambda_o \left(\frac{\partial P_o}{\partial x} - \frac{\partial P_w}{\partial x}\right), \tag{4.108}$$

and apply Eqs. (4.107) and (4.108) to obtain

$$Q = Q_{wx}\left(1 + \frac{\Lambda_o}{\Lambda_w}\right) + \Lambda_o \Delta \rho g \sin \vartheta + \Lambda_o \Delta \rho g \cos \vartheta \frac{\partial h(x,t)}{\partial x} \quad \text{or}$$

$$Q_{wx} = Q \frac{\Lambda_w}{\Lambda_w + \Lambda_o} - \frac{\Lambda_w \Lambda_o}{\Lambda_w + \Lambda_o} \Delta \rho g \left(\sin \vartheta + \cos \vartheta \frac{\partial h}{\partial x}\right). \tag{4.109}$$

Substitution of Eq. (4.109) into Eq. (4.104) leads to

$$\varphi(1 - S_{wc} - S_{or})\frac{\partial h(x,t)}{\partial t} + \frac{\partial}{\partial x}\left(Q\frac{\Lambda_w}{\Lambda_w + \Lambda_o} - \frac{\Lambda_w \Lambda_o}{\Lambda_w + \Lambda_o} \Delta \rho g \left(\sin \vartheta + \cos \vartheta \frac{\partial h}{\partial x}\right)\right) = 0, \tag{4.110}$$

and therefore leads to a convection-diffusion equation, which we shall call the Dietz-Dupuit equation. This nonlinear equation in $h(x,t)$ must be solved with numerical methods. This is a straightforward task with today's desk computers.

4.8.5   QUASI STATIONARY SOLUTION OF THE DIETZ-DUPUIT EQUATION
FOR $M < G + 1$

Eq. (4.110) follows after twice using that viscous forces in the X-dip direction can be disregarded, and thus that we have both above and below the interface hydrostatic equilibrium. All the same, we expect to obtain the stationary solution given in Eq. (4.97) also for this equation as the uniform flow field indeed implies that the viscous forces in the X-dip direction are zero. To prove this mathematically, we rearrange Eq. (4.110) to obtain

$$\varphi \Delta S \frac{\partial h}{\partial t} + \frac{\partial}{\partial x}\left(\frac{Q\Lambda_w}{\Lambda_o + \Lambda_w} - \frac{\Lambda_w \Lambda_o}{\Lambda_o + \Lambda_w}\Delta\rho g \sin\vartheta\right) = \frac{\partial}{\partial x}\left(\frac{\Lambda_w \Lambda_o}{\Lambda_o + \Lambda_w}\Delta\rho g \cos\vartheta \frac{\partial h}{\partial x}\right). \quad (4.111)$$

We use moving coordinates, i.e., $\xi = x - \frac{Qt}{\varphi \Delta SH}$ to rearrange Eq. (4.111) to (note that $\frac{\partial h}{\partial t} = -\frac{\partial h}{\partial \xi}\frac{Q}{\varphi \Delta SH}$)

$$\frac{\partial}{\partial \xi}\left(\frac{Q\Lambda_w}{\Lambda_o + \Lambda_w} - \frac{\Lambda_w \Lambda_o}{\Lambda_o + \Lambda_w}\Delta\rho g \sin\vartheta - \frac{Qh}{H}\right) = \frac{\partial}{\partial \xi}\left(\frac{\Lambda_w \Lambda_o}{\Lambda_o + \Lambda_w}\Delta\rho g \cos\vartheta \frac{\partial h}{\partial \xi}\right), \quad (4.112)$$

where we introduce quasi-stationarity by the assumption that $\left(\frac{\partial h}{\partial t}\right)_\xi = 0$. The equation can be integrated once to obtain (interchanging the left and right side of the equation). We obtain

$$\left(\frac{\Lambda_w \Lambda_o}{\Lambda_o + \Lambda_w}\Delta\rho g \cos\vartheta \frac{\partial h}{\partial \xi}\right) = \left(\frac{Q\Lambda_w}{\Lambda_o + \Lambda_w} - \frac{\Lambda_w \Lambda_o}{\Lambda_o + \Lambda_w}\Delta\rho g \sin\vartheta - \frac{Qh}{H}\right),$$

where the integration constant must be zero at $h = H$. This can be rearranged to

$$\frac{\Lambda_w \Lambda_o}{\Lambda_o + \Lambda_w}\Delta\rho g \sin\vartheta\left(1 + \frac{1}{\tan\vartheta}\frac{\partial h}{\partial \xi}\right) = \left(\frac{Q\Lambda_w}{\Lambda_o + \Lambda_w} - \frac{Qh}{H}\right),$$

$$(4.113)$$

and

$$\Lambda_o \Delta\rho g \sin\vartheta\left(1 + \frac{1}{\tan\vartheta}\frac{\partial h}{\partial \xi}\right) = \left(Q - \frac{Qh}{H}\frac{\Lambda_w + \Lambda_o}{\Lambda_w}\right).$$

$$(4.114)$$

This can be rewritten as

$$\frac{H\Lambda_w \Delta\rho g \sin\vartheta}{Q}\left(1 + \frac{1}{\tan\vartheta}\frac{\partial h}{\partial \xi}\right) = \left(\frac{\Lambda_w(H - h) - \Lambda_o h}{\Lambda_o}\right). \quad (4.115)$$

The gravity number is $G = \frac{H\lambda_w \Delta\rho g \sin\vartheta}{Q}$, the mobility ratio λ_w/λ_o and it follows that

$$Gh\left(1 + \frac{1}{\tan\vartheta}\frac{\partial h}{\partial \xi}\right) = \frac{\lambda_w h(H - h) - \lambda_o h(H - h)}{\lambda_o(H - h)} = (M - 1)h,$$

from which we obtain the exactly correct (within an interface model but without the VE-approximation) equation

$$\frac{dh}{dx} = \frac{M - 1 - G}{G}\tan\vartheta.$$

The VE approximation in the interface model for water oil-displacement leads for the situation that $M < G + 1$ to the exactly correct steady-state solution that has been obtained without invoking the VE approximation.

4.8.6   EXERCISE, SHOCK SOLUTION VERSUS INTERFACE ANGLE SOLUTION

About comparison of the shock solution with the stationary profile that uses interface angle calculations for $M < G + 1$. Start with the same conditions as in exercise (4.7.3)

- Plot the fractional flow function Φ_w versus the height h. Disregard the term containing $\frac{\partial h}{\partial x}$,

$$\Phi_w(h) = \frac{\Lambda_w}{\Lambda_o + \Lambda_w} - \frac{\Lambda_w \Lambda_o}{\Lambda_o + \Lambda_w} \frac{\Delta \rho g \sin \vartheta}{Q},$$

where

$$\Lambda_w(h) = \frac{kk'_{rw}}{\mu_w} h := \lambda_w h, \ \Lambda_o(h) = \frac{kk'_{ro}}{\mu_o}(H - h) := \lambda_o(H - h).$$

- Draw the shock solution both in the fractional flow function Φ_w and in the height of interface $h(x,t)$ as a function of distance (x) for various times, i.e.,

$$x_{shock}(t) = \frac{Qt}{\varphi \Delta S} \frac{\mathcal{F}_w(H) - \mathcal{F}_w(h=0)}{H} = \frac{Qt}{\varphi \Delta S H}$$

- How does this height of interface $h(x,t)$ as a function of distance (x) solution relate to the solution obtained for the quasi-stationary interface using

$$\frac{dh}{dx} = \frac{M - 1 - G}{G} \tan \vartheta.$$

4.8.7   ANALYTICAL SOLUTIONS

In the time of Dietz (in 1953), further approximations were required to find the analytical solution of Eq. (4.110). He disregarded the terms containing a gravity term $\Delta \rho g$. In this way, the same solution procedure as used in the Buckley–Leverett theory could also be applied. Dake maintains one of the gravity terms but disregards the term with $\partial h / \partial x$. Also for Dake's case, the "Buckley–Leverett" solution procedure can be used. Both approximations, i.e., the one used by Dietz and the one used by Dake lack a physical foundation. They are "convenience" approximations, of which the only virtue is that it allows to obtain an analytical solution. In spite of all this, the analytical solutions help to improve our understanding. We distinguish:

Dietz's approximation

$$Q_{wx}(h) \approx Q \frac{\Lambda_w}{\Lambda_w + \Lambda_o} := Q \mathcal{F}_w(h). \tag{4.116}$$

We substitute Eq. (4.116) in Eq. (4.104) and find

$$\varphi(1 - S_{wc} - S_{or}) \frac{\partial h(x,t)}{\partial t} + Q \frac{\partial \mathcal{F}_w(h)}{\partial x} = 0.$$

The second term depends only on h and we can write

$$\varphi(1 - S_{wc} - S_{or}) \frac{\partial h(x,t)}{\partial t} + Q \frac{d\mathcal{F}_w(h)}{dh} \frac{\partial h}{\partial x} = 0.$$

In the same way as for the Buckley–Leverett theory, we find a constant state solution and a rarefaction solution.

For $M > 1$, we obtain a combination of a

- Constant state solution,

$$h(x,t) = \text{constant} \tag{4.117}$$

- and a rarefaction wave solution, which reads for Dietz's-approximation

$$\eta = \frac{Q}{\varphi(1 - S_{wc} - S_{or})} \frac{d\mathcal{F}_w}{dh(x,t)}. \tag{4.118}$$

For $M < 1$, we obtain also a constant-state solution and a

- shock solution, i.e., the interface drops from the value $h = H$ to the value $h = 0$. The shock speed for Dietz's-approximation is given by

$$v_s = \frac{x_{shock}}{t} = \frac{Q}{\varphi(1 - S_{wc} - S_{or})} \frac{\mathcal{F}_w(h = H) - \mathcal{F}_w(h = 0)}{H} = \frac{Q}{\varphi(1 - S_{wc} - S_{or})H} \tag{4.119}$$

The theory is completely analogous to the Buckley–Leverett theory with a fractional flow function $\mathcal{F}_w(h)$ versus h. For $M > 1$ the function is concave and we use Eq. (4.118) to obtain the solution. Constant state solutions are used to connect the rarefaction solution to the injection and production side. For $M < 1$, the function is convex, implying from the mathematical point of view that we get a shock solution. If we were to use Eq. (4.118), we would obtain a solution where water is above the oil. This is not physically realistic. The shock solution is connected to the injection and production point by constant-state solutions. The interface between water and oil is perpendicular to the layer direction. This occurs because we disregarded the term with $\partial h / \partial x$. If we would have included the term eventually the steady state solution (4.97) would have been obtained. However, we note that the shock solution is often not at all a bad approximation of the steady-state solution, which we derived above for $M < G + 1$.

Dake's approximation Here we need to substitute

$$Q_{wx}(h) \approx Q \frac{\Lambda_w}{\Lambda_w + \Lambda_o} - \frac{\Lambda_w \Lambda_o}{\Lambda_w + \Lambda_o} \Delta \rho g \sin \vartheta := Q \Phi_w(h).$$

in Eq. (4.104). and obtain the rarefaction solution.
For $M > G + 1$ we obtain a combination of a

- Constant-state solution

$$h(x,t) = constant \tag{4.120}$$

- and a rarefaction wave solution, which reads for Dake's-approximation.

$$\eta = \frac{Q}{\varphi(1 - S_{wc} - S_{or})} \frac{d\Phi_w}{dh(x,t)} \tag{4.121}$$

For $M < G + 1$ we obtain also a constant-state solution and a

- Shock solution, i.e., the interface drops from the value $h = H$ to the value $h = 0$. The shock speed for Dake's-approximation is given by

$$v_s = \frac{x_{shock}}{t} = \frac{Q}{\varphi(1 - S_{wc} - S_{or})} \frac{\Phi_w(h = H) - \Phi_w(h = 0)}{H} = \frac{Q}{\varphi(1 - S_{wc} - S_{or})H}, \tag{4.122}$$

i.e., the same solution as obtained for Dietz's approximation.

4.8.8   ANALYTICAL EXPRESSIONS FOR THE INTERFACE AS A FUNCTION OF POSITION IN THE RESERVOIR

To benefit fully from our convenience approximations, we like to derive the analytical expressions. It requires some tedious but straightforward algebra. To simplify the notation, we introduce again dimensionless quantities: the mobility ratio M and the dimensionless gravity number G (Only required in Dake's approximation). Moreover, we introduce $b(x,t) = h(x,t)/H$ as the dimensionless position of the interface. The mobility ratio $M = \frac{k'_w/\mu_w}{k'_o/\mu_o}$ is the mobility of displacing fluid divided by the mobility of the displaced fluid. The end point permeabilities k'_w and k'_o are permeabilities for water at residual oil saturation and the oil permeability at connate water saturation respectively. In a water-wet medium is $k'_w < k'_o$ and this leads to a more favorable mobility ratio. The gravity number $G_x = G = H\Delta\rho g\sin\vartheta/(Q\mu_w/k'_w)$ is the gravity force divided by the viscous force. We give the result of the algebraic manipulations only for Dake's approximation as the result of Dietz's approximation can be obtained by putting $G = 0$. We use the integrated mobilities $\Lambda_w = Hb(x,t)\lambda_w$ and $\Lambda_o = H(1 - b(x,t))\lambda_o$. First, we obtain the following expression for $\Phi(b)$

$$\Phi_w(b) = \frac{Mb}{Mb + (1-b)} - \frac{Gb(1-b)}{Mb + (1-b)} = \frac{Mb - Gb(1-b)}{Mb + (1-b)}. \tag{4.123}$$

After differentiation we find the following result for the velocity of the velocity $[m/s]$ of the interface.

$$\eta = \frac{Q}{H\varphi(1 - S_{wc} - S_{or})}\frac{d\Phi_w(b)}{db(x,t)} = \frac{Q}{H(\varphi\Delta S)}\frac{M[b^2G + 1] - (b-1)^2G}{(Mb + (1-b))^2}. \tag{4.124}$$

This equation has the following interpretation. The right side is the velocity of the interface at the dimensionless height b. It is interesting to derive from this equation the velocity of the tip of the tongue (bottom part of the interface with ($b = 0$) and the top of the interface ($b = 1$). Substitution in Eq. (4.124) leads to

- Top of the interface

$$(\eta)_{b=1} = \frac{Q}{H(\varphi\Delta S)}\frac{G+1}{M},$$

- bottom of the interface (the water tongue)

$$(\eta)_{b=0} = \frac{Q}{H(\varphi\Delta S)}(M - G).$$

For typical Schoonebeek conditions ($M >> G$) and $M = 100$ one observes that the tongue moves about $10000\times$ times as fast as the top part. The situation is extremely unfavorable because for the tip of the tongue capillary effects can not be disregarded and also the convenience approximations lead to inaccurate results. Again the results for Dietz's approximation are obtained by putting $G = 0$.

4.8.9   ANALYTICAL EXPRESSIONS FOR THE PRODUCTION BEHAVIOR

The derivation of the production behavior for segregated flow is completely analogous to the derivation for the Buckley–Leverett theory. We rewrite Eq. (4.124). We use the dimensionless cumulative water injection $W_{iD} = Qt/(\varphi\Delta SHL)$, where the capital "$D$" is used to indicate that the reference volume refers to movable pore volumes $HL\varphi(1 - S_{wc} - S_{or})$ instead of the small "d", which refers to porevolumes $HL\varphi$. We rewrite Eq. (4.124)

$$\frac{x}{L} = \frac{Qt}{HL\varphi(1 - S_{wc} - S_{or})} \frac{M[b^2G + 1] - (b-1)^2 G}{(Mb + (1-b))^2}$$

$$= W_{iD} \frac{M[b^2G + 1] - (b-1)^2 G}{(Mb + (1-b))^2}. \tag{4.125}$$

For $x = L$ we denote the dimensionless interface position by b_e where the subscript e means evaluation at the end point $x = L$. We find

$$W_{iD} = \left(\frac{1}{\frac{d\Phi_w}{db}}\right)_e = \frac{(Mb_e + (1 - b_e))^2}{M(b_e^2 G + 1) - (b_e - 1)^2 G}. \tag{4.126}$$

From this equation, we can obtain b_e in terms of W_{iD}. We obtained the solution by application of MAPLE. For notational convenience here, below we use W instead of W_{iD}. The solution of the equation is

$$b_e = \frac{-GW + M - 1(+-)\sqrt{MW(G - M + 1)(GW - M + 1)}}{(M - 1)(GW - M + 1)}$$

$$= \frac{Y(+-)\sqrt{(MW(M - G - 1)Y)}}{(1 - M)Y} \tag{4.127}$$

where we have introduced $Y := M - 1 - GW$ for reasons of notational convenience.

The dimensionless oil production dN_{pD} is equal to the product of the fractional flow function of oil $\Phi_o = 1 - \Phi_w$ and the dimensionless injection (=total fluid) rate dW. From this we can calculate the dimensionless cumulative oil production. We denote with subindex e the end of the layer

$$N_{pD} = \int_0^W (1 - \Phi_{we}) dW. \tag{4.128}$$

We use Eq. (4.121) and write

$$\eta = \frac{x}{t} = \frac{Q}{H\varphi(1 - S_{wc} - S_{or})} \frac{d\Phi_w(b)}{db(x,t)} \tag{4.129}$$

and find

$$\frac{x}{L} = \frac{Qt}{HL\varphi(1 - S_{wc} - S_{or})} \frac{d\Phi_w(b)}{db(x,t)} = W \frac{d\Phi_w(b)}{db(x,t)}. \tag{4.130}$$

from which it follows that

$$W = \left(\frac{db(x,t)}{d\Phi_w(b)}\right)_{b=b_e}, \tag{4.131}$$

where b_e is the dimensionless height at the production (end) point. For evaluating the integral (4.128) we use integration by parts and write W instead of W_{iD} for notational convenience.

$$N_{pD} = (1 - \Phi_w)_e W_{iD} + \int_0^{\Phi_{we}} W d\Phi_{we}$$

$$= (1 - \Phi_w)_e W + \int_0^{b_e} \frac{db}{d\Phi_{we}} d\Phi_{we} = (1 - \Phi_w)_e W + b_e$$

$$= \frac{1 - b_e + Gb_e(1 - b_e)}{Mb_e + (1 - b_e)} W + b_e. \tag{4.132}$$

We use Eq. (4.127) and obtain

$$N_{pD} = 2 \frac{\begin{array}{c} \left(\left(-\frac{1}{2} + \left(\frac{G}{2} - \frac{1}{2} \right) W \right) M + \frac{1}{2} + \left(\frac{G}{2} + \frac{1}{2} \right) W \right) \sqrt{MW \left(G - M + 1 \right) \left(GW - M + 1 \right)} \\ + MW \left(G - M + 1 \right) \left(GW - M + 1 \right) \end{array}}{\sqrt{MW \left(G - M + 1 \right) \left(GW - M + 1 \right)} \left(M - 1 \right)^2}$$

$$= 2 \frac{\begin{array}{c} \left(\left(-\frac{1}{2} + \left(\frac{G}{2} - \frac{1}{2} \right) W \right) M + \frac{1}{2} - \left(\frac{G}{2} + \frac{1}{2} \right) W \right) \sqrt{MW \left(G - M + 1 \right) \left(GW - M + 1 \right)} \\ + MW \left(G - M + 1 \right) \left(GW - M + 1 \right) \end{array}}{\sqrt{MW \left(G - M + 1 \right) \left(GW - M + 1 \right)} \left(M - 1 \right)^2}$$

$$(4.133)$$

The algebra needed for the case that $G = 0$ is much simpler but completely analogous. For $G = 0$, Eq. (4.126) reduces to

$$W = \frac{(Mb_e + (1 - b_e))^2}{M}. \qquad (4.134)$$

Eq. (4.123) can be simplified to

$$\Phi_w|_{G=0} = \frac{Mb}{Mb + (1 - b)}. \qquad (4.135)$$

With Eq. (4.127) one can calculate b_e: $b_e = \frac{\sqrt{MW} - 1}{M - 1}$. Substitution into Eqs. (4.135) and (4.134) leads to an expression for the dimensionless oil production.

$$N_{pD} = \frac{2\sqrt{WM} - W - 1}{M - 1}. \qquad (4.136)$$

As water breaks through, we have that $W = N_{pD}$. Substitution into Eq. (4.136) leads to

$$N_{pD,bt} = \frac{1}{M}.$$

When all oil has been recovered, we would have that $N_{pD} = 1$. Substitution into Eq. (4.136) leads to

$$W_{iD,max} = M, \qquad (4.137)$$

where $W_{iD,max}$ is the dimensionless pore volume of water needed to recover all movable oil. When $G \neq 0$, we need to solve Eq. (4.133) for $N_{pD} = 1$, i.e., the expression stated is equal to $W = 1$ to obtain $N_{pD,bt}$ and to obtain $W = N_{pD}$ to obtain $W_{iD,max}$, i.e.,

$$N_{pD,bt} = \frac{1}{M - G},$$

$$W_{iD,max} = \frac{M}{G + 1}. \qquad (4.138)$$

The analytical theory used hereafter the introduction of, e.g., Dake's approximation leads to a shock solution if $M < G + 1$. If one retains in Eq. (4.109), the term $\frac{\partial h}{\partial x}$, then the interface will have a finite slope given by Eq. (4.97). In most cases, for $M < G + 1$, the shock solution is a reasonable approximation. For $M < 1$, in Dietz's approximation and for $M < G + 1$ for Dake's approximation, we obtain that before water breakthrough the cumulative dimensionless oil production equals the dimensionless water injection $N_{pD} = W$. After breakthrough the cumulative dimensionless oil production remains constant $N_{pD} = N_{pD,bt}$.

4.8.10   SUMMARY OF ANALYTICAL PROCEDURE FOR INTERFACE MODELS

Here we back introduce the notation W_{iD} instead of W

Determine the mobility ratio and gravity number

$$M = \frac{k'_w/\mu_w}{k'_o/\mu_o}, \; G = \frac{g\Delta\rho_{wo}\sin\vartheta}{u\mu_w/k'_w}.$$

If M< G+1

The slope $\frac{dy}{dx}$ of the interface is constant and given by

$$\frac{dy}{dx} = \frac{M-1-G}{G}\tan(\vartheta).$$

We use geometric considerations to determine the oil recovery curves

For the situation shown in Figure 4.25, the volume swept by water can be expressed by the dimensionless production related to the total flowing pore volume in the reservoir. It is indeed equal to the total movable oil volume minus the movable pore volume of oil still present in the upper rectangular triangle in the reservoir

$$N_{pD} = 1 - \frac{(H-y_e)^2}{2HL\tan\beta}.$$

The amount of water that was injected up till now can be determined by ignoring the presence of the production well. The volume of water that flows beyond the well is given by $\frac{y_e}{2}\frac{y_e}{\tan\beta}w\varphi(1 - S_{wc} - S_{or})$. The cumulative water injection is therefore

$$W_{iD} = N_{pD} + \frac{y_e^2}{2HL\tan\beta}.$$

At breakthrough $(y_e = 0)$ we have

$$N_{pD,bt} = 1 - \frac{H}{2L\tan\beta}.$$

In this derivation, we have assumed that $\tan\beta > H/L$.

If M>G+1

The slope is not constant. The interface can be sketched with the help of Eq. (4.124).

$$\eta = \frac{Q}{H(\varphi\Delta S)}\frac{M[b^2G+1]-(b-1)^2G}{(Mb+(1-b))^2}. \tag{4.139}$$

This equation has the following interpretation. The right side is the velocity of the interface at dimensionless height $b = h/H$.

The cumulative oil production at water breakthrough $N_{pD,bt}$ $(N_{pD} = W_{iD})$ and the cumulative water injection needed for total recovery $W_{iD,max}$ $(N_{pD} = 1)$ are given by

$$N_{pD,bt} = \frac{1}{M-G},$$

$$W_{iD,max} = \frac{M}{G+1}. \tag{4.140}$$

4.8.11   EXERCISE, ADVANTAGE OF $M \leq G + 1$.

About an inventory of the advantage of the condition $M \leq G + 1$.

A high mobility ratio, i.e., $M > 1$ has adverse effects on the recovery efficiency. The adverse effect may be partly compensated by stabilizing gravity effects so that the mobility ratio can be higher than $M > 1$, i.e., $M > G + 1$ for the adverse effect to clearly express itself. It is, however, important to keep the displacing fluid upstream (below for the heavier displacing fluid or above for the lighter displacing fluid). Explain the adverse effects of the high mobility ratio for

- Viscous fingering,
- Buckley–Leverett displacement in terms of the saturation jump at the shock,
- vertical sweep efficiency for interface model both for $M \leq G + 1$ and $M \geq G + 1$,
- the areal sweep efficiency [140].

Do capillary effects have a stabilizing or destabilizing effect?

4.A   NUMERICAL APPROACH FOR INTERFACE MODELS

The numerical approach is the same as given for the situation $M > G + 1$ but is given here for easy reference.

We now turn to the finite difference approach. The equation we need to solve reads

$$\varphi \Delta S \frac{\partial h}{\partial t} + Q \frac{\partial \mathcal{F}_w}{\partial x} = 0.$$

In this case, the fractional flow function $\mathcal{F}_w$ is given by the relation

$$\mathcal{F}_w\left(h, \frac{\partial h}{\partial x}\right) = \frac{\Lambda_w}{\Lambda_o + \Lambda_w} - \frac{\Lambda_w \Lambda_o}{\Lambda_o + \Lambda_w} \frac{\Delta \rho g}{Q} \left(\sin \vartheta + \cos \vartheta \frac{\partial h}{\partial x}\right),$$

where again

$$\Lambda_w(h) = \frac{kk'_{rw}}{\mu_w} h, \quad \Lambda_o(h) = \frac{kk'_{ro}}{\mu_o}(H - h). \tag{4.141}$$

We integrate over one grid block, which can be visualized as a cell of your spreadsheet. Therefore, we obtain

$$\varphi \Delta S \int\limits_{in}^{out} \frac{\partial h}{\partial t} dx + Q \int\limits_{in}^{out} \frac{\partial \mathcal{F}_w}{\partial x} dx = 0, \tag{4.142}$$

$$\varphi \Delta S \frac{d}{dt} \int\limits_{in}^{out} h(x,t) dx + Q \int\limits_{in}^{out} \frac{\partial \mathcal{F}_w}{\partial x} dx = 0.$$

We define the average height $\bar{h} = \frac{1}{\Delta x} \int\limits_{in}^{out} h(x,t)\, dx$, where Δx is the length of the cell. Therefore, in any grid cell labeled i, we obtain

$$\bar{h}_i(t + \Delta t) = \bar{h}_i(t) + \frac{Q}{\varphi \Delta S} \frac{\Delta t}{\Delta x}(\mathcal{F}_{w,in} - \mathcal{F}_{w,out}). \tag{4.143}$$

We now use the so-called upstream weighting approximation, which implies that $\mathcal{F}_{w,in}$ flows with the height value upstream, i.e., of cell $(i-1)$ and that $\mathcal{F}_{w,out}$ flows with the height properties of the cell considered, i.e., of cell i. For the situation considered here it is obvious where the flow is coming from but for the $2-D$ case or other situations this may have to be determined in advance. A system that flows with the properties of the cell where it is flowing to becomes very unstable (you may want to try it). The flow out of cell i is equal to the flow flowing into cell $i+1$. Therefore, we only specify $\mathcal{F}_{w,in}$ over the boundary between cell $i-1$ and cell i

$$\mathcal{F}_{w,in} = \frac{\Lambda_{w,u}}{\Lambda_{o,u}+\Lambda_{w,u}} - \frac{\Lambda_{w,u}\Lambda_{o,u}}{\Lambda_{o,u}+\Lambda_{w,u}} \frac{\Delta\rho g}{Q}(\sin\vartheta + \cos\vartheta\frac{h_i - h_{i-1}}{\Delta x}),$$

where $\Lambda_{w,u} = \Lambda_{w,i-1}$ if the water is flowing out of cell $i-1$ into cell i, but $\Lambda_{w,u} = \Lambda_{w,i}$ if the water is flowing out of cell i into cell $i-1$. In the same way $\Lambda_{o,u} = \Lambda_{o,i-1}$ if the oil is flowing out of cell $i-1$ into cell i, but $\Lambda_{o,u} = \Lambda_{o,i}$ if the oil is flowing out of cell i into cell $i-1$. In gravity dominated cases oil and water can flow in opposite directions.

4.A.1 EXERCISE. BEHAVIOR FOR $M > G+1$

About interface profile calculations for $M > G+1$

We consider a layer with a distance between the row of injection and the row of production wells of 400 m. The reservoir dip is $5°$. The permeability k of the reservoir is 2 Darcy and the height H is 20 m. The porosity φ is 0.3 [-]. The density difference $\Delta\rho$ between water and oil is $[200\,\mathrm{kg/m^3}]$, the acceleration due to gravity g is $9.81[\mathrm{m/s^2}]$. We use for the viscosity of oil $\mu_o = 0.002[\mathrm{Pa.s}]$ and for the water viscosity $\mu_w = 0.0005[\mathrm{Pa.s}]$. The interfacial tension σ_{ow} between oil and water is $30 \times 10^{-3}\mathrm{N/m}$.
The connate water saturation $S_{wc} = 0.15$ and the residual oil saturation S_{or} is also 0.15. The total Darcy velocity $u = 1m/d$. The end point relative permeabilities are 0.5 and 0.95 for water and oil, respectively. Make a named list of variables in EXCEL by using the "Formulas-Create from Selection" command.

- Calculate the mobility ratio M and the gravity number G.

$$M = \frac{kk'_{rw}}{\mu_w} \frac{\mu_o}{kk'_{ro}} \text{ and } G = \frac{kk'_{rw}g\Delta\rho\sin\vartheta}{u\mu_w}.$$

- Plot the fractional flow function $\mathcal{F}_w$ versus height; vary the oil viscosity between 10^{-4} and $0.16[Pa.s]$. Disregard the term containing $\frac{\partial h}{\partial x}$,

$$\mathcal{F}_w(h) = \frac{\Lambda_w}{\Lambda_o + \Lambda_w} - \frac{\Lambda_w\Lambda_o}{\Lambda_o + \Lambda_w} \frac{\Delta\rho g\sin\vartheta}{Q},$$

where

$$\Lambda_w(h) = \frac{kk'_{rw}}{\mu_w}h, \Lambda_o(h) = \frac{kk'_{ro}}{\mu_o}(H-h).$$

- Perform a numerical calculation on a separate sheet called, e.g., "Dietz" and compare to the analytical results.

4.B NUMERICAL APPROACHES FOR BUCKLEY LEVERETT AND INTERFACE MODELS IMPLEMENTED WITH EXCEL

4.B.1 SIMPLE SHEET FOR BUCKLEY–LEVERETT MODEL

This spreadsheet for the numerical simulation of the Buckley-Leverett problem also calculates the pressure drop ΔP and the oil production rate. The product $Q\Delta P$, where Q is the flow rate calculates the net pumping energy, which must be divided by the conversion efficiencies (0.45 $\times$ 0.8 $\times$ 0.9 = 0.324), respectively from fuel $\rightarrow$ to electricity, the mechanical driver energy and the electrical driver energy to obtain the practical pumping exergy. The outflow is expressed in $m^3/m^2/s$. The energy of hydrocarbons is 10.7 kWh/L = 10.7 MWh / m^3 [159]. In this way we can calculate the exergy of the hydrocarbons divided by the pumping exergy. When this number drops below one, more exergy is spent to circulate the fluids than recovered and the oil project must be stopped.

- We first define the input data (Figure 4.28),
- then we use the input data to find computed input data (Figure 4.29),
- then we write the equations (Figure 4.30)
- we convert to manual according to the settings shown in Figure 4.31,
- we insert the control variables as in Figure 4.32
- put the statements in the spreadsheet for the EXCEL computation; the first cell is as in Figure 4.33, which includes the left boundary condition for which the fractional flow function is one; the second cell (right of the first cell defines a general cell, which does not include the boundary condition; the last cell is the cell at the far right, which includes the boundary condition stating that the capillary pressure at the end is zero.

   The numerical calculation of the production curves, though in principle possible, will be considered outside the scope of these notes.

◢	A	B	C	D
1		**data**		
2	sinit		0.15 [-]	initial water saturation
3	diameter		0.04 [m]	diameter core
4	length		1 [m]	length core
5	theta		0.00 [rad]	inj./prod. line with horizontal
6	phi		0.38 [-]	porosity
7	k		1.00E-12 [m^2]	permeability
8	swc		0.15 [-]	connate water
9	sor		0 [-]	residual oil
10	Sco		0.16 [-]	cross-over saturation
11	kroo		1 [-]	end point oil permeability
12	krwo		0.5 [-]	end point water permeability
13	utot		0.0001 [m/s]	total input Darcy velocity
14	gamma		0.5 [-]	scaling parameter capillary pressure
15	sow		0.03 [N/m]	interfacial tension
16	lambda		7 [-]	parameter for Pc curve
17	fracdim		2.5 [-]	fractal surface dimension
18	rho		800 [kg/m^3]	density oil
19	rhow		1000 [kg/m^3]	density water
20	muo		1.00E-02 [Pa s]	viscosity oil
21	muw		1.00E-03 [Pa s]	viscosity water
22	g		9.81 [m/s^2]	acc. gravity

Figure 4.28   input data for Buckley–Leverett model, $S_{co} = 0.16$

25		computed		
26				
27	nw	=2/lambda+3	[-]	saturation exponent water
28	no	=2/lambda+1	[-]	saturation exponent oil
29	area	=PI()*diameter*diameter/4	[m^2]	crossection core
30	Pcb	=gamma*sog*SQRT(phi/k)*((0.5-sor)/(1-sor-swc))^(1/lambda)	[Pa]	bubbling pressure
31	Pcba	=Pcb*Sco^(1/fracdim)*((Sco-swc)/(1-swc-sor))^(-1/lambda)	[Pa]	surface bubbling pressure
32	time	=time	[s]	time

Figure 4.29 Computed input data

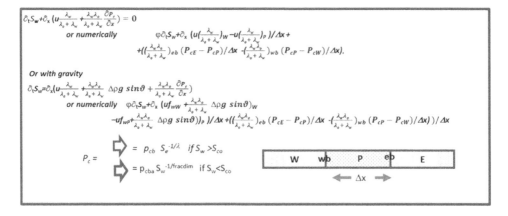

Figure 4.30 Basic equations for Buckley–Leverett model.

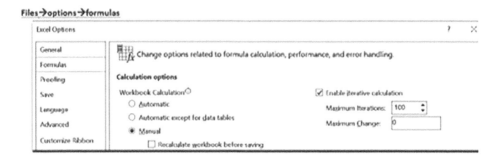

Figure 4.31 Manual mode for iterative EXCEL calculation.

4.C NUMERICAL DIFFUSION FOR FIRST ORDER UPSTREAM WEIGHTING SCHEME

A first-order upstream weighting scheme introduces (artificially) a diffusion term in the model equation. We need to check whether this is going to upset our results as we require them. The derivation of the numerical diffusion constant always runs along the lines given below. For convenience, we show here the artificially introduced diffusion term into our interface model.

5		control variables			
6					
7	start	1		[-]	0=initialize,1=run
8	counter	=IF(start=0,0,H8+1)		[-]	# sweeps through spreadsheet
9	dsmax	=0.01		[-]	allowed change per time step
10	maxds	=MAX(MAX(Del_sw),-MIN(Del_sw))		[-]	observed change per time step
11	nz	100		[-]	# of grid cells to be counted manually
12	dz	=length/nz		[m]	grid size
13	dt	=IF(counter<4,0.00001,MIN(dt*(dsmax/maxds)		[s]	time step
14	time	=IF(start=0,0,H14+H13)		[s]	time
15	iterations	100		[-]	# of iterations; keep in this cell!
16	ncalc	10		[-]	# calculations / iteration

Figure 4.32 Control variables for iterative spreadsheet.

cell no	1	2
se	=(Sw-swc)/(1-swc-sor)	=(Sw-swc)/(1-swc-sor)
lrw	=krwo*se^nw/muw	=krwo*se^nw/muw
lro	=kroo*(1-se)^2*(1-se^no)/muo	=kroo*(1-se)^2*(1-se^no)/muo
fw	=lrw/(lro+lrw)	=lrw/(lro+lrw)
lw*lo/(lw+lo)=lof	=k*fw*lro	=k*fw*lro
u*fw	=fw*(utot+k*lro*(rhow-rho)*g*SIN(theta))	=fw*(utot+k*lro*(rhow-rho)*g*SIN(theta))
Pc	=IF(Sw>Sco,Pcb*se^(-1/lambda),Pcba*Sw^(-1/fracdim))	=IF(Sw>Sco,Pcb*se^(-1/lambda),Pcba*Sw^(-1/fracdim))
fluxoutt	=(ufw+lofw*(C43-Pc)/dz+ABS(ufw+lofw*(C43-Pc)/dz))/2+(C42+(	=(ufw+lofw*(D43-Pc)/dz+ABS(ufw+lofw*(D43-Pc)/dz))/2+(D4
Del sw	=(utot-fluxoutt)*dt/(phi*dz)	=(B44-fluxoutt)*dt/(phi*dz)
dist	=dz/2	=B46+dz
Sw	=IF(start=0,sinit,Sw+Del_sw)	=IF(start=0,sinit,Sw+Del_sw)
∆P	=(-utot/k/(lrw+lro)+(rho*g*(1-fw)+rhow*g*fw)*SIN(theta))*dz	=(-utot/k/(lrw+lro)+(rho*g*(1-fw)+rhow*g*fw)*SIN(theta))*dz
presdrop	=SUM(DP)	

Figure 4.33 First and second column (which can be copied to the right until the last cell). The row with the entry "se", se is the name of the entire row. This has been accomplished by painting B37:cw50, create from selection, left column, ok. Now the entire row has the name of the entry indicated in the left column. The fluxoutt cells consists of the term = (ufw+lofw*(C43-Pc)/dz+ABS(ufw+lofw*(C43-Pc)/dz))/2, which is the sum of the convective flux and the capillary flux. Addition of the same term with an absolute sign causes that the term is zero if its contribution would be negative. The entry with fluxoutt is partially concealed. It should read = (ufw+lofw*(D43-Pc)/dz+ABS(ufw+lofw*(D43-Pc)/dz))/2+(D42+D41*(D43-Pc)/dz-ABS(D42+D41*(D43-Pc)/dz))/2. This term in full reads as the convective term inclusive the capillary pressure term (C42+C41*(C43-Pc)/dz-ABS(C42+C41*(C43-Pc)/dz))/2. If the term (C42+C41*(C43-Pc)/dz) is positive due to the absolute sign it becomes zero; if it is negative the flow if from right to left. In summary with the absolute sign this formulation automatically implements upstream weighting.
The capillary pressure at the production side is zero.

Consider the VE-approximation for the interface model in which we have omitted the diffusional term, i.e., we use the volume balance equation

$$\frac{\partial h}{\partial t} + \frac{Q}{\varphi \Delta S}\frac{\partial \mathcal{F}_w}{\partial x} = 0,$$
$$\frac{\partial h}{\partial t} + q\frac{\partial \mathcal{F}_w}{\partial x} = 0, \tag{4.144}$$

and we use the fractional flow function $\mathcal{F}_w$ in which we disregard the term containing $\frac{\partial h}{\partial x}$. We note that q has units $[m^2/s]$

$$\mathcal{F}_w(h) = \frac{\Lambda_w}{\Lambda_o + \Lambda_w} - \frac{\Lambda_w \Lambda_o}{\Lambda_o + \Lambda_w}\frac{\Delta \rho g \sin \vartheta}{Q}.$$

The numerical equation, which we solve reads

$$\frac{h(t+\Delta t)-h(t)}{\Delta t} = q\frac{\mathcal{F}_w(x_{i-1})-\mathcal{F}_w(x_i)}{\Delta x}.$$

We use Taylor expansion to show the second-order terms that are present in the numerical scheme

$$\mathcal{F}_w(x_{i-1}) = \mathcal{F}_w(x_i) - \frac{\partial \mathcal{F}_w}{\partial x}\Delta x + \frac{1}{2}\frac{\partial^2 \mathcal{F}_w}{\partial x^2}(\Delta x)^2, \tag{4.145}$$

$$h(t+\Delta t) = h(t) + \frac{\partial h}{\partial t}\Delta t + \frac{1}{2}\frac{\partial^2 h}{\partial t^2}(\Delta t)^2. \tag{4.146}$$

This means that to second-order accuracy, we solve

$$\frac{\partial h}{\partial t} + q\frac{\partial \mathcal{F}_w}{\partial x} = \frac{q}{2}\frac{\partial^2 \mathcal{F}_w}{\partial x^2}\Delta x - \frac{1}{2}\frac{\partial^2 h}{\partial t^2}\Delta t := \frac{\partial}{\partial x}\left(D\frac{\partial h}{\partial x}\right),$$

where D is the numerical diffusion constant. Therefore, we want to express the second derivative $\frac{\partial^2 h}{\partial t^2}$ in terms of a derivative toward x. For this, we use the original model equation (4.144) and we take the derivative toward t, i.e.,

$$\frac{\partial^2 h}{\partial t^2} = -q\frac{\partial}{\partial t}\frac{\partial \mathcal{F}_w}{\partial x} = -q\frac{\partial}{\partial x}\left(\frac{\partial \mathcal{F}_w}{\partial t}\right)$$

$$= -q\frac{\partial}{\partial x}\left(\left(\frac{d\mathcal{F}_w}{dh}\right)\frac{\partial h}{\partial t}\right) = q^2\frac{\partial}{\partial x}\left(\frac{d\mathcal{F}_w}{dh}\frac{\partial \mathcal{F}_w}{\partial x}\right) = q^2\frac{\partial}{\partial x}\left(\left(\frac{d\mathcal{F}_w}{dh}\right)^2\frac{\partial h}{\partial x}\right). \tag{4.147}$$

Therefore, we obtain for the numerical diffusion constant the term in square brackets.

$$\frac{\partial}{\partial x}\left(D\frac{\partial h}{\partial x}\right) = \frac{q}{2}\frac{\partial^2 \mathcal{F}_w}{\partial x^2}\Delta x - \frac{q^2\Delta t}{2}\frac{\partial}{\partial x}\left(\frac{d\mathcal{F}_w}{dh}\right)^2\frac{\partial h}{\partial x}$$

$$= \frac{\partial}{\partial x}\left(\left[\left(\frac{q\Delta x}{2}\frac{d\mathcal{F}_w}{dh} - \frac{q^2\Delta t}{2}\left(\frac{d\mathcal{F}_w}{dh}\right)^2\right)\right]\frac{\partial h}{\partial x}\right). \tag{4.148}$$

5 Dispersion in Porous Media

OBJECTIVE OF THIS CHAPTER

- Be able to distinguish the Darcy velocity, interstitial velocity and "true" velocity due to tortuosity (Figure 2.1),
- to understand the effect of no mixing, physical (realistic) mixing, and exaggerated mixing (numerical diffusion) in miscible displacement (Figure 5.1),
- explain the occurrence of local and macroscopic dispersion in mass transport problems in porous media,
- explain the occurrence of longitudinal and transverse dispersion,
- to derive the average velocity and longitudinal dispersion from an effluent profile,
- to understand the Gelhar relation that describes the longitudinal dispersion in a heterogeneous medium and shows that it is proportional to the variance of the logarithm of the permeability (related to the Dykstra-Parsons coefficient) times the correlation length.

5.1 INTRODUCTION

Reactive transport in porous media plays an important role in environmental hydrology [18], petroleum engineering, agricultural engineering and chemical engineering. In the most common application in environmental engineering, one of the fluids is freshwater and one of the fluids is water in which a contaminant is dissolved. In reactive transport, in porous media, mixing of components plays a central role. The most common application in petroleum engineering is tracer flow [143], i.e., a component that does not affect the transport properties, e.g., the viscosity. Miscible displacement [206] where considerable fractions of the components of the injected fluids dissolve in the oil are outside the scope of the theory discussed here [174]. Mixing in porous media can also be caused by different sorption behavior (heterogeneous sorption) of the components [198]. However, this chapter considers the mixing, in the absence of heterogeneous sorption and nonlinear transport mechanisms.

Mixing of fluids in porous media is very slow. Diffusion coefficients of small molecules in liquids with a viscosity of 10^{-3} [Pa.s] are of the order of 10^{-9} m^2/s. For larger molecules, they are even smaller. To diffuse over a distance of one meter costs about 30 years. However, without diffusion, there would be no mixing at all. Chemical reactions occur when fluids mix. Consequently, diffusion is very important for reactive transport in porous media. In porous media, enhanced mixing occurs at the pore scale because of fluid flow as explained in this chapter. Even if this enhances the diffusion, the effective diffusion coefficient remains small. In numerical flow simulators, no matter how sophisticated the numerical scheme, the mixing effect is exaggerated for all practical choices of numerical grid sizes. This is caused by the fact that all numerical schemes introduce mixing, usually in the form of numerical diffusion. With reference to Figure 5.1, we can distinguish between no diffusion, which means that reactants do not mix and no chemical reaction occurs, realistic diffusion with correct chemical reaction terms and with numerical diffusion, which overestimates the mixing of reactants and thus chemical reaction. The formulation of transport problems with analytical models often completely disregards diffusion. This means that the "truth" lies somewhere between the analytical and numerical methods. Due to the fact that diffusional mixing effects are so small, it is also very difficult to obtain reliable experimental data. This chapter also shows the derivation of the diffusional terms in the transport processes

DOI: 10.1201/9781003168386-5

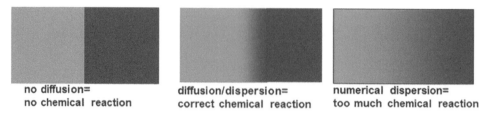

no diffusion=
no chemical reaction

diffusion/dispersion=
correct chemical reaction

numerical dispersion=
too much chemical reaction

Figure 5.1 No miscibility, realistic miscibility and exaggerated miscibility due to numerical diffusion and its influence on chemical reactions.

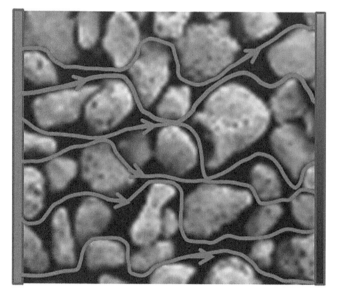

Figure 5.2 Mechanism of local longitudinal dispersion, the horizontal axis is denoted by x, the vertical axis is denoted by y. Particles starting at the same x-location, will be at a different x-location depending on whether they have to travel more or less around grains. Particles that have to travel much around the grains will have penetrated less far (be at smaller x-locations).

in porous media. We distinguish between molecular diffusion, local dispersion and macroscopic dispersion. Molecular diffusion in porous media is reduced because fluids move along tortuous flow paths. This reduction is usually small typically of the order of 70%.

Local dispersion is caused by heterogeneities on the pore scale, which forces fluids to move around the grains. The reason for *local dispersion* due to pore-level heterogeneity can be qualitatively understood from Figure 5.2. *Longitudinal dispersion* describes the spreading *in the direction of flow*. To understand longitudinal dispersion, we consider a planar front of a concentration (tracer) vertical line pulse entering a porous medium. The tracer flows with the fluid along the streamlines. The flow in the porous medium must be around the pore grains (or around areas with low permeability). Some flow paths are longer than others. This means that after some time not every part of the original line pulse has penetrated equally far in the porous medium. Different tracer concentrations enter the same pores and here true molecular diffusive mixing occurs. As the mixing is influenced by the flow and leads to spreading in the flow direction, we use the word *longitudinal hydrodynamic dispersion*. As a result, the effective diffusion will be enhanced. This

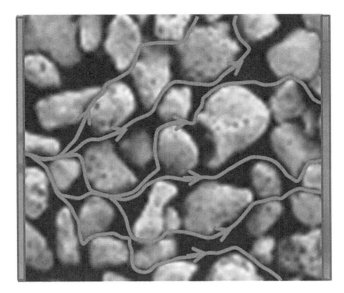

Figure 5.3 Mechanism of local transverse dispersion. The splitting is exaggerated with respect to Figure 5.2 for reasons of illustration.

convective enhanced diffusion coefficient is always calculated from an upscaling procedure supported by experimental data. We distinguish between upscaling from the pore scale to the core scale (local dispersion) and from the core-scale to the field scale (macroscopic dispersion) [97]. We ignore all the scales below the pore scale (see however, [24]).

Transverse dispersion is spreading in a direction perpendicular to the direction of flow. For *transverse dispersion*, we consider a horizontal line pulse. Transverse dispersion can be qualitatively understood from Figure 5.3. The splitting is exaggerated for illustration purposes. As the stream tube (bundle of neighboring streamlines) that runs through this line pulse enters the porous medium it may hit a grain (or low permeability region) and split up, one part going to the right and the other part going to the left. Thus, the split-up stream tubes split up at the next grain and the process repeats itself. The analogy to the inebriated sailor is again clear as there is little chance for a particular streamline to go right at all steps. There is again the analogy to a diffusional process and the concentration spreads in a direction *perpendicular* to the flow. Again we distinguish between local dispersion when the dispersion occurs at the grain level [129] and actual mixing occurs or macroscopic dispersion when we deal with permeability regions and we want to calculate average concentrations.

For interpretation of laboratory experiments, core-scale equations are relevant. However, for upscaling to the field scale, macroscopic dispersion needs to be considered. The macroscopic dispersion is due to the same mechanism as local dispersion, except that now the fluids mainly flow in high permeability streaks around less permeable parts before being recombined. On the macroscale, this recombination occurs by mathematical averaging. Indeed, the behavior of spatially averaged concentrations can also often be described by a diffusion process, which is called macroscopic dispersion. If the scale extends to the reservoir scale instead of macroscopic dispersion, sometime the word "gigascopic dispersion" is used. It hardly causes actual mixing of fluids (see, however, [129]). When the flow will be retarded in low permeability regions and enhanced in high permeable regions, different concentrations will enter finite regions of the reservoir small enough to represent a single grid block in a simulator. Mixing by molecular diffusion will not occur, but the average concentration in the relevant region will be different from zero or one. We can use the average concentration in the calculations, but we have to remember that this does not indicate true mixing, e.g., when thermodynamic equilibrium calculations are carried out. The word *hydrodynamic dispersion* is used for both *local dispersion* and *macroscopic dispersion*. The ensuing concentration profile is therefore rather related to the probability that one or the other fluid occurs than to the concentration at the pore level. Indeed, due to the large size of the heterogeneities, there is no actual mixing when the fluids are recombined. The mathematical mixing in upscaled flow equations can be large, but we are now not talking about true mixing. However, some partial mixing occurs [23, 129]. As it turns out for convective-enhanced diffusion coefficients, we distinguish longitudinal dispersion, meaning that the mixing occurs in the direction of flow and transverse dispersion [206], which is much smaller, is concerned with mixing perpendicular to the main flow direction.

It should be noted that experiments to obtain accurate data [70, 71] are not trivial due to the low values of the dispersion coefficients. Experimental data, however, may be incorrect due to streamline splitting at the entrance and production point, i.e., if special fluid distributors at the injection and production point were not used. Small entrapped air bubbles can also cause an apparent increase of the dispersion coefficient. Finally, also the sand pack must be homogeneous, which requires special experimental preparation techniques [253]. Figures 5.4 and 5.5 show experimental data of longitudinal and transverse dispersion coefficients, respectively.

In summary, we distinguish

1. dispersion due to *molecular diffusion,*
2. dispersion due to the heterogeneity of the reservoir.
   a. *local dispersion* due to mixing at the pore level,
   b. *macroscopic dispersion* due to large scale heterogeneities leading to mixing in the mathematical sense.
3. *Longitudinal* versus *transverse* dispersion

The reason to introduce a macroscopic dispersion coefficient is to avoid detailed simulation of the concentration profile at the heterogeneity scale. As a simple example, we consider Figure 5.6. Say that we wish to upscale the situation where we have a high permeable layer at the bottom and

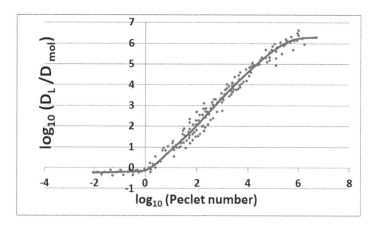

Figure 5.4 Log_{10} (Longitudinal dispersion coefficient divided by molecular diffusion coefficient) (y) versus $\log_{10}$ (Peclet number) (x), retrieved from [175] (flow velocity × characteristic length / molecular diffusion coefficient), see also [30, 52, 81, 86, 131, 133, 175, 199, 210, 222].

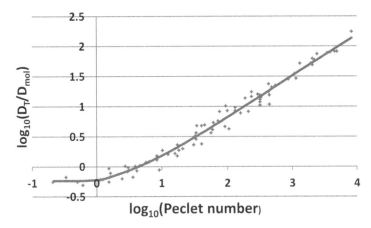

Figure 5.5 $\log_{10}$ (Transverse dispersion coefficient divided by molecular diffusion coefficient) y versus $\log_{10}$ (Peclet number) (x) [56] (flow velocity $\times$ grain diameter / molecular diffusion coefficient) [30, 56, 103, 112, 171, 207, 209].

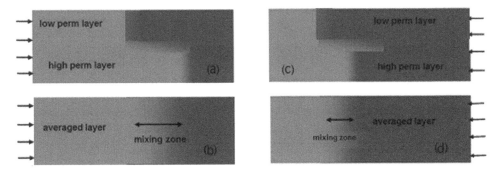

Figure 5.6 Illustration with schematic pictures of occurrence of partial miscibility with macroscopic dispersion and flow reversal (arrows indicate flow direction): panel (a) detailed modeling, panel (b) using a macroscopic dispersion coefficient, panel (c) after reversing the flow by detailed modeling, showing actual mixing between the layers, panel (d) modeling reverse flow with a negative macroscopic dispersion coefficient.

a lower permeable layer at the top. The result will be as in panel a of Figure 5.6. Flow is from left to right and the contaminant moves faster in the high perm layer. There will be some true mixing at the boundary between the layers. In panel (b), instead, we upscale to a single layer and describe the mixing by a macroscopic dispersion coefficient. This dispersion coefficient may depend on time to deal with the convective nature of the "mixing" (see Section 5.6). When we would reverse the flow (from right to left) [23, 129], we observe, as in panel c, that the contaminant will flow back except in a region between the layer where actual mixing has occurred. If we would like to describe this by a dispersion coefficient as in panel d, this dispersion coefficient has to be negative. Figure 5.6 illustrates some of the problems if one wishes to upscale, using enhanced mixing by introducing a macroscopic dispersion coefficient.

Another way of modeling mixing processes is by random walk [229, 230, 231, 232]. A particular streamline will enter randomly distributed regions of high and low permeability. Therefore, the tracer pulse in that streamline moves sometimes ahead and some times lags behind. With respect to its neighboring streamlines, it is highly unlikely that following a specific streamline the tracer is always lucky and only passes through high permeable parts. It is equally unlikely that tracer has only tough luck and never enters a high permeable region. Most likely, the tracer enters as many low as high permeable regions. This is considered analogous to the problem of the inebriated sailor [165], still able to move but unable to navigate, but fortunately confined to the sidewalk (1D movement). He tries to move away from the lantern post. After N steps, he has moved on average $\sqrt{N}$ steps from the lantern post. This is typical of a diffusional process.

We will discuss the following aspects. In Section 5.2, we will review the basic ideas for two-component Fickian diffusion. In Section 5.3, we will give a solution of the convection-diffusion equation and we will discuss Taylor's problem of "dispersion" in a capillary tube. In Section 5.4, we will average the accumulation convection equation to obtain the convection-dispersion equation. In Section 5.5, we discuss the relation between statistics and dispersion. We also show the experimental scale dependence of measured dispersion coefficients. In Section 5.6, we derive the dispersion coefficient for a layered reservoir. In Section 5.7, we derive the famous "Gelhar relation" between the dispersion coefficient and the statistical properties of a reservoir that has a lognormally distributed permeability field with a small correlation length with respect to the reservoir dimensions.

5.2   MOLECULAR DIFFUSION ONLY

The "conditio sine qua non" of actual mixing is *molecular diffusion.* In porous media, diffusion on the pore scale can interact with flow on the pore scale to enhance mixing. In the absence of flow or at very small flow rate, mixing due to diffusion occurs in the absence of these enhancement effects. This pure diffusion aspect will be discussed in this section.

Molecular diffusion is generally described by *Fick's law.* In the class of linear laws (e.g., Newton's law of viscosity, Ohm's law, the law of Darcy, which are admittedly all more complex than they appear), Fick's law is the most complicated. This has to do with the difference between the velocity of species and the diffusional velocity. In a diffusing mixture, various chemical species are moving with different velocities. This velocity can be defined as the average molecular velocity of that species. Let each species with mass density ρ_i move with velocity v_i. We can define the mass average velocity [26] as

$$v = \frac{\sum \rho_i v_i}{\sum \rho_i}. \tag{5.1}$$

This leads to the definition of the diffusion velocity $(v_i - v)$ of a species or rather the velocity of species i with respect to v. The mass flux of species i is denoted by n_i, i.e.,

$$n_i = \rho_i v_i. \tag{5.2}$$

The mass flux relative to the mass average velocity is denoted by

$$j_i = \rho_i(v_i - v). \tag{5.3}$$

We note that for a two-component mixture consisting of species A and B, the $j_A = -j_B$. Fick's law of diffusion states that

$$j_A = -D_{AB} \, \mathbf{grad} \, \rho_A \approx -D_{AB} \bar{\rho} \, \mathbf{grad} \, \omega_A, \tag{5.4}$$

where $\bar{\rho}$ is the average density and ω_A is the mass fraction of species A, and D_{AB} is the binary diffusion coefficient [m^2 / s].

In dilute solutions of a mixture of contaminants, each of the contaminants can, as an approximation, be considered as a two-component mixture with water, without considering the other contaminants.

For estimating the molecular diffusion coefficient [m²/s] in gases, we recommend to use the formula proposed by Slattery and Bird [212]

$$\frac{p\mathcal{D}_{AB}}{(p_{cA}p_{cB})^{1/3}(T_{cA}T_{cB})^{5/12}(1/M_A+1/M_B)^{1/2}} = a\left(\frac{T}{\sqrt{T_{cA}T_{cB}}}\right), \tag{5.5}$$

where the pressure p and the critical pressure p_{cA} are in atmospheres, the temperature T and critical temperature T_{cA} are in [K]. The diffusion coefficient $\mathcal{D}_{AB}$ is in [m²/s]. For non polar gases, excluding nitrogen and helium, $a = 2.745 \times 10^{-8}$ and $b = 1.823$. For pairs, consisting of H_2O and a nonpolar gas $a = 3.640 \times 10^{-8}$ and $b = 2.334$. These equations are between 6% and 8% accurate at atmospheric pressure.

For estimating the molecular diffusion coefficient [m²/s] in liquids, we recommend the Wilke-Chang (1955) [251] equation (see also [191]), which is in essence, an empirical modification of the Stokes-Einstein relation and which reads

$$D_{AB} = 7.4 \times 10^{-12}\frac{\sqrt{\Phi}\sqrt{M_A}T}{\mu_A V_B^{0.6}}, \tag{5.6}$$

where M_A [g/mole] is the molecular weight of solvent, T [K] is the temperature, μ_A [m Pa.s] is the viscosity of the solvent, V_B [cm³ /mole] is the molar volume of the solute and Φ is the association factor of the solvent, which we take equal to 1 if the solvent is oil as Wilke and Chang (1955) recommended for unassociated materials, whereas the recommendation for water as solvent is $\Phi = 2.6$. As an example we obtain for CO_2 in water with $V_B = 35$ [cm³ /mole] [180] at $T = 295$ K, $\mu_A = 0.96$ [m Pa.s], $M_A = 18$ [g/mole], $D_{AB} = 1.84 \times 10^{-9}$ [m² /s], which can be compared to a measured value of 2.3×10^{-9}. This is outside the range of accuracy of 10% claimed for Eq. (5.6) [191].

We will apply Fick's law (5.4) to flow in a porous medium. For this, it is important to understand the meaning of the *interstitial velocity* v_A of a species A and its *Darcy velocity* u_A. To understand these aspects, it is convenient to contemplate the schematic picture shown in Figure 2.1. The interstitial velocity is not the average velocity in the pore because it does not include the effect of tortuosity. It can be verified that the porosity φ in Figure 2.1 is $\varphi = \frac{h}{H\cos\vartheta}$. In the tube we assume that the fluid moves with a single average velocity, i.e., we ignore a parabolic velocity profile in the tube. The *interstitial velocity or pore velocity* is the time it takes for the fluid to travel a length L (from the inlet to the outlet) per unit time (ignoring the tortuous path). Let the flow through the tubes be given by q. Therefore, the "*actual velocity*" in the pores is $v_{\text{real}} = \frac{q}{h}$. The Darcy velocity (taking the width of the porous medium (perpendicular to the paper) equal to unity is $u = \frac{q}{H}$. The interstitial velocity v

$$v = \frac{u}{\varphi} = \frac{q}{h}\cos\vartheta = v_{\text{real}}\cos\vartheta := \frac{v_{\text{real}}}{\tau}, \tag{5.7}$$

where τ is the tortuosity factor.

Our procedure is to state the mass conservation for a volume element of the porous medium exclusive the grains and subsequently use some averaging argument. We assume incompressibility of the fluids. We consider such a volume element $\tilde{V}_\varphi$ where the subindex φ means integration over only the pores. Therefore, the rate of change of mass of A is equal to minus the flux of species A integrated over the surface $\tilde{S}_\varphi$ of the control volume that has pores, i.e.,

$$\frac{dm}{dt} = \frac{d}{dt}\int_{\tilde{V}_\varphi}\rho_A dV = -\oint_{\tilde{S}_\varphi}\rho_A v_{An}dS. \tag{5.8}$$

We use the v_{An} to denote that we mean the interstitial velocity perpendicular to the outflowing surface (Figure 2.1). We apply the divergence theorem (Gauss's integral theorem) and convert from interstitial velocities to real velocities $\mathbf{v}_A = \mathbf{v}_{A,real}/\tau$ and assume that the tortuosity is a constant to obtain

$$\frac{d}{dt}\int_{\tilde{V}_\varphi} \rho_A dV = -\frac{1}{\tau}\int_{\tilde{V}_\varphi} \mathbf{div}\,(\rho_A \mathbf{v}_{A,real})\,dV. \tag{5.9}$$

Then, we subtract and add the mass average velocity $\mathbf{v}_{real}$ and obtain

$$\frac{d}{dt}\int_{\tilde{V}_\varphi} \rho_A dV = -\int_{\tilde{V}_\varphi} \mathbf{div}\frac{\rho_A}{\tau}(\mathbf{v}_{A,real} - \mathbf{v}_{real})dV - \frac{1}{\tau}\int_{\tilde{V}_\varphi} \mathbf{div}\rho_A \mathbf{v}_{real}dV. \tag{5.10}$$

We substitute Fick's law and obtain

$$\frac{d}{dt}\int_{\tilde{V}_\varphi} \rho_A dV = \int_{\tilde{V}_\varphi} \mathbf{div}\frac{1}{\tau}\left(\frac{D_{AB}}{\tau}\mathbf{grad}\,\rho_A\right)_{real} dV$$

$$-\frac{1}{\tau}\int_{\tilde{V}_\varphi} \mathbf{div}\rho_A \mathbf{v}_{real}\,dV. \tag{5.11}$$

We note that $\frac{-D_{AB}}{\tau}\,\mathbf{grad}\,\rho_A$ is the actual flux in the pores as we take the gradient as the concentration at the outlet minus the concentration at the inlet divided by L. Indeed, the gradient (Figure 2.1) must be corrected for the tortuous path. In other words, as we measure the concentration between points in the porous medium, the actual concentration gradient must be corrected for the tortuosity factor τ. We must, however, convert to interstitial values because they are directly related to the mass fluxes that are used in the conservation laws.

We set the volume $\tilde{V}_\varphi$ equal to the *representative elementary volume (REV)*, i.e., a volume large enough such that the integral becomes proportional to $\tilde{V}_\varphi$, but small enough with respect to the total volume of the domain to allow for smooth concentration variations. We divide the integrals by the REV $\tilde{V}_\varphi$ to obtain the averages. We will denote the averages just in the same way as the variables used in the integrand of Eq. (5.12), i.e.,

$$\frac{\partial \rho_A}{\partial t} = +\mathbf{div}\frac{1}{\tau}\left(\frac{D_{AB}}{\tau}\mathbf{grad}\,\rho_A\right)_{real} - \frac{1}{\tau}\mathbf{div}(\rho_A \mathbf{v}_{real}).$$

Subsequently, we reconvert from real to interstitial values, i.e., we multiply by τ both the diffusional term and the convection term and drop the word real. The final result is thus in terms of interstitial variables.

$$\frac{\partial \rho_A}{\partial t} - \mathbf{div}\left(\frac{\mathcal{D}_{AB}}{\tau}\mathbf{grad}\rho_A\right) + \mathbf{div}\,(\rho_A \mathbf{v}) = 0. \tag{5.12}$$

The equation is usually written in terms of molar concentration c instead of mass concentrations. This can be achieved by using the molar average velocity v^* instead of a mass average velocity [26], i.e.,

$$v^* = \frac{\sum c_i v_i}{\sum c_i}. \tag{5.13}$$

This leads to the definition of the diffusion velocity $(v_i - v^*)$ of a species or rather the velocity of species i with respect to v^*. The molar flux relative to the molar average velocity is denoted by

$$J_i = c_i(v_i - v^*). \tag{5.14}$$

We note again that for a two-component mixture consisting of species A and B, the $J_A = -J_B$. Fick's law of diffusion states that

$$J_A = -D_{AB}\,\mathbf{grad}\,c_A \approx -D_{AB}\bar{c}\,\mathbf{grad}\,x_A, \tag{5.15}$$

where $\bar{c}$ is the average concentration and x_A is the mole fraction of species A, and D_{AB} is the binary diffusion coefficient $[\text{m}^2/\text{s}]$.

Analogously to the derivation of Eq. (5.12), we can write in porous media

$$\frac{\partial c}{\partial t} - \mathbf{div}\left(\mathcal{D}_{\text{eff}}\,\mathbf{grad}\,c\right) + \mathbf{div}\left(c\mathbf{v}^*\right) = 0, \tag{5.16}$$

where we also defined the effective diffusion constant $D_{\text{eff}} = D_{AB}/\tau \approx 0.7 D_{AB}$.

In the case that we deal with ideal mixtures, i.e.,

$$c_A \bar{V}_A + c_B \bar{V}_B = 1, \tag{5.17}$$

it is also possible to use volume averaged velocities [26], pp 541. Here $\bar{V}_A$ is the partial molar volume of A, and C_A is molar concentration of A. We write the volume averaged velocities as

$$v^{\blacksquare} = c_A \bar{V}_A v_A + c_B \bar{V}_B v_B \tag{5.18}$$

and it is possible [27] to show for binary mixtures that

$$j^{\blacksquare} = \rho_A\left(v_A - v^{\blacksquare}\right) = \rho\frac{\bar{V}_B}{M_B}j_A = -D_{AB}\,\mathbf{grad}\,\rho_A. \tag{5.19}$$

When the partial molar volumes, $(\bar{V}_A, \bar{V}_B)$, are constant, we can show that

$$\frac{\partial \upsilon}{\partial t} - \mathbf{div}\left(\mathcal{D}_{\text{eff}}\,\mathbf{grad}\,\upsilon\right) + \mathbf{div}\left(\upsilon\mathbf{v}^{\blacksquare}\right) = 0, \tag{5.20}$$

where υ is the volume fraction of one of the components.

5.3 SOLUTIONS OF THE CONVECTION-DIFFUSION EQUATION

5.3.1 INJECTION IN A LINEAR CORE

We consider a linear core in which we displace the original fluid that is colored purple with the injection fluid colored red. Our conditions are such that the mixing is dominated by molecular diffusion. As an initial condition, the right part of the core is filled with purple fluid and the left part with red fluid. At time zero, we withdraw a partitioning screen between the fluids and start to flow at a velocity v. At the far upstream side ($x \to -\infty$ the concentration c (of red fluid) is one and at the far downstream side ($x \to \infty$) it is zero. Eq. (5.12) reduces in one dimension to

$$\frac{\partial c}{\partial t} - \frac{\partial}{\partial x}\left(D_{\text{eff}}\frac{\partial c}{\partial x}\right) + \frac{\partial v c}{\partial x} = 0, \tag{5.21}$$

with initial and boundary conditions

$$\begin{array}{ll} \text{IC} & c(x, t = 0) = c_o(1 - H(x)) \\ \text{BC} & c(x \to -\infty, t) < \infty \\ \text{BC} & c(x \to \infty, t) < \infty, \end{array} \tag{5.22}$$

where $H(x)$ is Heaviside's unit step function

$$H(x) = \begin{array}{ll} 0 & x < 0 \\ [0,1] & x = 1 \\ 1 & x > 0 \end{array} \tag{5.23}$$

The solution of the equation is

$$c = \frac{c_o}{2}\left(1 - \text{erf}\frac{x - vt}{2\sqrt{\mathcal{D}_{\text{eff}}t}}\right),$$ (5.24)

where the errorfunction $\text{erf}(x)$ and the complementary error function $\text{erf}c(x)$ are defined as [3]

$$\text{erf}(x) = \frac{2}{\sqrt{\pi}}\int_0^x \exp(-u^2)du$$

$$\text{erfc}(x) = 1 - \text{erf}(x).$$

Very often, this equation is also used to describe a situation with a different boundary condition [99] namely when the red fluid at the origin is set equal to c_o, i.e., $(c(x = 0,t) = c_o)$. Even if this is formally not correct, the solution (5.24) is very similar to the correct solution

$$c = \frac{c_o}{2}\left(2 - \text{erf}\frac{x - vt}{2\sqrt{\mathcal{D}_{\text{eff}}t}} - \exp\frac{vx}{\mathcal{D}_{\text{eff}}}\text{erf}\frac{x + vt}{2\sqrt{\mathcal{D}_{\text{eff}}t}}\right),$$ (5.25)

except near the injection point.

In most of the cases, we are interested in the injection of a fluid at a constant concentration at the injection point, usually taken as the origin. The correct boundary condition at the origin is $(vc(x = 0+,t) = vc_0 + D_{\text{eff}}\frac{dc}{dx}$. This equation implies that the concentration at the origin is in general lower than the injection concentration. In this case, we obtain

$$c = \frac{c_o}{2}\left(\text{erfc}\left(\frac{x - vt}{2\sqrt{\mathcal{D}t}}\right) - \exp\frac{vx}{\mathcal{D}}\text{erfc}\left(\frac{x + vt}{2\sqrt{\mathcal{D}t}}\right)\left(1 + \frac{v(x + vt)}{\mathcal{D}}\right)\right)$$
$$+ c_o\sqrt{\frac{v^2t}{\pi\mathcal{D}}}\exp\left(\frac{vx}{\mathcal{D}} - \frac{(x + vt)^2}{4\mathcal{D}t}\right).$$ (5.26)

Also the solution (5.24) is very similar to the solution (5.26) except at small times. Equations (5.25) and (5.26) are difficult to evaluate numerically. This has to do in Eq. (5.25) with the fact that a product of a very high and very small number is involved and in Eq. (5.26) that it involves subtraction of very large terms of almost equal magnitude. You may want to use Abramowitz to solve this problem, but, in practice, we always use Eq. (5.24) because it is an excellent approximation. Therefore, in experiments, we determine the concentration profile at the outflow end and use equation (5.24) and plot the argument of the error function versus the inverse error function of $1 - 2c/c_o$.

$$\text{erf}^{-1}\left(1 - 2\frac{c}{c_o}\right) = \frac{x - vt}{2\sqrt{\mathcal{D}_{\text{eff}}t}}.$$ (5.27)

For practical purposes, it is more convenient to use the relation between the error function and the cumulative distribution function in its standard normal form (see next subsection) because it allows plots on probability paper and inverse functions of the normal distribution functions are more readily available (see, for instance, Abramowitz and Stegun [3] or EXCEL).

Van Genughten was also able to get analytical solutions when a decay term is included [239].

Comparison to the Standard Normal Form

The cumulative normal distribution in its standard normal form, i.e., when the average is zero and the standard deviation is equal to one is given by

$$F(z) = \frac{1}{\sqrt{2\pi}} \int_{-\infty}^{z} \exp\left(\frac{-t^2}{2}\right) dt. \tag{5.28}$$

This can be converted to an error function by substituting $y = \frac{t}{\sqrt{2}}$ and $dy = \frac{dt}{\sqrt{2}}$ and thus we find

$$F(z) = \frac{\sqrt{2}}{\sqrt{2\pi}} \int_{-\infty}^{z/\sqrt{2}} \exp\left(-y^2\right) dy$$

$$= \frac{1}{2}\frac{2}{\sqrt{\pi}} \int_{-\infty}^{0} \exp\left(-y^2\right) dy + \frac{1}{2}\frac{2}{\sqrt{\pi}} \int_{0}^{z/\sqrt{2}} \exp\left(-y^2\right) dy$$

$$= \frac{1}{2} + \frac{1}{2}\operatorname{erf}\frac{z}{\sqrt{2}}.$$

This can be rearranged to

$$2F(z) - 1 = \operatorname{erf}\frac{z}{\sqrt{2}},$$

$$2F(\frac{x - vt}{\sqrt{2\mathcal{D}_{\mathrm{eff}}t}}) - 1 = \operatorname{erf}\left(\frac{1}{\sqrt{2}}\frac{x - vt}{\sqrt{2\mathcal{D}_{\mathrm{eff}}t}}\right) = 1 - \frac{2c}{c_o}.$$

as the solution of the equation was

$$\frac{c}{c_0} = \frac{1}{2}\left(1 - \operatorname{erf}\frac{x - vt}{2\sqrt{\mathcal{D}_{\mathrm{eff}}t}}\right). \tag{5.29}$$

and this can be written as

$$F(\frac{x - vt}{\sqrt{2\mathcal{D}_{\mathrm{eff}}t}}) = 1 - \frac{c}{c_o} \quad \text{or}$$

$$\frac{x - vt}{\sqrt{2\mathcal{D}_{\mathrm{eff}}t}} = F^{-1}\left(1 - \frac{c}{c_o}\right).$$

The coordinate x at the outlet is $x = L$. Therefore, if we plot $\frac{L-vt}{\sqrt{2\mathcal{D}_{\mathrm{eff}}t}}$ versus the inverse of the cumulative distribution function $z = F^{-1}\left(1 - \frac{c}{c_o}\right)$ (in EXCEL, $z = \mathrm{NORMINV}\left(\left(1 - \frac{c}{c_o}\right), 0, 1\right)$) we get a straight line and can read the diffusion constant from the plot. Note that the equation is in its standard normal form, i.e., $z = \frac{L-vt}{\sqrt{2\mathcal{D}_{\mathrm{eff}}t}}$ is the standard normal form of the variable. Therefore, the time $t = t_o$ value for which $\left(1 - \frac{c}{c_o}\right) \approx 0.5$ leads to $\frac{L-vt_o}{\sqrt{2\mathcal{D}_{\mathrm{eff}}t_o}} = 0$ and the coordinate in its standard normal form $z = 0$, and consequently $v = L/t_o$. In the same way, it leads the $t = t_1$ value for which $\left(1 - \frac{c}{c_o}\right) \approx 0.84$ to $\frac{L-vt_1}{\sqrt{2\mathcal{D}_{\mathrm{eff}}t_1}} = 1$ and the coordinate in its standard normal form $z = 1$ and the dispersion coefficient can be obtained.

5.3.2 TAYLOR'S PROBLEM IN A CYLINDRICAL TUBE

We consider a cylindrical tube with radius R originally filled with purple fluid. We inject a red fluid from the left at an average velocity $\bar{v}$. There will be a parabolic velocity profile $v(r) = 2\bar{v}\left(1 - (r/R)^2\right)$ in the tube if we disregard entrance effects. Molecular diffusion occurs both axially (in the flow direction) as radially. We want to describe the average concentration profile (averaged over the cross-section of the tube) in terms of a convection-diffusion equation in $1D$ (one dimension). It turns out that for long times, the effective diffusion constant in the $1D$ equation consists of a diffusional part and a dispersion part proportional to the average flow rate in the tube-squared. Here, we state only the results of the equations, i.e.,

$$\frac{\partial c}{\partial t} - D_{AB}\left(\frac{\partial^2 c}{\partial x^2} + \frac{1}{r}\frac{\partial}{\partial r}\left(r\frac{\partial c}{\partial r}\right)\right) + v(r)\frac{\partial c}{\partial x} = 0. \tag{5.30}$$

The solution for the average concentration approaches

$$\frac{\partial \bar{c}}{\partial t} - \frac{\partial}{\partial x}\left(D_{ct}\frac{\partial \bar{c}}{\partial x}\right) + \bar{v}\frac{\partial \bar{c}}{\partial x} = 0, \tag{5.31}$$

where D_{ct} is given in references [79, 203]

$$D_{ct} = \frac{R^2\bar{v}^2}{48\mathcal{D}_{AB}}. \tag{5.32}$$

We note that for a cylindrical model, the dispersion shows a quadratic dependence on the average velocity value. Like so often with cylindrical tube models the porous medium chooses to behave differently, i.e., shows dispersion coefficients which are rather proportional to the average velocity to the power one.

5.4   DERIVATION OF THE DISPERSION EQUATION

It has been suggested that the dispersion equation can be derived from the averaged mass conservation law in a porous medium. This is not the case [79], but we can show that by the averaging an extra term evolves. Experimentally it appears that this term can be cast in a diffusional form with the "diffusion" constant termed the hydrodynamic dispersion coefficient.

Let us start with the equation we have derived above.

$$\frac{\partial \rho_A}{\partial t} - \mathbf{div}\left(\frac{\mathcal{D}_{AB}}{\tau}\mathbf{grad}\rho_A\right) + \mathbf{div}\left(\rho_A\mathbf{v}\right) = 0. \tag{5.33}$$

Now we want to use this equation on a larger scale and consequently average the equation over a volume V_g of interest. One may think of V_g as the volume of a grid cell in a computer simulation. In order to obtain an averaged equation, we will also divide by the volume. Therefore, we obtain

$$\frac{d}{dt}\frac{1}{V_g}\int_{V_g}\rho_A dV = +\frac{1}{V_g}\int_{V_g}\mathbf{div}\left(\frac{\mathcal{D}_{AB}}{\tau}\mathbf{grad}\rho_A\right)dV$$

$$-\frac{1}{V_g}\int_{V_g}\mathbf{div}\left(\rho_A\mathbf{v}\right)dV. \tag{5.34}$$

We may define the averaged quantities

$$\bar{\rho}_A = \frac{1}{V_g}\int_{V_g}\rho_A dV$$

$$\bar{\mathbf{v}} = \frac{1}{V_g}\int_{V_g}\mathbf{v}dV.$$

We split the velocity $\mathbf{v} = \bar{\mathbf{v}} + \mathbf{v}'$ where $\bar{\mathbf{v}}$ is the average velocity and $\mathbf{v}'$ is the deviation from the average velocity. In addition, we split the concentration ρ_A into an average and a fluctuation, i.e., $\rho_A = \bar{\rho}_A + \rho_A'$ and substitute into Eq. (5.34) and average. We obtain for the term

$$\overline{\rho_A \mathbf{v}} = \frac{1}{V_g} \int_{V_g} \mathbf{div}(\rho_A \mathbf{v})\, dV = \bar{\rho}_A \bar{\mathbf{v}} + \overline{\rho_A' \mathbf{v}'}, \tag{5.35}$$

because $\overline{\bar{\rho}_A \mathbf{v}'} = 0$ and $\overline{\rho_A' \bar{\mathbf{v}}} = 0$. Therefore, we obtain (using that the order of differentiation and averaging can be reversed), i.e.,

$$\frac{\partial \bar{\rho}_A}{\partial t} - \mathbf{div}\left(\frac{\mathcal{D}_{AB}}{\tau} \mathbf{grad} \bar{\rho}_A\right) + \mathbf{div}(\bar{\rho}_A \bar{\mathbf{v}}) + \mathbf{div}\left(\overline{\rho_A' \mathbf{v}'}\right) = 0. \tag{5.36}$$

As always, with averaging, we get a closure problem: we do not know how to handle the term $\mathbf{div}\left(\overline{\rho_A' \mathbf{v}'}\right)$. It is, however, reasonable to assume that the concentration fluctuation is proportional to the average applied concentration gradient. The most general form for this would be

$$\rho_A' = \chi \cdot \mathbf{grad} \bar{\rho}_\mathbf{A}, \tag{5.37}$$

where χ is an arbitrary vector, which can be obtained using homogenization [13]. It follows that [79]

$$\overline{\rho_A' \mathbf{v}'_\mathbf{A}} = \overline{\mathbf{v}' \otimes \chi} \cdot \mathbf{grad}\, \bar{\rho}_A := -\mathbf{D} \cdot \mathbf{grad}\, \bar{\rho}_A, \tag{5.38}$$

where we use the notation $(\mathbf{v} \otimes \chi)$ for the product between the two vectors $\mathbf{v}$ and χ. This product is called the dyadic product. The dyadic product is a tensor with as elements, e.g., on the $(1,2)$ position the x-component of one vector with the y-component of the other vector.

$\overline{\mathbf{v}' \otimes \chi}$ defines the hydrodynamic dispersion tensor [17] $\mathbf{D}$, which can be written in it diagonalized form (assuming symmetry between the two transverse directions as

$$\mathbf{D} = \begin{pmatrix} \mathcal{L} & 0 & 0 \\ 0 & \mathcal{T} & 0 \\ 0 & 0 & \mathcal{T} \end{pmatrix}, \tag{5.39}$$

where $\mathcal{L}$ is the longitudinal convective dispersion coefficient and $\mathcal{T}$ is the transverse convective dispersion coefficient. Using homogenization [13], the closure relation (5.38) can be derived.

Therefore, we obtain the final equation (we use $\mathbf{I}$ to denote the unit matrix and omit the bars to calculate the averages)

$$\frac{\partial \rho_A}{\partial t} - \mathbf{div}\left(\left(\frac{\mathcal{D}_{AB}\mathbf{I}}{\tau} + \mathbf{D}\right) \cdot \mathbf{grad}\rho_A\right) + \mathbf{div}(\rho_A \mathbf{v}) = 0, \tag{5.40}$$

which can also be written in terms of molar comcentrations

$$\frac{\partial c}{\partial t} - \left(\mathbf{div}\left(\mathcal{D}_{\text{eff}}\mathbf{I} + \mathbf{D}\right) \cdot \mathbf{grad}c\right) + \mathbf{div}(c\mathbf{v}) = 0. \tag{5.41}$$

We also again defined the effective diffusion constant $\mathcal{D}_{\text{eff}} = \mathcal{D}_{AB}/\tau \approx 0.7 \mathcal{D}_{AB}$.

In the dispersion equation, the concentration has a different meaning. It is an average concentration in a large grid cell. It would be possible to make the same computations but now using more grid cells. I would see in the results of the fine gridded computations some concentration variations within the large grid cell. Therefore, the concentration indicates the probability that I find contaminated water in the grid cell and it has certainly nothing to do with the actual concentration. Therefore, if one is including nonlinear chemical reaction terms into the equation such a description in terms of dispersion can only be used, if one remains aware of the fact that one is using some ill-defined model. (Engineers use the adage "Any answer is better than no answer"; but this approach has the drawback that eventually you are starting to believe the general validity of "any answer".)

The effective diffusion constant comprises the effect of molecular diffusion D_m and dispersion. For the measurement of transverse dispersion, a fluid labeled with a tracer is injected in the lower half of the tube and pure solvent in the upper half. On the axis, we have the 50% line. We can use an error function to describe the concentration profile in the transverse direction, thus deducing the transverse (effective) dispersion coefficient. Usually, the transverse dispersion coefficient is rather small, and we confine interest to the longitudinal dispersion coefficient.

For small rates, the measured effective longitudinal dispersion coefficient is the effective molecular diffusion constant. For high rates, the hydrodynamic dispersion mechanism dominates (Figures 5.4 and 5.3). We distinguish

- D_{eff} the apparent molecular diffusion constant $D_{eff} = \frac{D_m}{\tau}$,
- K_l is the longitudinal dispersion coefficient (total),
- $\mathcal{L}$ is the longitudinal convective dispersion coefficient,
- K_t is the transverse dispersion coefficient (total).
- T is the transverse convective dispersion coefficient.

In carefully designed experimental setups in the laboratory, we obtain the following empirical relation for the longitudinal dispersion coefficient

$$K_l = D_{eff} + \mathcal{L} = D_{eff} + c_l \, v \, \sigma \, d_p, \tag{5.42}$$

where d_p is the grain diameter, c_l a constant (0.5), v the interstitial velocity, and σ a heterogeneity factor which equals 3.5 for a random pack. In the same way we have an empirical relation for the transverse dispersion coefficient

$$K_t = D_{eff} + T = D_{eff} + c_t v \sigma d_p, \tag{5.43}$$

where c_t is a constant (0.0157). Therefore, we note that the transverse hydrodynamic dispersion coefficient is smaller than its longitudinal counter part by a factor of 30. The effective heterogeneity factors often depend on the grain size, which is indicative of packing problems [187]. The heterogeneity factor in outcrop material is still much larger.

5.5 STATISTICS AND DISPERSION

5.5.1 RANDOM WALK MODELS

The random walk models [165] are traditionally explained by following the path of an inebriated sailor [165], still able to walk but unable to navigate. In the one- dimensional example (he is fortunately confined to the sidewalk), he is as likely to do a step west as to do a step east. The probability that he has done $\frac{1}{2}(N+m)$ steps east and $\frac{1}{2}(N-m)$ steps west is denoted by $p(m,N)$ and given by

$$p(m,N) = \frac{N!}{[\frac{1}{2}(N+m)]![\frac{1}{2}(N-m)]!} \left(\frac{1}{2}\right)^N. \tag{5.44}$$

As a result of this total of N steps, he ends up m steps to the east of his starting point. The expression above can be evaluated by means of the Stirling formula, i.e.,

$$\log N! \approx (N + \frac{1}{2}) \log N - N + \frac{1}{2} \log(2\pi). \tag{5.45}$$

When we substitute Eq. (5.44), we find

$$\log p(m,N) = (N + \frac{1}{2}) \log N - \frac{1}{2}(N + m + 1) \log[\frac{N}{2}(1 + \frac{m}{N})]$$
$$- \frac{1}{2}(N - m + 1) \log[\frac{N}{2}(1 - \frac{m}{N})] - \frac{1}{2} \log(2\pi) - N \log 2. \tag{5.46}$$

Since $m \ll N$, we can expand $\log(1 \pm \frac{m}{N} = \pm\frac{m}{N} - \frac{m^2}{2N^2} \pm \ldots$ and substitute into Eq. (5.46), i.e.,

$$\log p(m,N) = (N + \frac{1}{2}) \log N - \frac{1}{2} \log(2\pi) - N \log 2$$
$$- \frac{1}{2}(N + m + 1)(\log N - \log 2 + \frac{m}{N} - \frac{m^2}{2N^2})$$
$$- \frac{1}{2}(N - m + 1)(\log N - \log 2 - \frac{m}{N} - \frac{m^2}{2N^2})$$
$$\approx -\frac{1}{2} \log N + \log 2 - \frac{1}{2} \log 2\pi - \frac{m^2}{2N}. \tag{5.47}$$

In other words, we can write

$$p(m,N) \approx \sqrt{\frac{2}{\pi N}} \exp -\frac{m^2}{2N}. \tag{5.48}$$

We note here that

$$\int_{-\infty}^{\infty} p(m,N)dm = \int_{-\infty}^{\infty} \sqrt{\frac{2}{\pi N}} \exp -\frac{m^2}{2N} dm =$$
$$\sqrt{\frac{4N\pi}{\pi N}} = 2.$$

The reason that the probability function is not normalized to one is that m can only take even or odd values depending on whether N is even or odd. When we introduce the step length of each step l, so that the distance $x = ml$, we can use this to express the probability $p(x,N)dx = p(m,N)dm$. Now we want to get rid of this even and odd number business, and we use that $dx = 2\, l\, dm$ (instead of the usual procedure $dx = ldm$), but keep $x = lm$ to get the probability distribution function properly normalized. Therefore, we obtain

$$p(x,N)dx = \sqrt{\frac{1}{2\pi l^2 N}} \exp -\frac{x^2}{2Nl^2} dx := \sqrt{\frac{1}{2\pi\sigma^2}} \exp -\frac{x^2}{2\sigma^2} dx, \tag{5.49}$$

where σ^2 is the standard deviation. Let us compare the solution with the solution of the diffusion equation (disregarding the convection term), i.e.,

$$\frac{\partial c}{\partial t} - \frac{\partial}{\partial x}\left(D_{\text{eff}} \frac{\partial c}{\partial x}\right) = 0. \tag{5.50}$$

When the initial condition is that the concentration is a delta function, which can be expressed as a normal distribution function in which the standard deviation $(\chi \to 0)$ goes to zero, i.e.,

$$c(x, t \to 0) = \lim_{\chi \to 0} \frac{c_0}{\sqrt{2\pi \chi^2}} \exp -\frac{x^2}{2\chi^2}, \qquad (5.51)$$

and we obtain a solution that remains finite for $x = \pm\infty$, i.e., $c(x \to \infty, t) < \infty$ and $c(x \to -\infty, t) < -\infty$ and

$$\frac{c}{c_0} = \frac{1}{2\sqrt{\pi \mathcal{D}_{\text{eff}} t}} \exp -\frac{x^2}{4\mathcal{D}_{\text{eff}} t}. \qquad (5.52)$$

We define n as the number of steps N per unit time $(n = N/t)$. From the above we can identify $2\sigma^2 = 4D_{\text{eff}} t = 2Nl^2 = 2nl^2 t$. We can now use some intuitive arguments to come up with the famous relation of Gelhar. This equation relates the hydrodynamic dispersion coefficient $\mathcal{L}$ in the same way as the effective diffusion constant above, but now with the number of steps per unit time $n = \frac{v}{\lambda}$ where v is the velocity of the fluid and λ is the correlation length of cells with a specific permeability. The length of a step depends on the product of the incremental permeability $\frac{\Delta k}{k}$ and the correlation length λ, and thus, we can relate $D_{\text{eff}} = nl^2/2 \sim \frac{1}{2}\frac{v}{\lambda}\frac{\Delta k^2}{k^2}\lambda^2 = \frac{1}{2}(\exp \sigma^2 - 1)v\lambda \approx \frac{1}{2}\sigma^2 v\lambda$. Here we have used that the square of the coefficient of variation of lognormal fields is $(\exp \sigma^2 - 1)$ where σ^2 is the variance of the logarithm of the permeability field. We will further discuss how the relation of Gelhar can be derived. Here a factor of $\frac{1}{2}$ comes in, which will not be present in the full derivation. Still the equation gives an order of magnitude and cannot be considered accurate within such a factor of 2.

5.6 VARIANCE OF CONCENTRATION PROFILE AND DISPERSION

By comparison of Eq. (5.49) and (5.52), one can consider the variance of a concentration profile that started out as a delta function as a measure of the dispersion coefficient $\mathcal{L}$, i.e., $2\sigma^2 = 4\mathcal{L}t$. We can generalize $\mathcal{L} = \frac{1}{2}\frac{\sigma^2}{t}$ to $\mathcal{L} = \frac{1}{2}\frac{d\sigma^2}{dt}$ to obtain a local definition of the dispersion coefficient, i.e.,

$$\mathcal{L} = \frac{1}{2}\frac{\sigma^2}{t} =>$$

$$\mathcal{L} = \frac{1}{2}\frac{d\sigma^2}{dt} = \frac{1}{2}\frac{d\sigma^2}{d\bar{x}}\frac{d\bar{x}}{dt} = \frac{1}{2}v\frac{d\sigma^2}{d\bar{x}} := Av, \qquad (5.53)$$

where A [m] is the dispersivity and $\bar{x}$ is the average position of the concentration profile. Indeed, we consider the concentration profile as a probability density function for the position of a fluid element that departed at time zero from x=0, i.e.,

$$\int_{-\infty}^{\infty} c\,dx = constant$$

$$E[X] = \bar{x} = \frac{\int_{-\infty}^{\infty} xc\,dx}{\int_{-\infty}^{\infty} c\,dx}$$

$$\sigma^2 = E[(x - \bar{x})^2] = \frac{\int_{-\infty}^{\infty} (x - \bar{x})^2 c\,dx}{\int_{-\infty}^{\infty} c\,dx}. \qquad (5.54)$$

We now reverse the idea of this statistical approach. Instead of calculating the variance with the concentration profile, we calculate the average position and the variance of the position in a given setting. In a layered system, we define x-position of the concentration following a concentration pulse at the inlet as $x(y)$, where y denotes the coordinate in the cross-layer direction. The thickness of the layered reservoir is H. We define the average and the variance as

$$\bar{x} = E(X) = \frac{1}{H} \int_0^H x(y) dy,$$

$$\text{var}(X) = E\left(X - E(X))^2\right) = \frac{1}{H} \int_0^H \left(x^2(y) - \bar{x}^2\right) dy,$$

where the capital letter, e.g., X denotes the stochastic variable. In this case, the stochastic variables are uniformly distributed in y and $p(y) = \frac{1}{H}$. Indeed, we use $v = \frac{d\bar{x}}{dt}$ is the velocity of the concentration profile or the average interstitial velocity. We can also follow the position of a fluid element to determine the dispersivity. The position x of a fluid element that travels with the interstitial velocity v is equal to $x = vt$. Therefore, we obtain for layered reservoir (in which the pressure gradient in the layer direction $\frac{\partial p}{\partial x}$ is assumed not to depend on the cross-layer direction coordinate (y)) (vertical equilibrium approximation), i.e.,

$$v(y) = \frac{k(y)}{\mu \phi} \frac{\partial p}{\partial x}$$

$$\bar{x} = E[X] = t \frac{E[K(y)]}{\mu \phi} \frac{\partial p}{\partial x} = \frac{\bar{k}t}{\phi \mu} \frac{\partial p}{\partial x}$$

$$\sigma_x^2 = E[(X - E(X))^2] = \left(\frac{1}{\phi \mu} \frac{\partial p}{\partial x}\right)^2 E[(K(y) - \bar{k})^2] t^2 =$$

$$= \frac{E[(K - \bar{k})^2]}{\bar{k}^2} \bar{x}^2. \tag{5.55}$$

The dispersivity for a layered reservoir (excluding the molecular diffusion part) is thus given as $(2\mathcal{L} = d\sigma_x^2/dt = \frac{d\sigma_x^2}{dx}\frac{dx}{dt} = vd\sigma_x^2/dx =: 2Av$

$$A = \frac{1}{2}\frac{d\sigma_x^2}{d\bar{x}} = \frac{E[(K - \bar{k})^2]}{\bar{k}^2} \bar{x} = C_v^2 \bar{x}, \tag{5.56}$$

where C_v is the coefficient of variation and A is the dispersivity [m]. This is the reason for the observed scale dependence of the dispersivity (Figure 7 from [175] shown in Figure 5.4).

5.7 DISPERSIVITY AND THE VELOCITY AUTOCORRELATION FUNCTION

We like to generalize the result of the previous paragraph to an expression for an arbitrary (not necessarily layered) permeability structure. We use all the time the commutative property between differentiation and expectation (integration with the probability density function). The way we do it is formally not very elegant, but a completely correct derivation is very lengthy and does not add to the understanding of concepts. We start with calculating the variance $\sigma^2 = E[(X - \bar{x})^2] = E[X^2] - (E[X])^2$, i.e.,

$$\sigma^2 = E[X^2] - (E[X])^2 = E\left[\left(\int_0^t V(t') dt'\right)^2\right] - \left(E\left[\int_0^t dt' V(t')\right]\right)^2, \tag{5.57}$$

where we use t' to distinguish from the integration boundary t. We write the interstitial velocity in the x-direction $V(t) = E[V] + \widetilde{V} := \bar{v} + \widetilde{V}$ as the sum of an average part $E(V) = \bar{v}$ and a fluctuating part $\widetilde{V}$ for which $E(\widetilde{V}) = 0$, and we obtain

$$\sigma^2 = E\left[\left(\int_0^t \widetilde{V}(t')dt'\right)^2\right], \tag{5.58}$$

where we put all terms denoting average fluctuations equal to zero.

We use this equation in $\mathcal{L}(t) = \frac{1}{2}\frac{d\sigma^2}{dt}$ and obtain by noting that we can interchange differentiation and integration and taking the expectation;

$$\mathcal{L} = \frac{1}{2}\frac{d\sigma^2}{dt} = E\left[\left(\int_0^t \widetilde{V}(t')dt'\right)\widetilde{V}(t)\right]$$

$$= E\left[\int_0^t \widetilde{V}(t)\widetilde{V}(t')dt'\right] = \int_0^t E\left[\widetilde{V}(t)\widetilde{V}(t')\right]dt', \tag{5.59}$$

where we used Leibnitz's rule for differentiation of integrals.

For second-order stationary fields, we have that $E[\widetilde{V}(t)\widetilde{V}(t')]$ only depends in the time difference $(t - t')$. We can now derive the longitudinal convective dispersion coefficient

$$\mathcal{L} = \frac{1}{2}\frac{d\sigma^2}{dt} = \int_0^t E\left[\widetilde{V}(t)\widetilde{V}(t')\right]dt'$$

$$= -\int_t^0 E\left[\widetilde{V}(t - t')\widetilde{V}(0)\right]d(t - t') = \int_0^t E\left[\widetilde{V}(t - t')\widetilde{V}(0)\right]d(t - t')$$

$$= \int_0^t E\left[\widetilde{V}(\tau)\widetilde{V}(0)\right]d(\tau) := \int_0^t d\tau R_{vv}(\tau). \tag{5.60}$$

$R_{vv}(\tau)$ is the velocity autocorrelation function (Figure 5.7). For small times $\mathcal{L} = \int_0^t d\tau R_{vv}(\tau) \approx \sigma_v^2 t := A\bar{v}$. Therefore, it follows that

$$A = \frac{\sigma_v^2}{\bar{v}^2}\bar{v}t \approx \frac{\sigma_k^2}{k^2}\bar{x} \quad \text{for} \quad t \ll \tau_{cor}. \tag{5.61}$$

For long times $\mathcal{L} = \int_0^t d\tau R_{vv}(\tau) \approx \sigma_v^2 \int_0^t \rho_{vv}(\tau)d\tau := A\bar{v}$, where $\rho_{vv}(\tau)$ is the normalized covariance function. The integral over the normalized covariance function $\int_0^\infty \rho_{vv}(\tau)d\tau$ is by definition

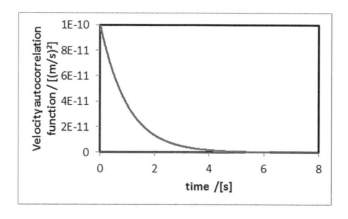

Figure 5.7   The velocity autocorrelation function.

the characteristic time, i.e., the correlation time τ_{cor}, and

$$A = \frac{\sigma_v^2}{\bar{v}^2}\bar{v}\tau_{cor} = \frac{\sigma_v^2}{\bar{v}^2}\lambda \approx \frac{\sigma_k^2}{\bar{k}^2}\lambda, \tag{5.62}$$

where $\lambda = \bar{v}\tau_{cor}$ is the correlation length. For lognormal fields the coefficient of variation squared $\frac{\sigma_k^2}{\bar{k}^2} = \exp(\sigma^2) - 1 \approx \sigma^2$ when σ^2 = the variance of the logarithm of the permeability is small. We obtain the well-known Gelhar relation

$$A = (\exp\sigma^2 - 1)\lambda \approx \sigma^2\lambda. \tag{5.63}$$

5.8 EXERCISE, NUMERICAL/ANALYTICAL 1D DISPERSION

About numerical and analytical solution of the 1D dispersion equation.

We consider a linear core in which we displace the water by a fluid containing a contaminant. The contaminant behaves as a tracer, i.e., it does not affect the viscosity and other fluid properties of the water. The diffusion constant $D_{AB} = 10^{-9}$ [m^2/s] of the contaminant has been measured in the laboratory. A typical value of the tortuosity factor would be 1.4. The interstitial velocity v is 1[m/d]. The time is 1/2 day. Without loss of generality we take the injection concentration c_0 equal to one. The core length is one meter. For the calculation, we need to determine the errorfunctions (see chapter 7, "Error functions and Fressnel Integrals" in Abramowitz and Stegun [3]). For this, we need the following parameters $aa = 0.254829592$, $ab = -0.284496736$, $ac = 1.421413741$, $ad = -1.453152027$, $ae = 1.061405429$ and $p = 0.3257911$. Make a named list of variables in EXCEL by using the insert-name-create command. Be careful with the analytical solutions. Sometimes, huge numbers are subtracted from other huge numbers so that one needs to use approximate procedures to get it right. If such a thing happens, do not worry about it. I have been spending once over 2 weeks to solve such a problem.

1. Use *Tools-Macro-Visual Basic-Editor* to enter in the visual basic editor. Use in the visual basic editor *Insert-Module* to type the function that calculates the complementary error function,

   a. type:
      Function ERFC(x)
      $aa = 0.254829592$
      $ab = -0.284496736$
      $ac = 1.421413741$
      $ad = -1.453152027$
      $ae = 1.061405429$
      $p = 0.3257911$
      $t = 1/(1 + p * \text{Abs}(x))$
      erh $= t * (aa + t * (ab + t * (ac + t * (ad + t * ae)))) * \text{Exp}(-x * x)$
      If $(x > 0)$ Then ERFC = erh
      If $(x < 0)$ Then ERFC = 2 - erh
      End Function

2. We assume that the diffusion convection equation describes this situation, i.e.,

$$\varphi\frac{\partial c}{\partial t} + u\frac{\partial c}{\partial x} = \frac{\partial}{\partial x}\left(D\varphi\frac{\partial c}{\partial x}\right), \tag{5.64}$$

   subject to the boundary and initial conditions:

$$c(x = 0, t) = c_0$$
$$c(x, t = 0) = 0,$$

which has the solution ($u = v\varphi$)

$$c = \frac{c_o}{2}\left(\text{erfc}\left(\frac{x-vt}{2\sqrt{Dt}}\right) + \exp\frac{vx}{D}\text{erfc}\left(\frac{x+vt}{2\sqrt{Dt}}\right)\right). \tag{5.65}$$

Make a plot of the concentration profile at after one half day by using both the first part of the above equation $\frac{c_o}{2}\text{erfc}\left(\frac{x-vt}{2\sqrt{Dt}}\right)$ and the full equation. It turns out that this equation is difficult to evaluate numerically, which is exactly the purpose of asking this question.

3. A better boundary condition would be

$$vc(x=0+,t) = vc_0 + D\left(\frac{\partial c}{\partial x}\right)_{x=0+},$$

where we use again the initial condition $c(x,t=0) = 0$.
Now we obtain

$$c = \frac{c_o}{2}\left(\text{erfc}\left(\frac{x-vt}{2\sqrt{Dt}}\right) - \exp\frac{vx}{D}\text{erfc}\left(\frac{x+vt}{2\sqrt{Dt}}\right)\left(1 + \frac{v(x+vt)}{D}\right)\right)$$
$$+ c_o\sqrt{\frac{v^2t}{\pi D}}\exp\left(\frac{vx}{D} - \frac{(x+vt)^2}{4Dt}\right).$$

Compare to the plot above. It turns out that this equation is almost impossible to evaluate numerically.

4. Freshwater at relative concentration $\frac{c}{c_0} = 0$ is introduced into a salt column with contaminated water at relative concentration $\frac{c}{c_0} = 1$. As the contaminated water is displaced, the following measurements are made. The average velocity is $v = 1.6$, [19].
(a) At what position were the readings given in Table 5.1 taken?
(b) Determine the dispersion coefficient and the velocity. Hint use Eq. (5.24) and divide the argument (both in the numerator and denominator) of the complementary error function by v.

$$c = \frac{c_o}{2}\text{erfc}\left(\frac{x/v - t}{2\sqrt{D_{\text{eff}}/(v^2t)}}\right), \tag{5.66}$$

Use x/v and D_{eff}/v^2 as parameters for minimizing the difference using the solver option in EXCEL.

5. We defined n as the number of steps N per unit time ($n = N/t$) and identified $2\sigma^2 = 4D_{\text{eff}}t = 2Nl^2 = 2nl^2t$. Therefore, we can use $4D_{\text{eff}}t = 4D_{\text{eff}}N\Delta t = 2Nl^2$. It follows that $l = \sqrt{2D_{\text{eff}}\Delta t}$. Take the average velocity $v = 10^{-5}$ [m/s]. We define a particle that moves every time step by $x(N) = x(N-1) + v\Delta t + \sqrt{12}\left(\mathcal{R} - \frac{1}{2}\right)l$ with $x(0) = 0$ and $\mathcal{R}$ is a random number uniformly distributed between $[0,1]$. The factor $\sqrt{12}$ corrects for the fact that the variance of a uniformly distributed variable is $1/12$.

6. Explain the transition from Eq. (5.58) to Eq. (5.59) and give a more detailed reason how the autocorrelation function is used to obtain the difference between Eqs. (5.61) and (5.62).

Use EXCEL. Define a column of some 1,000 cells say the C-column. Write, e.g., in cell C3:= if(start=0,0,c3+v*dt+$\sqrt{}$(12)((rand()-0.5))*l) and copy to cells c3:c1002. After performing a reasonable number of steps say 400,000, copy/paste special (values) the C-column into the H-column. Sort and estimate the cumulative distribution function $F = (i - \frac{1}{2})/1,000$, where i is the

Table 5.1
Time versus Concentration

time /s	concentration [fraction]
2450	0.002940577
2470	0.005206637
2490	0.00887362
2510	0.014574326
2530	0.02309635
2550	0.035358049
2570	0.052353755
2590	0.075067233
2610	0.104359134
2630	0.140841549
2650	0.184758665
2670	0.235895056
2690	0.293531127
2710	0.356458451
2730	0.423057526
2750	0.491429134
2770	0.559560675
2790	0.625502877
2810	0.687531423
2830	0.744272184
2850	0.79477659
2870	0.838543107
2890	0.875489712
2910	0.905888868
2930	0.930279954
2950	0.949374329
2970	0.963965849
2990	0.97485582
3010	0.982797066
3030	0.988457971
3050	0.992404472

position in the sorted column and put the result in the F-column. Plot versus the inverse cumulative distribution function in the standard normal form *(norminv(F,0,1)* in the G-column. Plot the G-column versus the H-column and use the insert trendline option to determine the (linear) slope. Draw some conclusions.

5.9   EXERCISE, GELHAR RELATION

About showing that the dispersion coefficient $\mathcal{L} = \left(\exp(\sigma^2) - 1\right)\lambda v$ by solving the concentration equation in a heterogeneous permeability field.

- Use the permeability field in exercise 2.5.7 to show that concentration profile averaged over the height behaves as the solution of the convection-diffusion equation with a longitudinal dispersion coefficient [98, 97]

$$\mathcal{L} = \left(\exp(\sigma^2) - 1\right)\lambda v, \tag{5.67}$$

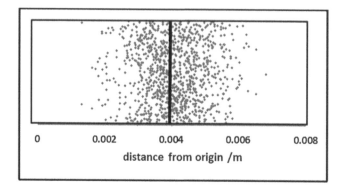

Figure 5.8 Random walk result with velocity $v = 1.0 \times 10^{-5}$ [m/s], diffusion coefficient $D = 1.0 \times 10^{-9}$, [m²/s], time step $\Delta t = 1.0 \times 10^{-3}$ [s], total time $t = 400$ [s]. Each particle moves moves every time step by $x(N) = x(N-1) + v\Delta t + (\mathcal{R} - \frac{1}{2})l$ with $x(0) = 0$ and $\mathcal{R}$ is a random number uniformly distributed between $[0, 1]$. The vertical axis assigns to each point a random uniformly distributed random displacement and is only added to make the points visible.

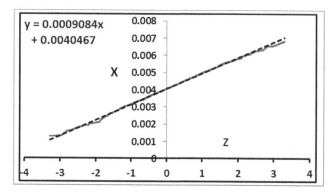

Figure 5.9 Plot of X-values versus standard normal variable Z (inverse of cumulative probability in standard normal form). The cumulative distribution function can be obtained after sorting the X-values and assign to the ith value in the sorted column the value $(i - 1/2)/N$, where N are the number of data points. The variable Z can be obtained with EXCEL with norminv(F, 0,1), where F is the estimated value of the cumulative distribution function. If X is normally distributed we have that $X = \sigma Z + X_{av}$. The estimated average value $X_{av} = vt = 0.0040467$ (Figure 5.8), compared to an expected value of 0.004. The standard deviation $\sigma = 0.0009084$ corresponding to a variance of $\sigma^2 = 8.25 \times 10^{-7} \approx 2Dt = 8.0 \times 10^{-7}$.

where v is the interstitial velocity, σ^2 is the variance of the natural logarithm of the permeablity and λ is the correlation length (distance between points). The standard variation of the variance [46] is about 0.1 can therefore be ignored,

- show that for a standard deviation of s = 1.60944, the Dykstra-Parsons coefficient $V_{DP} = 1 - \exp(-s) = 0.8$, the variance of the $\ln k$ field is 2.59 and the dispersion coefficient $\mathcal{L} \approx 12.33\lambda v$. As the average velocity for the example chosen by us is $5.06/20 = 0.253$[m/s] and the correlation length is the grid size (0.05), leading to a dispersion coefficient of 0.155[m²/s],

- use the nomenclature shown in Figure 2.23,

- calculate the (cast) velocity field across the east boundary of every cell as $v_x^{eb} = \lambda_{PE}(\Phi_P - \Phi_E))/\Delta x$. Note that the velocity on the eb of the W cell is the velocity across the wb of the P cell ,
- calculate the (south) velocity field across the south boundary of every cell as $v_y^{sb} = \lambda_{PS}(\Phi_P - \Phi_S))/\Delta y$,
- denote the concentration by c and solve the equation

$$
\begin{aligned}
c_P(t+\Delta t) = c_P(t) &- (v_x^{eb} + abs(v_x^{eb}))/\Delta x/2)c_P - (v_x^{eb} - abs(v_x^{eb}))/\Delta x/2)c_E \\
&+ (v_x^{wb} + abs(v_x^{wb}))/\Delta x/2)c_W + (v_x^{wb} - abs(v_x^{wb}))/\Delta x/2)c_P \\
&+ (v_y^{nb} + abs(v_y^{nb}))/\Delta y/2)c_N - (v_y^{nb} - abs(v_y^{nb}))/\Delta y/2)c_P \\
&- (v_y^{sb} + abs(v_y^{sb}))/\Delta y/2)c_P - (v_y^{sb} - abs(v_y^{sb}))/\Delta y/2)c_S,
\end{aligned}
\tag{5.68}
$$

- Put in cell C10"=IF(startt=0,cinit,C10+((vx!B10*B10+ABS(vx!B10)*B10)/(2*dx) +(vx!B10*C10-ABS(vx!B10*C10))/(2*dx)-(vx!C10*C10+ABS(vx!C10)*C10)/(2*dx) -(vx!C10*D10-ABS(vx!C10*D10))/(2*dx)+(vy!C9*C9+ABS(vy!C9)*C9)/(2*dy) +(vy!C9*C10-ABS(vy!C9*C10))/(2*dy)-(vy!C10*C10+ABS(vy!C10)*C10)/(2*dy) -(vy!C10*C11-ABS(vy!C10*C11))/(2*dy))*dt)".
- Use concentration c = 1 at the injection side cells B10: B30 ,
- compare the numerically calculated concentration profile and the theoretical concentration profile

$$
c = \frac{c_o}{2}\left(\text{erfc}\left(\frac{x-vt}{2\sqrt{\mathcal{L}t}}\right) + \exp\frac{vx}{\mathcal{L}}\,\text{erfc}\left(\frac{x+vt}{2\sqrt{\mathcal{L}t}}\right)\right),
\tag{5.69}
$$

- draw conclusions (Figure 5.9).

5.10   NUMERICAL ASPECTS

By our way of discretization of the convection term, we introduce a diffusion term in the equation. This is called numerical dispersion. We try to solve

$$
\frac{\partial c}{\partial t} + v\frac{\partial c}{\partial x} = 0.
\tag{5.70}
$$

We convert to the numerical representation with the following substitutions (Figure 5.10)

- contaminant into cell $(i) = vc_{i-1}(t)$,
- contaminant out of cell $(i) = vc_i(t)$
- contaminant accumulation in cell $(i) = \frac{(c_i(t+\Delta t) - c_i(t))\Delta x}{\Delta t}$.

and use $W = i-1$, $P = i$. Therefore, our numerical equation reads for an equidistant grid

$$
\frac{c_i(t+\Delta t) - c_i(t)}{\Delta t} = \frac{v(c_{i-1}(t) - c_i(t))}{\Delta x}.
\tag{5.71}
$$

To understand that this gives numerical dispersion, we use the Taylor expansions up to second order

$$
c_i(t+\Delta t) = c_i(t) + \frac{\partial c_i}{\partial t}\Delta t + \frac{1}{2}\frac{\partial^2 c_i}{\partial t^2}(\Delta t)^2 ,
\tag{5.72}
$$

and

$$
c_{i-1}(t) = c_i(t) - \frac{\partial c_i}{\partial x}\Delta x + \frac{1}{2}\frac{\partial^2 c_i}{\partial x^2}(\Delta x)^2.
\tag{5.73}
$$

Substitution of (Eq. 5.72) and (Eq. 5.73) into (Eq. 5.71) leads to

$$\frac{\partial c_i}{\partial t} + v\frac{\partial c_i}{\partial x} = \frac{1}{2}v\frac{\partial^2 c_i}{\partial x^2}\Delta x - \frac{1}{2}\frac{\partial^2 c_i}{\partial t^2}\Delta t \ . \tag{5.74}$$

Differentiation of (Eq. 5.70) toward t leads to

$$\frac{\partial^2 c}{\partial t^2} + v\frac{\partial^2 c}{\partial x \partial t} = 0, \tag{5.75}$$

while differentiation of (Eq. 5.70) toward x leads to

$$\frac{\partial^2 c}{\partial x \partial t} + v\frac{\partial^2 c}{\partial x^2} = 0. \tag{5.76}$$

From (Eq. 5.76) and (Eq. 5.75), it follows that

$$\frac{\partial^2 c}{\partial t^2} = v^2\frac{\partial^2 c}{\partial x^2} \ . \tag{5.77}$$

We obtain by substitution of (Eq. (5.77) in (Eq. (5.74))

$$\frac{\partial c_i}{\partial t} + v\frac{\partial c_i}{\partial x} = D_{num}\frac{\partial^2 c}{\partial x^2} \ , \tag{5.78}$$

where D_{num} is the numerical diffusion coefficient, i.e.,

$$D_{num} = \frac{1}{2}v\Delta x - \frac{1}{2}v^2\Delta t. \tag{5.79}$$

It turns out that when we apply the same reasoning to the implicit formulation

$$\frac{c_i(t+\Delta t) - c_i(t)}{\Delta t} = \frac{v(c_{i-1}(t+\Delta t) - c_i(t+\Delta t))}{\Delta x}, \tag{5.80}$$

that we obtain a numerical diffusion coefficient of

$$D_{num} = \frac{1}{2}v\Delta x + \frac{1}{2}v^2\Delta t. \tag{5.81}$$

5.A   HIGHER-ORDER FLUX FUNCTIONS FOR HIGHER-ORDER SCHEMES

Use *Tools-Macro-Visual Basic-Editor* to enter in the visual basic editor. Use in the visual basic editor *Insert-Module* to type the function for the following upstream weighted flow functions. One may prefer to program these functions as shown below. Exit the visual basic program when you are done. This speeds up the program. In the schemes below, put "lambda" = 0 in nonlinear problems [108, 160, 202]. In the linear diffusion convection type of equation $\frac{\partial c}{\partial t} + v\frac{\partial c}{\partial x} = \frac{\partial}{\partial x}\left(D\frac{\partial c}{\partial x}\right)$, we take $\lambda = \frac{v\Delta t}{\Delta x}$. We do not deal here with the diffusive flux, which needs to be separately added. The higher-order schemes are only applicable for equidistant grids. Input in the programs are the fluxes over the east boundaries of the cells. Respectively $fluxup, flux, fluxdown$ are the upstream weighted fluxes over the east boundary over the cells $W(est), P(oint)$ and $E(ast)$.

For the convection-diffusion problem, where the flow is from west to east we have $fluxup = uc_W, flux = uc_P$ and $fluxdown = uc_E$.

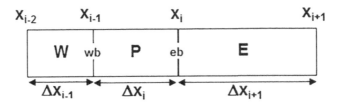

Figure 5.10 Three adjacent grid cells. The middle cell is labeled $P(oint)$, the cell to the left $W(est)$ and the cell to the right $E(ast)$. The coordinates of the grid cell boundaries are given at the top. Below we indicate the gridcell lengths. The left and right boundary of grid cell are indicated as wb and eb, i.e., (west-boundary) and (east-boundary) respectively.

One Point Upstream Weighting

Function QWD(flux)
QWD = flux 'this is the flux crossing the east boundary of the grid cell P (Figure 5.10)
End Function

Two Point Upstream Weighting

Function QWTPD(fluxup, flux, fluxdown, lambda)
QWTPD = flux: 'this is the flux crossing the east boundary of grid cell P(oint)
If (fluxdown - flux <> 0)
Then: 'fluxdown is the flux crossing the east boundary of cell E(ast),
r = (flux - fluxup) / (fluxdown - flux): 'fluxup is the flux crossing the east boundary of cell W(est)
phi = r: 'ratio of flux gradients
If (phi > 2) Then phi = 2: 'flux limiter
If (phi < = 0) Then phi = 0
QWTPD = flux + 0.5* phi * (fluxdown - flux) * (1 - lambda)
End If
End Function

Two Point Upstream Weighting for Engineers

Function QWTWOPD(fluxup, flux, fluxdown)
fluxhulp = 2.5 * flux - 1.5 * fluxup
If (fluxhulp < 0) Then fluxhulp = 0
QWTWOPD = fluxhulp
End Function

Mid Point Upstream Weighting

Function QWMPD(fluxup, flux, fluxdown, lambda)
QWMPD = flux
If (fluxdown - flux <> 0) Then r = (flux - fluxup) / (fluxdown - flux)
phi = r + Abs(r): 'limits the gradient of flux ratios as being larger than zero
If (phi > 1) Then phi = 1: ' limits the gradient of flux ratios as being smaller than one
QWMPD = flux + 0.5* phi * (fluxdown - flux) * (1 - lambda)
End If
End Function

Third Order Upstream Weighting

Function QWLND(fluxup, flux, fluxdown, lambda) ' third-order leonard scheme
lambda = v Dt / Dx ' can be put equal to zero for a nonlinear case
QWLND = fluxdown
If (fluxdown - flux <> 0) Then r = (flux - fluxup) / (fluxdown - flux)
phi = (2 - lambda) / 3 + (1 + lambda) / 3 * r
If (phi > 2) Then phi = 2
If (phi <= 0) Then phi = 0
If (phi > 2 * r And phi <= 2) Then phi = 2 * r
QWLND = flux + phi / 2 * (fluxdown - flux) * (1 - lambda)
End If
End Function

5.B NUMERICAL MODEL WITH THE FINITE VOLUME METHOD

The numerical model gives a discretized version of the model Eq. (5.64) as shown in Figure 5.10. One defines n grid cells of which three grid cells are shown, which are labeled $W(est), P(oint)$, and $E(ast)$. One may prefer to number the grid cells, i.e., $W = i-1, P = i$ and $E = i+1$. The boundaries of cell $P = i$ have distances from the origin equal to x_{i-1} and x_i and one may assume that $x_0 = 0$. These boundaries are also labeled as $wb = $ west boundary and $eb = $ east boundary respectively. The grid cell lengths are indicated for the cells W, P, E as $\Delta x_{i-1}, \Delta x_i$ and Δx_{i+1}, respectively. When all grid cell lengths are equal, we talk about an equidistant grid. We will mostly use an equidistant grid in all our examples. For all our one$-D$ conservation law examples, we will use this grid cell notation.

The numerical model can be derived by integrating the partial differential equation over the grid block cell P, i.e.,

$$\int_{x_{i-1}}^{x_i} \varphi \frac{\partial c}{\partial t} dx + \int_{x_{i-1}}^{x_i} \frac{\partial (uc)}{\partial x} dx = \int_{x_{i-1}}^{x_i} \frac{\partial}{\partial x} \left(D_{eff} \varphi \frac{\partial c}{\partial x} \right) dx. \tag{5.82}$$

It is usually assumed that the accumulation term is about constant in the grid cell. Therefore, we obtain

$$\varphi \frac{\partial c}{\partial t} \Delta x_i = (uc)_{wb} - (uc)_{eb} + \left(D_{eff} \varphi \frac{\partial c}{\partial x} \right)_{eb} - \left(D_{eff} \varphi \frac{\partial c}{\partial x} \right)_{wb}. \tag{5.83}$$

All quantities on the right side of the equation are considered to be evaluated at time t, i.e., at the present time. This is called the explicit numerical scheme. A scheme in which the right side is evaluated at the future time $t + \Delta t$ is called an implicit scheme. If we were to evaluate the concentrations in the convection term at the west and east boundary, we introduce negative diffusion and the numerical solution becomes unstable. Therefore, we must evaluate this term at the grid cell where the flow is coming from. We also treat the accumulation term in terms of a forward difference scheme. The concentration is considered to change linearly with x in for the evaluation of the diffusion term. The result is

$$\varphi \frac{\Delta x_i}{\Delta t} (c(t + \Delta t) - c(t)) = \frac{1}{2} (u_{wb} + abs(u_{wb})) c_W$$

$$+ \frac{1}{2} \left(u_{wb} - abs\left(u_{wb} \right) \right) c_P - \frac{1}{2} \left(u_{eb} + abs\left(u_{eb} \right) \right) c_P$$

$$- \frac{1}{2} \left(u_{eb} - abs\left(u_{eb} \right) \right) c_E + \left(2 D_{eff} \varphi \left(\frac{c_E - c_P}{\Delta x_P + \Delta x_E} \right) \right)$$

$$- \left(2 D_{eff} \varphi \left(\frac{c_P - c_W}{\Delta x_W + \Delta x_P} \right) \right)$$

where we have used $\Delta x_{i-1} = \Delta x_W$, $\Delta x_i = \Delta x_P$, and $\Delta x_{i+1} = \Delta x_E$. As the flux over the east boundary of cell i is equal to the flow over the west boundary of the next cell $i+1$ it is convenient to identify flux over the east boundary of cell P and cell W respectively

$$(flux(P))_{eb} = + \frac{1}{2} \left(u_{eb} + abs\left(u_{eb} \right) \right) c_P + \frac{1}{2} \left(u_{eb} - abs\left(u_{eb} \right) \right) c_E$$

$$- \left(2 D_{eff} \varphi \left(\frac{c_E - c_P}{\Delta x_P + \Delta x_E} \right) \right)$$

$$(flux(W))_{eb} = + \frac{1}{2} \left(u_{eb} + abs\left(u_{eb} \right) \right) c_W + \frac{1}{2} \left(u_{eb} - abs\left(u_{eb} \right) \right) c_P$$

$$- \left(2 D_{eff} \varphi \left(\frac{c_P - c_W}{\Delta x_W + \Delta x_P} \right) \right)$$

Therefore, we can write the numerical scheme as

$$\Delta c_P(t) = \frac{\Delta t}{\varphi \Delta x_P} \left((flux(W))_{eb} - (flux(P))_{eb} \right) \tag{5.84}$$

$$c_P(t + \Delta t) = c_P(t) + \Delta c_P(t) \tag{5.85}$$

For an equidistant grid ($\Delta x_P = \Delta x$), one can use the so-called Courant-Friedrichs-Lewy (CFL) condition, which states that $\Delta t << \frac{\varphi \Delta x}{u}$. In practice, one often chooses Δt to limit the concentration change $\Delta c_P(t)$ per time step. To get a more uniform programming style, we prefer to use the latter option in all the programs.

Petroleum engineers like to write

$$\varphi \frac{d}{dt} \int_{\tilde{V}_{\varphi}} \rho_A dV = - \int_{\tilde{V}_{\varphi}} \mathbf{div}\, \rho_A \mathbf{u}_A dV, \tag{5.86}$$

which is preferable.

Glossary

capillary pressure: difference between wetting and non-wetting phase pressures. Units [Pa].

conservation laws: in iso-thermal transport processes, we must use the law of mass conservation and the equation of motion (Darcy's law). That all there is to it for isothermal displacement processes.

Darcy velocity of phase u_α: volumetric flux of phase α, i.e., $u_\alpha = Q_\alpha/A$, where A is the surface area of the porous medium (not the pores) through which the flow is occurring.

Darcy's Law for two-phase flow: relates the Darcy velocity to the phase potential, i.e., $u_\alpha = -k_\alpha/\mu_\alpha \, \mathbf{grad}(P_\alpha + \rho_\alpha gz)$ where k_α is the permeability of phase α. It has units [m^2]. In practice, the unit Darcy is used. The conversion is 1 Darcy $= 0.987 \times 10^{-12}$m^2. Apart from the usual assumptions for one-phase flow (disregarding inertia effects), Darcy's law of two-phase flow assumes that the flow of one phase is only affected by the presence and not by the flow of the other phase.

density of phase ρ_α : mass per unit volume of phase α. Units [kg/m$^3_\alpha$].

dispersion coefficients: coefficients that describe various type of mixing processes in porous media. Units [m^2/s].

- D_{eff} the apparent molecular diffusion constant $D_{\text{eff}} = \frac{\mathcal{D}_m}{\tau}$, where $\mathcal{D}_m$ is the molecular diffusion coefficient outside the porous medium. There is no porosity dependence because the convective flux and the volume are all reduced by a factor of φ,
- K_l is the longitudinal dispersion coefficient (total). It describes the convective mixing in the flow direction. K_l adds the convective mixing and the contribution due to molecular diffusion coefficient $\frac{\mathcal{D}_m}{\tau}$,
- $\mathcal{L}$ is the longitudinal convective dispersion coefficient without the molecular diffusion contribution,
- K_t is the transverse dispersion coefficient (total), with the molecular diffusion contribution
- T is the transverse convective dispersion coefficient without the molecular diffusion contribution.

drainage: a process where the non-wetting phase saturation is increasing.

end point permeability k'_α: permeability of one-phase at irreducible saturation of the other phase. It has units [m^2]. In the same way we define the relative end point permeability $k'_{r\alpha} = k'_\alpha/k$ as the relative permeability of one phase at irreducible saturation of the other phase.

ERoEI: Exergy return on Exergy invested: a bookkeeping procedure that calculates the main contributors to the exergy invested, like circulation costs, drilling costs to the exergy recovered (10.7 kWh/L), 40–55 MJ/kg for coal, oil, gas.

exergy: energy that can be converted into work.

imbibition: a process where the wetting phase saturation is increasing.

irreducible oil saturation S_{or}: oil saturation at which and below which oil is not able to flow if subjected to a potential gradient in the oil- phase $grad(P_o + \rho_o gz)$. Its units are [m^3_o/m$^3_{void}$], where the subindex o denotes fluid phase oil.

irreducible water saturation S_{wc}: water saturation at which and below which water is not able to flow if subjected to the potential gradient in the water-phase $grad(P_w + \rho_w gz)$. Water (in a water wet medium (see below)) will always flow but at such a low rate that it becomes practically negligible. Instead of irreducible water saturation also, the term connate water saturation or initial water saturation is used. Units $[m_w^3/m_{void}^3]$, where the subindex w denotes fluid phase *water*.

phase potential ϕ_α: potential of a phase α is the pressure + potential energy per unit volume of that phase, i.e., $P_\alpha + \rho_\alpha gz$ with units [Pa]. Groundwater people use the piezometric head defined by $(P_\alpha/\rho_\alpha g + z)$ with units [m], but for two-phase flow calculations, the definition of potential is preferred as it relates more directly to the Navier-Stokes equation.

porosity φ: void fraction, i.e., volume of voids divided by total volume. Units $[m_{void}^3/m_{pm}^3]$, where the subindex pm denotes porous medium. Groundwater people use the symbol "n" for porosity.

relative permeability $k_{r\alpha}$: phase permeability k_α divided by the one-phase permeability k.

saturation S_α: volume fraction of pores occupied by fluid (liquid or gas) α. Units $[m_\alpha^3/m_{void}^3]$, where the subindex α denotes fluid phase α. Groundwater people use the word retention to describe the moisture content; it denotes the volume fraction of porous medium occupied by liquids and this has units $[m_\alpha^3/m_{pm}^3]$.

total flow rate of phase Q_α: volume of fluid α that flows through an area A Units [m³/s]. s

viscosity of phase μ_α: defined by Newton's law (proportionality constant between stress and velocity gradient) $\sigma_{xz} = \mu_\alpha \frac{\partial v_z}{\partial x}$, and μ_α has units [Pa s].

wetting and non-wetting phase: phase for which the interfacial energy of the fluid and the grains is lower and higher, respectively.

worldwide (Dutch) energy consumption: 15,000-18,000 GW (110 GW)

List of Symbols[1]

Roman

$A_{\alpha\beta}$	Surface area between phase α and β	[m^2]
a	Wetted surface / volume bed	[m^{-1}]
B_o	Oil formation volume factor	[m^3/m^3]
c_α	Compressibility of phase $\alpha = o, w, f$	[Pa^{-1}]
C_A	Dietz shape factor (see Eq. (3.57))	[−]
CF	Carnot Factor $1 - T_0/T$, output work/ throughput energy	[−]
C_w	Heat capacity	[J/m^3/K]
Ex	Exergy	[J]
D_{ip}	Distance between injection and production well	[m]
D_p	Grain diameter	[m]
g	Acceleration due to gravity	[m/s^2]
$h(k)$	Probability density of permeability	[m^{-2}]
$E(k)$	Expected value of k	unit of k (see Eq. (2.68))
E_k	Fuel energy content	[J]
$F(k)$	Cumulative distribution function of k	[−]
f_p	Footprint	[kg CO$_2$ / MJ]
G	Gravity number $\frac{g \Delta \rho_{\alpha\beta} \sin \vartheta}{u \mu_w k'_{rw}}$	
h	Total head $p/(\rho g) + z + \frac{1}{2} v^2/g$	[m/s^2]
H	Height of reservoir	[m]
II	Injectivity index	[m^3/s/(Pa)]
k	Permeability	[m^2] or Darcy
K	Hydraulic conductivity	[m/s]
$k_{r\alpha}$	Relative phase α permeability	[−]
k_{ro}	Relative oil permeability	[−]
k'_{ro}	Relative oil end point permeability	[−]
k_{rw}	Relative water permeability	[−]
k'_{rw}	Relative water end point permeability	[−]
K	Hydraulic conductivity	[m/s]
L	Length of reservoir	[m]
LHV	Combustion energy excluding heat of condensation	[J]
$m(p)$	Generalized pressure for gases (see (Eq. 3.11))	[Pa]
M	Mobility ratio $\frac{\mu_o k'_{ro}}{\mu_w k'_{rw}}$	[−]
p, P	Pressure	[Pa]
P	80% Passing size for feed	[mm]
P_α	Pressure of phase α	[Pa]
P_c	Capillary pressure	[Pa]
P_{cb}	Capillary pressure at unit wetting phase saturation	[Pa]
P	Power to circulate fluids between wells	[J/s]
P	Central point for numerical calculations	[−]
P	80% passing size for product	[mm]
$P(k)$ or $p(k)$	Probability density function of k	[unit of k]$^{-1}$
PI	Productivity index	[m^3/s/(Pa)]
q	Flow rate	[m^3/s]
R	Gas constant	[J/mole/K]
R	Tube radius	[m]
R_e	Reynolds number	[−]
R_h	Area/ wetted perimeter	[m]

---

[1]locally used symbols not quoted, symbols may be duplicate to use most usual symbol i.

s	Standard deviation of $\ln(k)$	[$-$]
R	Tube radius	[m]
R	Tube radius	[m]
S_{or}	Residual oil saturation	[$-$]
S_{wc}	Connate water saturation	[$-$]
S_{we}	Effective saturation (Eq. (4.16))	[$-$]
S_{wi}	Initial water saturation	[$-$]
T	Temperature	[K]
T_0	Temperature of environment	[K]
T_{ref}	Reference Temperature	[$313\ K$]
u	Flow rate [m^3/s]/ A[m^2]	[m/s]
V	Volume	[m^3]
VE	Vertical (capillary-gravity) Equilibrium	[$-$]
V_{DP}	Dykstra Parsons coefficient ($V_{DP} = 1 - \exp(-s)$)	[$-$]
v^2/g	Velocity head	[m]
W	Dimensionless water production (Eq. 4.131)	[$-$]
W_V	Crushing energy	[J/m^3]
x, y	Coordinates	[m]
z	Elevation head, or coordinate	[m]

subindex

D	Dimensionless	[$-$]
E	East gridcell	[$-$]
f	Formation (see Eq. (3.9))	[$-$]
g	Gas	[$-$]
N	North gridcell	[$-$]
nb	North boundary	[$-$]
o	Oil	[$-$]
PE	Between central gridcell(P) and East gridcell (E)	[$-$]
PN	Between central gridcell(P) and North gridcell (N)	[$-$]
PS	Between central gridcell(P) and South gridcell (S)	[$-$]
PW	Between central gridcell(P) and West gridcell (W)	[$-$]
S	South gridcell	[$-$]
sb	South boundary	[$-$]
w	Water	[$-$]
W	West gridcell	[$-$]
wb	West boundary	[$-$]
$//$	x, y coordinate	[$-$]

Greek

β	Inertia factor	[m^{-1}]
γ	Geometry factor	[$-$]
γ	Inverse of percolation threshold	[$-$]
γ	Between 0.3 and 0.7 for capillary pressure function	[$-$]
δ	Wall thickness	[m]
$\Delta H_{combust}$	LHV (methane, oil, coal)	[(10.7 [kWh/l], 50 [MJ/kg]); 43.5 [MJ/kg] K]; 40.17 [MJ/kg]
ζ	X-dip coordinate	[m]
€GJ	Costs in euro of a GJ	59.7 [€/ GJ]
ϑ	Between 0 and π (180°) for contact angles	[$-$]
ϑ	Between 0 and π (180°) for dip angles	[$-$]
Ξ	Drilling energy	[$\approx 100\,\text{MJ/m}$]
λ	Sorting factor $0.1 \leqq \lambda \leqq 10$	[$-$]
λ	Mobility (permeabiliteit / viscosity)	[m^2/(Pa.s)]
μ	Viscosity	[Pa.s]
μ	Average of $\ln(k)$	[$-$]
ρ	Density	[kg/m^3]
ρ_s	Density of grains	[kg/m^3]
ρ_f	Density of porous medium	[kg/m^3]
$\sigma_{\alpha\beta}$	Interfacial tension between phase α and phase β	[N/m]

τ	Decay time for production	[years]
φ	Porosity	$[\mathrm{m}^3_{pore}/\mathrm{m}^3_{pm}]$
ϕ	Potential $\phi = p \pm \rho g z$	[Pa]
Φ	Mechanical energy $(P + \rho\, g\, z + \frac{1}{2}\rho\, v^2$	[Pa]
χ	Production rate	[bbl/day] or [m^3/s]
ψ	Stream function	[m^2/s]

References

1. J. Abate and W. Whitt. A unified framework for numerically inverting laplace transforms. *INFORMS Journal on Computing*, 18(4):408–421, 2006.
2. I. Thompson. *NIST Handbook of Mathematical Functions*, edited by F.W.J. Olver, D.W. Lozier, R.F. Boisvert, and C.W. Clark, Taylor & Francis, New York, 2011.
3. M. Abramowitz and I.A. Stegun. *Handbook of Mathematical Functions with Formulas, Graphs, and Mathematical Tables*, volume 55 of *National Bureau of Standards Applied Mathematics Series*. For sale by the Superintendent of Documents, U.S. Government Printing Office, Washington, DC, 1964.
4. M. Abramowitz and I.A. Stegun, editors. *Handbook of Mathematical Functions with Formulas, Graphs, and Mathematical Tables*. Dover Publications Inc., New York, 1992. Reprint of the 1972 edition.
5. A.W. Adamson and A.P. Gast. *Physical Chemistry of Surfaces*. Interscience, New York, 1967.
6. T. Ahmed. *Reservoir Engineering Handbook*. Gulf Professional Publishing, 2006.
7. R. Al-Hussainy, H.J. Ramey Jr., and P.B. Crawford. The flow of real gases through porous media. *Journal of Petroleum Technology or Transactions of the AIME*, 18:624–636, May 1966.
8. J.W. Amyx, D.M. Bass, and R.L. Whiting. *Petroleum Reservoir Engineering: Physical Properties*, volume 1. McGraw-Hill College, NewYork, 1960.
9. C.A.J. Appelo and D. Postma. *Geochemistry, Groundwater and Pollution*. Taylor & Francis, Boca Raton, FL, 2005.
10. C.A.J. Appelo and D. Postma. *Groundwater, Geochemistry and Pollution*. Balkema, Rotterdam, 1993.
11. K. Arnold, J.D. Clegg, E.D. Holstein, and H.R. Warner. *Petroleum Engineering Handbook: General Engineering*, volume 1. Society of Petroleum Engineers, Richardson, TX, 2006.
12. J.L. Auriault, C. Geindreau, and C. Boutin. Filtration law in porous media with poor separation of scales. *Transport in Porous Media*, 60(1):89–108, July 2005.
13. J.L. Auriault, C. Boutin, and C. Geindreau. *Homogenization of Coupled Phenomena in Heterogenous Media*. John Wiley & Sons, Hoboken, NJ, 2009.
14. J.J. Azar and G.R. Samuel. *Drilling Engineering*. PennWell Books, Tulsa, OK, 2007.
15. K. Aziz, L. Mattar, S. Ko, G.S. Brar, et al. Use of pressure, pressure-squared or pseudo-pressure in the analysis of transient pressure drawdown data from gas wells. *Journal of Canadian Petroleum Technology*, 15(02), 1976.
16. K. Aziz and A. Settari. *Petroleum Reservoir Simulation*, volume 476. Applied Science Publishers, London, 1979.
17. J. Bear. On the tensor form of dispersion in porous media. *Journal of Geophysical Research*, 66(4):1185–1197, 1961.
18. J. Bear. *Dynamics of Fluids in Porous Media*. Dover Publications, Inc., Dover, 1972.
19. J. Bear. Dynamics of fluids in porous media. Technical report, Elsevier, 1972.
20. J. Bear. *Dynamics of Fluids in Porous Media*. Courier Corporation, Chelmsford, MA, 2013.
21. P. Bedrikovetsky and J. Bruining. A percolation based upscaling technique for viscous force dominated waterflooding in uncorrelated heterogeneous reservoirs. In *IOR 1995-8th European Symposium on Improved Oil Recovery*, pages cp–107. European Association of Geoscientists & Engineers, 1995.
22. W.J. Beek, K.M.K. Muttzall, and J.W. Van Heuven. *Transport Phenomena*. Wiley-Blackwell, Hoboken, NJ, 1999.
23. C.W.J. Berentsen, M.L. Verlaan, and C. Van Kruijsdijk. Upscaling and reversibility of Taylor dispersion in heterogeneous porous media. *Physical Review E*, 71(4):46308, 2005.
24. R.N. Bhattacharya, V.K. Gupta, and H.F. Walker. Asymptotics of solute dispersion in periodic porous media. *SIAM Journal on Applied Mathematics*, 49(1):86–98, 1989.

25. M.A. Biot. General theory of three-dimensional consolidation. *Journal of Applied Physics*, 12(2):155–164, 1941.

26. R.B. Bird, W.E. Stewart, and E.N. Lightfoot. *Transport Phenomena*. John-Wiley, New York etc., 1960.

27. R.B. Bird, W.E. Stewart, and E.N. Lightfoot. Transport phenomena, 2002.

28. R.B. Bird, R.C. Armstrong, and O. Hassager. *Dynamics of Polymeric Liquids. Vol. 1: Fluid mechanics*. John Wiley and Sons Inc, New York, 1987.

29. K.S. Birdi. *Surface and Colloid Chemistry: Principles and Applications*. CRC Press, Boca Raton, FL, 2009.

30. R.J. Blackwell. Laboratory studies of microscopic dispersion phenomena. *Old SPE Journal*, 2(1):1–8, 1962.

31. M.J. Blunt. *Multiphase Flow in Permeable Media: A Pore-Scale Perspective*. Cambridge University Press, Cambridge, 2017.

32. D. Bodansky. *Nuclear Energy: Principles, Practices, and Prospects*. Springer Science & Business Media, Berlin, 2007.

33. F.C. Bond. The third theory of comminution. *Transactions of the American Institute of Mining, Metallurgical, and Petroleum Engineers*, 193:484—494, 1952.

34. F.C. Bond. Crushing and grinding calculations, part i-ii. *British Chemical Engineering*, 6(part I-II):378—385, 543–548, 1961.

35. A. Bourgeat, M. Jurak, and A.L. Piatnitski. Averaging a transport equation with small diffusion and oscillating velocity. *Mathematical Methods in the Applied Sciences*, 26(2):95–117, 2003.

36. A. Bourgeat and A. Piatnitski. Approximations of effective coefficients in stochastic homogenization. *Annales de l'Institut Henri Poincare/Probabilites et statistiques*, 40(2):153–165, 2004.

37. A. Bourgeat, M. Quintard, and S. Whitaker. Comparison between homogenization theory and volume averaging method with closure problem. *Comptes Rendus De L' Academie Des Sciences Serie Ii*, 306(7):463–466, 1988.

38. G.E.P. Box. All models are wrong, but some are useful. *Robustness in Statistics*, 202, 1979.

39. A.R. Brandt. Oil depletion and the energy efficiency of oil production: The case of california. *Sustainability*, 3(10):1833–1854, 2011.

40. R.H. Brooks and A.T. Corey. Hydraulic properties of porous media and their relation to drainage design. *Transactions of the ASAE*, 7(1):26–0028, 1964.

41. S. Brown, J. Sathaye, M. Cannell, and P.E. Kauppi. Mitigation of carbon emissions to the atmosphere by forest management. *The Commonwealth Forestry Review*, 75:80–91, 1996.

42. S.L. Brown, P. Schroeder, and J.S. Kern. Spatial distribution of biomass in forests of the eastern usa. *Forest Ecology and Management*, 123(1):81–90, 1999.

43. H. Bruining, M. Darwish, and A. Rijnks. Computation of the longitudinal and transverse dispersion coefficient in an adsorbing porous medium using homogenization. *Transport in Porous Media*, 91(3):833–859, 2012.

44. J. Bruining. Modeling reservoir heterogeneity with fractals: Enhanced oil and gas recovery research program, rept. no. 92-5. *Center for Petroleum and Geosystems Engineering, Univ*, 1992.

45. J. Bruining, D.N. Dietz, A. Emke, G. Metselaar, and J.W. Scholten. Improved recovery of heavy oil by steam with added distillables. In *Proceedings of the Third European Meeting on Improved Oil Recovery, Rome*, pp. 371–378, 1985.

46. J. Bruining, D. van Batenburg, L.W. Lake, and A.P. Yang. Flexible spectral methods for the generation of random fields with power-law semivariograms. *Mathematical Geology*, 29(6):823–848, 1997.

47. J. Bruining and C.J. Van Duijn. Uniqueness conditions in a hyperbolic model for oil recovery by steamdrive. *Computational Geosciences*, 4(1):65–98, 2000.

48. J. Bruining, C.J. Van Duijn, and R.J. Schotting. Simulation of coning in bottom water-driven reservoirs. *Transport in Porous Media*, 6(1):35–69, 1991.

49. S.E. Buckley and M.C. Leverett. Mechanism of fluid displacement in sands. *Transactions of the AIME*, 146(01):107–116, 1942.

50. R.M Butler. *Thermal Recovery of Oil and Bitumen*, vol. 46. Prentice Hall, Englewood Cliffs, NJ, 1991.

51. Improved Oil Recovery by Surfactant and Polymer Flooding. Do shah and rs schechter, 1977.

52. J.J. Carberry and R.H. Bretton. Axial dispersion of mass in flow through fixed beds. *AIChE Journal*, 4(3):367–375, 1958.

53. N.L.S. Carnot. Reflections on the motive power of heat, accompanied by kelvin wt:an account of carnot theory, 1897.

54. H.S. Carslaw and J.C. Jaeger. *Conduction of Heat in Solids*. Clarendon Press, Oxford, 1959.

55. G. Chavent and J. Jaffre. *Mathematical Models and Finite Elements for Reservoir Simulation: Single Phase, Multiphase and Multicomponent Flows through Porous Media*. Elsevier, Amsterdam, 1986.

56. G. Chiogna, C. Eberhardt, P. Grathwohl, O.A. Cirpka, and M. Rolle. Evidence of compound-dependent hydrodynamic and mechanical transverse dispersion by multitracer laboratory experiments. *Environmental Science & technology*, 44(2):688–693, 2009.

57. S.L. Clegg and K.S. Pitzer. Thermodynamics of multicomponent, miscible, ionic solutions: generalized equations for symmetrical electrolytes. *The Journal of Physical Chemistry*, 96(8):3513–3520, 1992.

58. A.T. Corey. The interrelation between gas and oil relative permeabilities. *Producers Monthly*, 19(1):38–41, 1954.

59. M. Coronado, J. Ramírez-Sabag, O. Valdiviezo-Mijangos, and C. Somaruga. A new scheme to describe multi-well compressible gas flow in reservoirs. *Transport in Porous Media*, 79(1):67–85, 2009.

60. F.F. Craig. *The Reservoir Engineering Aspects of Waterflooding*. Henry L. Doherty Memorial Fund of AIME, Society of Petroleum Engineers, Richardson, TX, 1993.

61. G.A. Croes and N. Schwarz. Dimensionally scaled experiments and the theories on the water-drive process. *Transactions of the AIME*, 204:35–42, 1955.

62. C.W.A. Cryer. A comparison of the three-dimensional consolidation theories of biot and terzaghi. *The Quarterly Journal of Mechanics and Applied Mathematics*, 16(4):401–412, 1963.

63. L.P Dake. *Fundamentals of Reservoir Engineering*, vol. 8. Elsevier, Amsterdam, 1983.

64. L.P. Dake. *Fundamentals of Reservoir Engineering*, vol. 8. Elsevier Science, Amsterdam, 1978.

65. H. Darcy. Les fontaines publiques de la ville de dijon, 1856.

66. B. Davies and B. Martin. Numerical inversion of the Laplace transform: A survey and comparison of methods. *Journal of Computational Physics*, 33(1):1–32, 1979.

67. G. De Josselin De Jong. The simultaneous flow of fresh and salt water in aquifers of large horizontal extension determined by shear flow and vortex theory. In *Soil Mechanics and Transport in Porous Media*, pages 244–251. Springer, 2006.

68. G. De Josselin De Jong. The simultaneous flow of fresh and salt water in aquifers of large horizontal extension determined by shear flow and vortex theory. In *Soil Mechanics and Transport in Porous Media*, edited by R.J. Schotting, H.C.J. van Duijn, and A. Verruijt, pp. 244–251. Springer, Dordrecht, 2006.

69. M. Dekking, A. Elfeki, C. Kraaikamp, and J. Bruining. Multi-scale and multi-resolution stochastic modeling of subsurface heterogeneity by tree-indexed markov chains. *Computational Geosciences*, 5(1):47–60, 2001.

70. J.M.P.Q. Delgado. A critical review of dispersion in packed beds. *Heat and Mass Transfer*, 42(4):279–310, 2006.

71. J.M.P.Q. Delgado. Longitudinal and transverse dispersion in porous media. *Trans IChemE, Part A, Chemical Engineering Research and Design*, 85(A9):1245–1252, 2007.

72. C.V. Deutsch and A.G. Journel. *Gslib: Geostatistical Software Library and User's Guide*. Oxford university Press, New York, 1992.

73. D.N. Dietz. A theoretical approach to the problem of encroaching and by-passing edge water. *Proc. Akad. van Wetenschappen*, 56:83–92, 1953. Also available as external research report TU-Eindhoven.

74. I. Dincer and M.A. Rosen. *Exergy: Energy, Environment and Sustainable Development*. Newnes, London, 2012.

75. E.C. Donaldson, G.V. Chilingarian, and T.F. Yen. *Enhanced Oil Recovery, I: Fundamentals and Analyses*, vol. 17. Elsevier, Amsterdam, 1985.

76. E.C. Donaldson, G.V. Chilingarian, and T.F. Yen. *Enhanced Oil Recovery, II: Processes and Operations*, vol. 17. Elsevier, Amsterdam, 1989.

77. M. Dong, F.A.L. Dullien, L. Dai, and D. Li. Immiscible displacement in the interacting capillary bundle model part I. Development of interacting capillary bundle model. *Transport in Porous media*, 59(1):1–18, 2005.

78. M. Dong, F.A.L. Dullien, L. Dai, and D. Li. Immiscible displacement in the interacting capillary bundle model part ii. applications of model and comparison of interacting and non-interacting capillary bundle models. *Transport in Porous media*, 63(2):289–304, 2006.

79. F.A.L. Dullien. *Porous Media-Fluid Transport and Pore Structure*, 574 pp. Academic Press, Cambridge, MA, 1992.

80. H. Dykstra and R.L. Parsons. The prediction of oil recovery by waterflood. *Secondary Recovery of Oil in the United States*, pp. 160–174. API, Washington DC, 1950.

81. E.A. Ebach and R.R. White. Mixing of fluids flowing through beds of packed solids. *AIChE Journal*, 4(2):161–169, 1958.

82. E. Ebeltoft, F. Lomeland, A. Brautaset, and . Haugen. Parameter based scal-analysing relative permeability for full field application. In *Proceedings of International Symposium of the Society of Core Analysis*, pp. 8–11, 2014.

83. BP Energy Economics. Bp energy outlook 2018 edition, 2018.

84. J.H. Edelman. Strooming van zoet en zout water. *Rapport inzake de watervoorziening van Amsterdam, Bijlage*, 2:8–14, 1940.

85. J. Edwards. Improving energy efficiency in e&p operations. In *SPE International Conference on Health, Safety, and Environment in Oil and Gas Exploration and Production*. Society of Petroleum Engineers, 2004.

86. M.F. Edwards and J.F. Richardson. Gas dispersion in packed beds. *Chemical Engineering Science*, 23(2):109–123, 1968.

87. A.A. Eftekhari, H. Van Der Kooi, and H. Bruining. Exergy analysis of underground coal gasification with simultaneous storage of carbon dioxide. *Energy*, 45(1):729–745, 2012.

88. A. Elfeki and M. Dekking. A markov chain model for subsurface characterization: Theory and applications. *Mathematical Geology*, 33(5):569–589, 2001.

89. A.M.M. Elfeki, F.M. Dekking, J. Bruining, and C. Kraaikamp. Influence of fine-scale heterogeneity patterns on the large-scale behaviour of miscible transport in porous media. *Petroleum Geoscience*, 8(2):159, 2002.

90. R. Farajzadeh, C. Zaal, P. van Den Hoek, and J. Bruining. Life-cycle assessment of water injection into hydrocarbon reservoirs using exergy concept. *Journal of Cleaner Production*, 235:812–821, 2019.

91. F.J. Fayers and A.H. Muggeridge. Extensions to dietz theory and behavior of gravity tongues in slightly tilted reservoirs. *SPE Reservoir Engineering*, 5(04):487–494, 1990.

92. F. Ferroni and R.J. Hopkirk. Energy return on energy invested (eroei) for photovoltaic solar systems in regions of moderate insolation. *Energy Policy*, 94:336–344, 2016.

93. R.P. Feynman, R.B. Leighton, and M. Sands. *The Feynman Lectures on Physics*, vol. 2. Mainly Electromagnetism and Matter. Addison-Wesley, Boston, MA, 1979.

94. P. Forchheimer. Wasserbewegung durch boden. *Z Vereines Deutsch Ing*, 45(1782):1788, 1901.

95. B. Gates. *How to Avoid a Climate Disaster: The Solutions We Have and the Breakthroughs We Need*. Knopf, New York, 2020.

96. J. Geertsma. Estimating the coefficient of inertial resistance in fluid flow through porous media. *Old SPE Journal*, 14(5):445–450, 1974.

97. L.W. Gelhar. *Stochastic Subsurface Hydrology*, 390 pp. Printice Hall, Englewood Cliffs, NJ, 1993.

98. L.W. Gelhar and C.L. Axness. Three-dimensional stochastic analysis of macrodispersion in aquifers. *Water Resources Research*, 19(1):161–180, 1983.

99. N.D. Gershon and A. Nir. Effects of boundary conditions of models on tracer distribution in flow through porous mediums. *Water Resources Research*, 5(4):830–839, 1969.

100. British Petroleum Global. Bp statistical review of world energy june 2017. *Relatorio. Disponível em: http://www. bp. com/en/global/corporate/energy-economics/statistical-review-of-world-energy. html*, 2017.

101. R.R.G.G. Godderij, A.G. Chessa, J. Bruining, and E. Kreft. Conditional simulation of subseismic faults. In *SPE Annual Technical Conference and Exhibition*. Society of Petroleum Engineers, 1995.

102. I.S. Gradshteyn and I.M. Ryzhik. *Table of Integrals, Series, and Products*. Academic Press, Cambridge, MA, 2014.

103. F. Grane. Measurements of transverse dispersion in granular media. *Journal of Chemical and Engineering Data*, 6(2):283–287, 1961.

104. M. Granovskii, I. Dincer, and M.A. Rosen. Exergetic life cycle assessment of hydrogen production from renewables. *Journal of Power Sources*, 167(2):461–471, 2007.

105. Walter Grassi. *Heat pumps: fundamentals and applications*. Springer, 2017.

106. D. Grassian, M. Bahatem, T. Scott, and D. Olsen. Development of an energy efficiency improvement methodology for upstream oil and gas operations. In *Abu Dhabi International Petroleum Exhibition & Conference*. Society of Petroleum Engineers, 2017.

107. A.O. Grindheim and J.O. Aasen. An evaluation of homogenisation techniques for absolute permeability. In *Lerkendal Petroleum Engineering Workshop*, 1991.

108. A.D. Gupta, L.W. Lake, G.A. Pope, K. Sepehrnoori, and M.J. King. High-resolution monotonic schemes for reservoir fluid flow simulation. *In Situ;(United States)*, 15(3):289–317, 1991.

109. F. Haces-Fernandez, H. Li, and D. Ramirez. Assessment of wind energy in the united states gulf of mexico area as power supply for offshore oil platforms. In *Offshore Technology Conference*, 2018.

110. J. Hagoort. *Fundamentals of Gas Reservoir Engineering*, vol. 23. Elsevier, Amsterdam, 1988.

111. C.A.S. Hall. *Energy Return on Investment: A Unifying Principle for Biology, Economics, and Sustainability*, vol. 36. Springer, Berlin, 2016.

112. D.R.F. Harleman and R.R. Rumer. Longitudinal and lateral dispersion in an isotropic porous medium. *Journal of Fluid Mechanics*, 16(03):385–394, 1963.

113. Z. Hashin and S. Shtrikman. A variational approach to the theory of the effective magnetic permeability of multiphase materials. *Journal of Applied physics*, 33(10):3125–3131, 1962.

114. A.M. Hassan, M. Ayoub, M. Eissa, T. Musa, H. Bruining, and R. Farajzadeh. Exergy return on exergy investment analysis of natural-polymer (guar-arabic gum) enhanced oil recovery process. *Energy*, 181:162–172, 2019.

115. H. Hassanzadeh and M. Pooladi-Darvish. Comparison of different numerical laplace inversion methods for engineering applications. *Applied Mathematics and Computation*, 189(2):1966–1981, 2007.

116. R.G. Hawthorne. Two-phase flow in two-dimensional systems-effects of rate, viscosity and density on fluid displacement in porous media. *Petroleum Transactions, AIME*, 219:81–87, 1960.

117. C.L. Hearn. Simulation of stratified waterflooding by pseudo relative permeability curves. *Journal of Petroleum Technology*, 23(7):805–813, 1971.

118. G.J. Hirasaki. Dependence of waterflood remaining oil saturation on relative permeability, capillary pressure, and reservoir parameters in mixed-wet turbidite sands. *SPE Reservoir Engineering*, 11(02):87–92, 1996.

119. G. Hon. Towards a typology of experimental errors: An epistemological view. *Studies in History and Philosophy of Science Part A*, 20(4):469–504, 1989.

120. M.M. Honarpour, F. Koederitz, and A. Herbert. *Relative Permeability of Petroleum Reservoirs*. CRC Press Inc, Boca Raton, FL, 1986.

121. R.N. Horne. *Modern Well Test Analysis*. Petroway Inc., Palo Alto, CA, 1995.

122. D.R. Horner. Pressure build-up in wells. In *3rd World Petroleum Congress*, 1951.

123. K.Z. House, A.C. Baclig, M. Ranjan, E.A. van Nierop, J. Wilcox, and H.J. Herzog. Economic and energetic analysis of capturing CO_2 from ambient air. *Proceedings of the National Academy of Sciences*, 108(51):20428–20433, 2011.

124. M.K. Hubbert. *Darcy's law and the field equations of the flow of underground fluids*. Shell Development Company, Exploration and Production Research Division, 1956.

125. R.J.M. Huijgens. *The Influence of interfacial tension on nitrogen flooding*. TU Delft, Delft University of Technology, 1994.

126. P. Ingsoy, R. Gauchet, and L.W. Lake. Pseudofunctions and extended dietz theory for gravity-segregated displacement in stratified reservoirs. *SPE Reservoir Engineering*, 9(01):67–72, 1994.

127. C.E. Jacob. On the flow of water in an elastic artesian aquifer. *Eos, Transactions American Geophysical Union*, 21(2):574–586, 1940.

128. J. Jensen, D. Hinkley, and L. Lake. A statistical study of reservoir permeability: distributions, correlations, and averages. *SPE Formation Evaluation*, 2(4):461–468, 1987.

129. R.K. Jha, A.K. John, S.L. Bryant, and L.W. Lake. Flow reversal and mixing. *SPE Journal*, 14:41–49, 2009.

130. M.B. Jones et al. Potential for carbon sequestration in temperate grassland soils. *Grassland Carbon Sequestration: Management, Policy and Economics*, 2010.

131. D. Kandhai, D. Hlushkou, A.G. Hoekstra, P.M.A. Sloot, H. Van As, and U. Tallarek. Influence of stagnant zones on transient and asymptotic dispersion in macroscopically homogeneous porous media. *Physical Review Letters*, 88(23):234501, 2002.

132. P.A. Kharecha and J.E. Hansen. Prevented mortality and greenhouse gas emissions from historical and projected nuclear power. *Environmental Science & Technology*, 47(9):4889–4895, 2013.

133. A.A. Khrapitchev and P.T. Callaghan. Reversible and irreversible dispersion in a porous medium. *Physics of Fluids*, 15:2649, 2003.

134. P.R. King. The use of renormalization for calculating effective permeability. *Transport in Porous Media*, 4(1):37–58, 1989.

135. S. Kirkpatrick. Percolation and conduction. *Reviews of Modern Physics*, 45(4):574, 1973.

136. L.J. Klinkenberg. The permeability of porous media to liquids and gases. In *Drilling and production practice*. American Petroleum Institute, 1941.

137. G.A. Korn and T.M. Korn. *Mathematical Handbook for Scientists and Engineers: Definitions, Theorems, and Formulas for Reference and Review*. Courier Corporation, Chelmsford, MA, 2000.

138. S.N. Krushkov. First-order quasilinear equations in several independent variables. *English Math. uSSR Sb.*, 10:217–273, 1970.

139. J.R. Kyte and D.W. Berry. New pseudo functions to control numerical dispersion. *Old SPE Journal*, 15(4):269–276, 1975.

140. L.W. Lake. *Enhanced Oil Recovery*. Prentice Hall Inc., Old Tappan, NJ, 1989.

141. L.W. Lake, R.T. Johns, W.R. Rossen, and G.A. Pope. *Fundamentals of Enhanced Oil Recovery*. Society of Petroleum Engineers, Richardson, TX, 2014.

142. L.W. Lake and P.B. Venuto. A niche for enhanced oil recovery in the 1990s. *Oil & Gas Journal*, 88(17):62–67, 1990.

143. L.W. Lake. *Enhanced Oil Recovery*. Prentice Hall, Hoboken, NJ, 1989.

144. R. Landauer. Electrical conductivity in inhomogeneous media. In *Electrical Transport and Optical Properties of Inhomogeneous Media*, vol. 40(1), pp. 2–45. AIP Publishing, Melville, Ny, 1978.

145. C. Laroche, M. Chen, Y.C. Yortsos, J. Kamath, et al. Determining relative permeability exponents near the residual saturation. In *SPE Annual Technical Conference and Exhibition*. Society of Petroleum Engineers, 2001.

146. P.D. Lax. *Hyperbolic Systems of Conservation Laws and the Mathematical Theory of Shock Waves*, vol. 11. SIAM, Philadelphia, PA, 1973.

147. K.M.S. Let and S. Kisoensingh. From pilot to commercial expansion of polymer flooding: a case study from the tambaredjo oilfield. In *2019 AAPG Latin America & Caribbean Region Geosciences Technology Workshop: Recent Discoveries and Exploration and Development Opportunities in the Guiana Basin*, 2019.

148. R.J. LeVeque. *Finite Volume Methods for Hyperbolic Problems*, vol. 31. Cambridge University Press, Cambridge, 2002.

149. M.C. Leverett. Capillary behavior in porous solids. *Transactions of the AIME*, 142(01):152–169, 1941.

150. Z. Li and K. Itakura. An analytical drilling model of drag bits for evaluation of rock strength. *Soils and Foundations*, 52(2):216–227, 2012.

151. P. Lingen, C.P.J.W. van Kruijsdijk, and J. Bruining. Capillary entrapment caused by small-scale wettability heterogeneities. *SPE Reservoir Engineering*, 11(2):93–100, 1996.

152. Y.P. Liu, J.W. Hopmans, M.E. Grismer, and J.Y. Chen. Direct estimation of air–oil and oil–water capillary pressure and permeability relations from multi-step outflow experiments. *Journal of Contaminant Hydrology*, 32(3):223–245, 1998.

153. F. Lomeland, E. Ebeltoft, and W.H. Thomas. A new versatile relative permeability correlation. In *International Symposium of the Society of Core Analysts, Toronto, Canada*, pp. 1–12, 2005.

154. F. Lomeland, E. Ebeltoft, and W.H. Thomas. A new versatile capillary pressure correlation. In *International Symposium of the Society of Core Analysts, Abu Dhabi, UAE*, volume 29, 2008.

155. F. Lomeland, B. Hasanov, E. Ebeltoft, and M. Berge. A versatile representation of upscaled relative permeability for field applications. In *SPE Europec/EAGE Annual Conference*. Society of Petroleum Engineers, 2012.

156. H.A. Lorentz. Grondwaterbeweging in de nabijheid van bronnen (groundwater movement in the neighbourhood of sources. *Ingenieur*, 2:24–26, 1913.

157. H.A. Lorentz. Opmerkingen bij het artikel van dr. a.h. borgesius, de ingenieur, volume 49 (1912). *De Ingenieur*, 2:24–26, 1913.

158. W.C. Lyons, T. Carter, and N.J. Lapeyrouse. *Formulas and Calculations for Drilling, Production, and Workover: All the Formulas You Need to Solve Drilling and Production Problems*. Gulf Professional Publishing, Houston, TX, 2015.

159. D. MacKay. *Sustainable Energy-Without the Hot Air*. UIT Cambridge, Cambridge, 2008.

160. B.T. Mallison, M.G. Gerritsen, K. Jessen, F.M. Orr, et al. High order upwind schemes for two-phase, multicomponent flow. *SPE Journal*, 10(03):297–311, 2005.

161. E.G. McPherson, Q. Xiao, and E. Aguaron. A new approach to quantify and map carbon stored, sequestered and emissions avoided by urban forests. *Landscape and Urban Planning*, 120:70–84, 2013.

162. B. Metz. *Carbon Dioxide Capture and Storage: IPCC Special Report. Summary for Policymakers, a Report of Working Group III of the IPCC; and, Technical Summary, a Report Accepted by Working Group III of the IPCC but not Approved in Detail*. World Meteorological Organization, 2006.

163. T. Miyazaki. *Water Flow in Soils*. CRC Press, Boca Raton, FL, 2005.

164. L. Moghadasi, A. Guadagnini, F. Inzoli, and M. Bartosek. Interpretation of two-phase relative permeability curves through multiple formulations and model quality criteria. *Journal of Petroleum Science and Engineering*, 135:738–749, 2015.

165. W.J. Moore. *Basic Physical Chemistry*. Prentice Hall, Hoboken, NJ, 1983.

166. J. Moreira, B. Gallinaro, and P. Carajilescov. Construction time of pwrs. *Energy Policy*, 55:531–542, 2013.

167. S. Morrell. A method for predicting the specific energy requirement of comminution circuits and assessing their energy utilisation efficiency. *Minerals Engineering*, 21(3):224–233, 2008.

168. N.R. Morrow. Physics and thermodynamics of capillary action in porous media. *Industrial & Engineering Chemistry*, 62(6):32–56, 1970.

169. M. Muskat. *The Flow of Homogeneous Fluids through Porous Media*. Springer, Berlin, 1982.

170. F. Oberhettinger and L. Badii. *Tables of Laplace Transforms*. Springer Science & Business Media, Berlin, 2012.

171. A. Olsson and P. Grathwohl. Transverse dispersion of non-reactive tracers in porous media: A new nonlinear relationship to predict dispersion coefficients. *Journal of contaminant hydrology*, 92(3-4):149–161, 2007.

172. T.N. Olsthoorn. The power of the electronic worksheet: modeling without special programs. *Groundwater*, 23(3):381–390, 1985.

173. T. Opolski. Speed of advance and power consumption in rotary drilling. *Bergbautechnik*, 12:654–658, 1956.

174. F.M. Orr Jr. *Theory of Gas Injection Processes*. Tie-Line Publications, Copenhagen, Denmark, 2007.

175. S. Ovaysi and M. Piri. Pore-scale modeling of dispersion in disordered porous media. *Journal of Contaminant Hydrology*, 2011.

176. J.T.G. Overbeek. *Colloid and Surface Chemistry*. MIT, Center for Advanced Engineering Study and Department of Chemical Engineering, 1970.

177. J.T.G. Overbeek, S.T. Mayr, R.G. Donnelly, and A. Vrij. *Colloid and Surface Chemistry, A Self-Study Course*. Department of Chemical Engineering and Center for Advanced Engineering Study Massachusetts Institute of Technology, 1971.

178. W.W. Owens and D.L.J. Archer. The effect of rock wettability on oil-water relative permeability relationships. *Journal of Petroleum Technology*, 23(07):873–878, 1971.

179. D.L. Parkhurst and C.A.J. Appelo. *Description of Input and Examples for PHREEQC Version Computer Program for Speciation, Batch-Reaction, One-Dimensional Transport, and Inverse Geochemical Calculations*. US Geological Survey, Denver, 2013.

180. W.J. Parkinson and N.J. De Nevers. Partial molal volume of carbon dioxide in water solutions. *Industrial & Engineering Chemistry Fundamentals*, 8(4):709–713, 1969.

181. S. Patankar. *Numerical Heat Transfer and Fluid Flow*. CRC Press, Boca Raton, FL, 1980.

182. T.W. Patzek. Thermodynamics of the corn-ethanol biofuel cycle. *Critical Reviews in Plant Sciences*, 23(6):519–567, 2004.

183. D.R. Pavone. A darcy's law extension and a new capillary pressure equation for two-phase flow in porous media. In *SPE Annual Technical Conference and Exhibition*. Society of Petroleum Engineers, 1990.

184. D.W. Peaceman. Interpretation of well-block pressures in numerical reservoir simulation. *Old SPE Journal*, 18(3):183–194, 1978.

185. D.W. Peaceman. Interpretation of well-block pressures in numerical reservoir simulation with non-square grid blocks and anisotropic permeability. *Old SPE Journal*, 23(3):531–543, 1983.

186. A.E. Peksa. Thesis tu-delft: Viability of storage options of co2 in ca silicates. *TU-Delft master thesis*, 2010.

187. T.K. Perkins and O.C. Johnston. A review of diffusion and dispersion in porous media. *Old SPE Journal*, 3(1):70–84, 1963.

188. G.E. Pickup, P.S. Ringrose, P.W.M. Corbett, J.L. Jensen, and K.S. Sorbie. Geology, geometry, and effective flow. In *SPE Annual Technical Conference and Exhibition*, 1994.

189. R.T. Pierrehumbert. *Principles of Planetary Climate*. Cambridge University Press, Cambridge,2010.

190. W.J. Plug, J. Bruining, E.C. Slob, and A.G. Goritti. Numerical validation of various mixing rules used for up-scaled geo-physical properties. In *Proc. of CMWR XVI, Copenhagen*, 2006.

191. B.E. Poling, J.M. Prausnitz, and O.C.John Paul. *The Properties of Gases and Liquids*, vol. 17. McGraw-Hill, New York, 2001.

192. W.H. Press, S.A. Teukolsky, W.T. Vetterling, and B.P. Flannery. *Numerical Recipes in C++*. Cambridge University Press, Cambridge, 2002. The art of scientific computing, Second edition, updated for C++.

193. W.H. Press, B.P. Flannery, S.A. Teukolsky, and W.T. Vetterling. *Numerical Recipes*, vol. 547. Cambridge University Press, Cambridge, 1986.

194. W.H. Press, S.A. Teukolsky, W.T. Vetterling, and B.P. Flannery. *Numerical Recipes*, Cambridge University Press, Cambridge, 1992.

195. W.H. Press, S.A. Teukolsky, W.T. Vetterling, and B.P. Flannery. *Numerical Recipes* 3rd edition: The Art of Scientific Computing. Cambridge University Press, Cambridge, 2007.

196. R.A. Prurapark. *Torque and Drag Calculations in Three-Dimensional Wellbores*. PhD thesis, Texas A & M University, 2010.

197. W.R. Purcell. Capillary pressures-their measurement using mercury and the calculation of permeability therefrom. *Journal of Petroleum Technology*, 1(02):39–48, 1949.

198. H.K. Rhee, R. Aris, and N.R. Amundson. *First-Order Partial Differential Equations*, vol. 2. Dover Pubns, Mineola, Ny, 2001.

199. M.N.E. Rifai, W.J. Kaufman, D.K. Todd, Berkeley. Sanitary Engineering Research Laboratory University of California, and Berkeley. Dept. of Civil Engineering University of California. *Dispersion phenomena in laminar flow through porous media*. University of California, 1956.

200. J. Rogelj, A. Popp, K.V. Calvin, G. Luderer, J. Emmerling, D. Gernaat, S. Fujimori, J. Strefler, T. Hasegawa, and G. Marangoni. Scenarios towards limiting global mean temperature increase below 1.5° c. *Nature Climate Change*, 8(4):325, 2018.

201. W. Rose, W.A. Bruce, et al. Evaluation of capillary character in petroleum reservoir rock. *Journal of Petroleum Technology*, 1(05):127–142, 1949.

202. B. Rubin, M.J. Blunt, et al. Higher-order implicit flux limiting schemes for black oil simulation. In *SPE Symposium on Reservoir Simulation*. Society of Petroleum Engineers, 1991.

203. P.G. Saffman. A theory of dispersion in a porous medium. *Journal of Fluid Mechanics*, 6(03): 321–349, 1959.

204. H. Salimi, H. Bruining, and V. Joekar-Niasar. Comparison of percolation theory to pore-network theory for relative permeabilities and capillary pressures. *submitted to water resources research*, pp. 1–18, 2018.

205. I. Sandrea and R. Sandrea. Global oil reserves-1: recovery factors leave vast target for eor technologies. *Oil and Gas Journal*, 105(41):44, 2007.

206. R.J. Schotting, H. Moser, and S.M. Hassanizadeh. High-concentration-gradient dispersion in porous media: experiments, analysis and approximations. *Advances in Water Resources*, 22(7):665–680, 1999.

207. F. Schwille. Dense chlorinated solvents in porous and fractured media: Model experiments. *Journal of Environmental Quality* , 19:158, 1990.

208. E. Sciubba. Extended exergy accounting applied to energy recovery from waste: The concept of total recycling. *Energy*, 28:1315–1334, 2003.

209. E.A. Seagren, B.E. Rittmann, and A.J. Valocchi. Quantitative evaluation of the enhancement of napl-pool dissolution by flushing and biodegradation. *Environmental Science & Technology*, 28(5): 833–839, 1994.

210. J.D. Seymour and P.T. Callaghan. Generalized approach to nmr analysis of flow and dispersion in porous media. *AIChE Journal*, 43(8):2096–2111, 1997.

211. A. Skauge and B. Ottesen. A summary of experimentally derived relative permeability and residual saturation on north sea reservoir cores. In *International Symposium of the SCA, Monterey, CA*, 2002.

212. J.C. Slattery and R.B. Bird. Calculation of the diffusion coefficient of dilute gases and of the self-diffusion coefficient of dense gases. *AIChE Journal*, 4(2):137–142, 1958.

213. J. Smoller. *Shock Waves and Reaction? Diffusion Equations*, vol. 258. Springer Science & Business Media, Berlin, 2012.

214. K.S. Sorbie. *Polymer-Improved Oil Recovery*. Springer Science & Business Media, Berlin, 2013.

215. K.S. Sorbie. Introduction to polymer flooding. In *Polymer-Improved Oil Recovery*, pp. 1–5. Springer, Dordrecht, 1991.

216. D. Sornette. Effective medium versus critical behavior of the failure threshold in percolation. *Physical Review B*, 36(16):8847, 1987.

217. M. Sperow. Estimating carbon sequestration potential on us agricultural topsoils. *Soil and Tillage Research*, 155:390–400, 2016.

218. E.J. Spiteri, R. Juanes, M.J. Blunt, and F.M. Orr. A new model of trapping and relative permeability hysteresis for all wettability characteristics. *Spe Journal*, 13(03):277–288, 2008.

219. F.I. Stalkup. *Miscible Displacement*. Society of Petroleum Engineers, Richardson, TX, 1983.

220. W. Stanek, L. Czarnowska, W. Gazda, and T. Simla. Thermo-ecological cost of electricity from renewable energy sources. *Renewable Energy*, 115:87–96, 2018.

221. H. Stehfest. Algorithm 368, numerical inversion of the Laplace transforms. *Communications of the ACM*, 1(13):47, 1970.

222. M. Stöhr, K. Roth, and B. Jähne. Measurement of 3d pore-scale flow in index-matched porous media. *Experiments in Fluids*, 35 (2):159–166, 2003.

223. N. Strauss. The role of the oil and gas industry in the transition toward a sustainable world. In *SPETT 2012 Energy Conference and Exhibition*. Society of Petroleum Engineers, 2012.

224. T.M. Tao and A.T. Watson. Accuracy of jbn estimates of relative permeability: Part 1-error analysis. *Society of Petroleum Engineers Journal*, 24(02):209–214, 1984.

225. T.M. Tao and A.T. Watson. Accuracy of jbn estimates of relative permeability: Part 2-algorithms. *Society of Petroleum Engineers Journal*, 24(02):215–223, 1984.

226. R. Teale. *The Concept of Specific Energy in Rock Drilling*, vol. 2,1. Elsevier, Amsterdam, 1965.

227. D. Teeuw. Prediction of formation compaction from laboratory compressibility data. *Society of Petroleum Engineers Journal*, 11(03):263–271, 1971.

228. D. Tiab and E.C. Donaldson. *Petrophysics: Theory and Practice of Measuring Reservoir Rock and Fluid Transport Properties*. Gulf Professional Publishing, Houston, TX, 2015.

229. G. Uffink, A. Elfeki, M. Dekking, J. Bruining, and C. Kraaikamp. Understanding the non-gaussian nature of linear reactive solute transport in 1d and 2d. *Transport in Porous Media*, 91(2):547–571, 2012.

230. G.J.M. Uffink. *Analysis of Dispersion by the Random Walk Method*. PhD thesis, Delft University of Technology, 1990.

231. A.J. Valocchi. Theoretical analysis of deviations from local equilibrium during sorbing solute transport through idealized stratified aquifers. *Journal of Contaminant Hydrology*, 2:191–207, 1988.

232. A.J. Valocchi and H.A.M. Quinodoz. Application of the random walk method to simulate the transport of kinetically adsorbing solutes. In *Groundwater Contamination*, volume 185 of *IAHS Publications*, pp. 35–42. IAHS, May 1989.

233. C. Van Dijk. Steam-drive project in the schoonebeek field, the netherlands. *Journal of Petroleum Technology*, 20(03):295–302, 1968.

234. C.J. Van Duijn, L.A. Peletier, and I.S. Pop. A new class of entropy solutions of the buckley–leverett equation. *SIAM Journal on Mathematical Analysis*, 39(2):507–536, 2007.

235. C.J. Van Duyn and L.A. Peletier. A class of similarity solutions of the nonlinear diffusion equation. *Nonlinear Analysis: Theory, Methods & Applications*, 1(3):223–233, 1977.

236. M. Van Dyke. *Perturbation Methods in Fluid Mechanics*, vol. 964. Academic Press, New York, 1964.

237. A.F. Van Everdingen and W. Hurst. The application of the laplace transformation to flow problems in reservoirs. *Transactions of the AIME*, 186(305):97–104, 1949.

238. A.F. Van Everdingen and W. Hurst. The application of the laplace transformation to flow problems in reservoirs. *Journal of Petroleum Technology*, 1(12):305–324, 1949.

239. R. Van Genuchten and M. Th. Analytical solutions for chemical transport with simultaneous adsorption, zero-order production and first-order decay. *Journal of hydrology*, 49(3-4):213–233, 1981.

240. P.P. Van Lingen. *Quantification and reduction of capillary entrapment in cross-laminated oil reservoirs*. PhD thesis, TU Delft, Delft University of Technology, 1998.

241. P.P. Van Lingen, J. Bruining, and C.P.J.W. Van Kruijsdijk. Capillary entrapment caused by small-scale wettability heterogeneities. *SPE Reservoir Engineering*, 11(02):93–100, 1996.

242. J. van Zalk and P. Behrens. The spatial extent of renewable and non-renewable power generation: A review and meta-analysis of power densities and their application in the us. *Energy Policy*, 123: 83–91, 2018.

243. E. Vanmarcke. *Random Fields: Analysis and Synthesis*. World Scientific, Singapore, 2010.

244. B. Verheggen and E.P. Weijers. *Climate change and the impact of aerosol: A literature review*. ECN, 2010.

245. W.W. van der Waal, D. Mikes, and J. Bruining. Inertia factor measurements from pressure-time decline curves obtained with probe-permeameters. *In Situ November*, 22(4):339–371, 1998.

246. A.J. Watson, U. Schuster, J.D. Shutler, T. Holding, I.G.C. Ashton, Peter Landschützer, David K Woolf, and Lonneke Goddijn-Murphy. Revised estimates of ocean-atmosphere CO2 flux are consistent with ocean carbon inventory. *Nature Communications*, 11(1):1–6, 2020.

247. H.J. Welge et al. A simplified method for computing oil recovery by gas or water drive. *Journal of Petroleum Technology*, 4(04):91–98, 1952.

248. H.J. Welge, E.F. Johnson, S.P. Ewing Jr., and F.H. Brinkman. The linear displacement of oil from porous media by enriched gas. *Journal of Petroleum Technology*, 13(08):787–796, 1961.

249. G.B. Whitham. Pure and applied mathematics. *Linear and Nonlinear Waves*, pp. 637–638, 1999.

250. O.H. Wiener. *Die theorie des mischkörpers für das feld der stationären strömung. 1. abhandlung: Die mittelwertsätze für kraft, polarisation und energie*. Teubner, B.G., Leipzig, 1912.

251. C. R. Wilke and P. Chang. Correlation of diffusion coefficients in dilute solutions. *AIChE Journal*, 1(2):264 – 270, 1955.

252. B.D. Wood. The role of scaling laws in upscaling. *Advances in Water Resources*, 32(5):723–736, 2009.

253. R.J. Wygal. Construction of models that simulate oil reservoirs. *Society of Petroleum Engineers Journal*, 3:281–286, 1963.

254. Y. Yang, D. Tilman, G. Furey, and C. Lehman. Soil carbon sequestration accelerated by restoration of grassland biodiversity. *Nature Communications*, 10(1):1–7, 2019.

255. Y.C. Yortsos. A theoretical analysis of vertical flow equilibrium. *Transport in Porous Media*, 18(2):107–129, 1995.

Index